T0259527

Explosion Protection

With gratitude to my wife Rosemarie to compensate for countless weekends writing this book

Explosion Protection

Electrical Apparatus and Systems
for
Chemical Plants
Oil and Gas Industry
Coal Mining

Dipl.-Phys. Dr. rer. nat. Heinrich Groh

ELSEVIER
BUTTERWORTH
HEINEMANN

AMSTERDAM • BOSTON • HEIDELBERG • LONDON • NEW YORK • OXFORD
• PARIS • SAN DIEGO • SAN FRANCISCO • SINGAPORE • SYDNEY • TOKYO

Elsevier Butterworth-Heinemann
Linacre House, Jordan Hill, Oxford OX2 8DP
200 Wheeler Road, Burlington, MA 01803

First published 2004

British Library Cataloguing in Publication Data
A catalogue record for this book is available from the British Library

Library of Congress Cataloguing in Publication Data
A catalogue record for this book is available from the Library of Congress

ISBN 0 7506 4777 9

For information on all Elsevier Butterworth-Heinemann publications
visit our website at http://books.elsevier.com

Typeset by Charon Tec Pvt Ltd., Chennai
Printed and bound by CPI Group (UK) Ltd, Croydon, CR0 4YY
Transferred to Digital Print 2012

Contents

Preface

Explosion protection, the prevention of ignition sources in areas endangered by combustible gases, vapours, mist or dusts in combination with the oxygen content of air, has been a main objective in engineering and physics for many years. Starting with Sir Humphrey Davy's lamp in 1815 and, ninety years later, Beyling's essential research in the field of flame transmission, considerable work has been done and produced safe as well as economic and reliable motor drives, process control and monitoring equipment, electric power control apparatus and lighting installations in hazardous zones. The focus of this book is on electrical apparatus and systems in hazardous areas in chemical plants, oil and gas industry and underground coal mining. The book intends to present a survey of this field, a snapshot demonstrating the 'state of the art'.

The inspiration for this book was a series of courses on 'Explosion Protection' given in chemical plants and the enthusiastic response to these courses, as well as the progress made in protection techniques, for example in pressurization, in more recent times.

This book is intended for engineers, scientists, plant safety personnel and for students in the field of electrical engineering to give an introduction to the basic principles of explosion protection and the relevant protection techniques.

The book is organized into three parts (Chapters 1–5, 6–8, 9–12) and a bibliography followed by an extensive index. The first chapter gives an introduction into basic physics – the determination of ignition temperatures and energies, of the maximum experimental safe gap and other safety related data, and summarizes these values into tables and diagrams. Chapter 2 deals with the classification of hazardous areas into 'zones' according to national and international standards or directives. In addition, a survey of standards for zones 0, 2 and 20–22 as well as for M1-equipment closes this chapter. In Chapter 3, national and international standardization for zone 1 and approval procedures are described, including European Harmonization, the 'New Approach' – ATEX 100a-Directive – and the IECEx-Scheme. Chapter 4 presents the grouping and classification of combustible substances according to their safety related data, e.g. ignition temperature and maximum experimental safe gap. Chapter 5 closes the first part with marking requirements according to international standards or directives and deals with selection aspects related to explosion protected apparatus.

Part two – as the main part of the book – starts with Chapter 6. Here, detailed descriptions of the different types of explosion protection are given, combined with illustrations of today's explosion protected apparatus. Chapter 7 deals with modern contents of 'pressurization'. According to those, an internal release of combustible substances can be handled in a safe way by dilution, and a draft standard for 'manned' pressurized enclosures has been published for voting. Closing the second part, Chapter 8 describes type testing procedures for 'flameproof enclosure, d' or 'intrinsic safety, i', accompanied by a description of partial discharge measurements, a highly efficient test method for monitoring the quality of insulation materials in electrical apparatus.

In the final part of the book, focal points are financial considerations (Chapter 9), inspection, maintenance and repair of explosion protected apparatus (Chapter 10), apparatus for installation in zone 0 and in zone 2 (Chapter 11) and cable protection systems, especially for underground installations in coal mines (Chapter 12).

The creation of such a work is impossible without competent help. So, first of all, I am indebted to Dipl.-Ing. Michael Hagen, Product Manager Instrumentation, R. STAHL Schaltgeräte GmbH, Waldenburg, Germany, for his sterling work as a co-author, namely his comprehensive representation of intrinsically safe electrical apparatus and systems. I have translated his part into English, and I must apologize for any stylistic incongruities and unclearness which may result therefrom.

Michael Hagen and I myself are greatly obliged to various companies, institutions and firms for providing very instructive photos and detailed diagrams.

I gratefully acknowledge the support of my publishers, Elsevier, in preparing this book, especially for the help in improving the style of my English!

A special expression of thanks is addressed to Dr Arnulf Krais, expert-Verlag GmbH, Renningen-Malmsheim, Germany, for his important part in launching this book project.

Heinrich Groh
Lünen Germany
September 2003

Acknowledgements

The authors are greatly obliged to companies, institutions and firms for very instructive photos and diagrams.
The following figures are by courtesy of:

Figure	Company, institution, firm
6.233	AB Allen Bradley Milwaukee, Wisconsin/USA
6.25, 6.124, 6.125, 6.126, 6.127, 6.128, 6.129	ABB Automation Products GmbH Frankfurt/Germany
6.69	ABB Calor Emag Mittelspannung GmbH Ratingen/Germany
6.26, 6.27, 6.33	ABB Industrie AG Birr/Switzerland
9.5, 9.6	ALSTOM Power Conversion GmbH Essen/Germany
6.7, 6.174	Auergesellschaft GmbH Berlin/Germany
6.103(a), (b) 6.104(a), (b)	BARTEC Sicherheits-Schaltanlagen GmbH Menden/Germany
6.37, 6.38, 6.70, 6.75, 6.120, 6.178, 6.179	BARTEC Componenten und Systeme GmbH Bad Mergentheim/Germany
6.20(a)–(d), 6.23, 6.24, 6.170(h)–(j), 6.173, 6.176(a), (b), 9.11, 9.12, 9.13(a), (b)	Becker Machaczek GmbH Friedrichsthal/Germany
6.187	BERNT GmbH, Düsseldorf/Germany and Drexelbrook Engineering Co., Horsham, Pennsylvania/USA
6.135	Bieler + Lang GmbH Achern/Germany
6.122, 6.123	Bopp & Reuther Messtechnik GmbH Mannheim/Germany and Josef Heinrichs GmbH & Co. Messtechnik KG Köln/Germany

(*continued*)

Figure	Company, institution, firm
6.91, 6.93, 6.94, 6.95, 6.96	BREUER-MOTOREN GmbH & Co. KG Bochum/Germany
6.97	BROOK CROMPTON Huddersfield/UK
6.109, 6.110, 6.111	BRUSH Transformers Ltd Loughborough, Leicestershire/UK
6.53(a)–(c), 6.61, 6.62, 6.65, 6.112	CEAG Sicherheitstechnik GmbH Soest/Germany
6.138, 6.139	DMT Deutsche Montan Technologie GmbH Essen/Germany
6.133, 6.134, 6.175	Dräger Sicherheitstechnik GmbH Lübeck/Germany
6.189, 6.190	ECOM Rolf Nied GmbH Assamstadt/Germany
6.3, 6.136, 6.137	Eickhoff Bergbautechnik GmbH Bochum/Germany
6.118, 6.119	ELMESS-Thermosystemtechnik GmbH & Co. Uelzen/Germany
6.202	Emerson Process Management Baar/Switzerland
6.177, 11.3, 11.4	Endress + Hauser Conducta GmbH & Co. Gerlingen/Germany
6.188	ESCAPELINE Ltd Luton, Bedfordshire/UK
6.28, 6.29	Federal Mogul RPB Ltd Shoreham-by-Sea/UK
6.243, 6.244	Fieldbus Int. AS Oslo/Norway
6.99	ITT Flygt Pumpen GmbH Langenhagen/Germany
6.170(a)–(g), 6.171(a), (b), 6.172	FHF Funke + Huster Fernsig Bergbautechnik GmbH Velbert/Germany
6.41, 6.42, 6.43, 6.44, 6.49, 6.74	El.-Ap., Elektro-Apparate Gothe & Co. GmbH Mülheim a. d. R./Germany
1.1, 1.2, 1.3(a)–(c), 1.4, 1.5 1.6, 1.7, 1.8, 1.9, 1.10(a)–(c), 1.11, 2.1, 2.2, 2.3, 2.4, 2.5, 2.6,	Dr. Heinrich Groh VDE Konstruktions-und Planungsbüro, Lünen/Germany

(continued)

Figure	Company, institution, firm
3.1(a), 3.1(b), 3.2, 4.1, 4.2, 4.3(a), (b), 5.1, 6.9, 6.11, 6.12, 6.13, 6.14, 6.15, 6.16, 6.17, 6.21, 6.22, 6.40(a)–(d), 6.56, 6.57, 6.58, 6.59, 6.76, 6.77, 6.78, 6.79, 6.82(a)–(c), 6.145, 6.146, 6.147, 6.148, 6.149, 6.150, 6.151, 6.152, 6.153(a), (b), 6.154(a), (b), 6.155, 6.156, 6.157, 6.158(a), (b), 6.159(a), (b), 6.160, 6.161, 6.162, 6.163, 6.164, 6.165, 6.166, 6.167, 6.168, 6.191, 6.194, 6.195, 6.237, 7.1, 7.2(a), (b), 8.1, 8.3, 8.4, 8.5(a), (b), 8.6, 8.7, 8.9, 8.10, 8.12, 9.1, 9.2, 9.3, 9.4, 9.8, 9.10, 10.1, 12.1, 12.2(a), (b), 12.3(a)–(d), 12.4(a), (b), 12.5, 12.6, 12.7, 12.8, 12.9, 12.10, 12.11, 12.12	
8.8	Haefely Test AG Basel/Switzerland
6.4, 6.66, 6.67, 6.68	HAGEN Batterie AG Soest/Germany
6.18, 6.19	H + R Hansen & Reinders GmbH & Co. KG Gelsenkirchen/Germany
6.208, 6.209	HART Communication Foundation Austin, Texas/USA
6.98	HÜBNER Elektromaschinen AG Berlin/Germany
6.140, 6.141	ISV Industrie Steck-Vorrichtungen GmbH Willstätt-Sand/Germany and MARÉCHAL Saint-Maurice/France
6.85(a), (b)	Leumann & Uhlmann AG Muttenz/Switzerland
6.86, 6.87, 6.88, 6.89, 6.90, 6.92	LOHER GmbH Ruhstorf/Germany
6.36	MAGNET-SCHULTZ GmbH & Co. KG Memmingen/Germany
6.130, 6.131, 6.132	MAIHAK AG Hamburg/Germany
6.64(b)	OSRAM GmbH München/Germany

(continued)

Figure	Company, institution, firm
6.8, 6.117, 11.1 11.2(a), (b)	F. H. PAPENMEIER GmbH & Co. KG Schwerte/Germany
6.180, 6.181, 6.182, 6.183, 6.184, 6.185, 6.196, 6.207, 6.216	Pepperl + Fuchs GmbH Mannheim/Germany
6.50, 6.51, 6.52	EMIL A. PETERS GmbH & Co. KG Iserlohn/Germany
6.116	Otto Pfannenberg Elektro-Spezialgerätebau GmbH Hamburg/Germany
6.54, 6.55	PFISTERER Kontaktsysteme GmbH & Co. KG Winterbach/Germany
6.39, 6.46	PHOENIX Contact GmbH & Co. Blomberg/Germany
6.226, 6.234, 6.235, 6.236, 8.13	PTB Physikalisch-Technische Bundesanstalt Braunschweig and Berlin/Germany
6.213, 6.214	PROFIBUS Nutzerorganisation Karlsruhe/Germany
6.81, 6.105(a), (b), 6.106 6.107, 6.108(a), (b), 6.142	PROMOS-Electronic GmbH Marl/Germany
6.6	VEM Sachsenwerk GmbH Dresden/Germany
8.2	Maschinenbau SCHOLZ GmbH & Co. KG Coesfeld/Germany
6.63, 6.64(a), 6.113, 6.114, 6.115	Adolf Schuch GmbH Worms/Germany
6.30, 6.60	Siemens AG Large Drives Div. Berlin/Germany
6.196	Siemens AG Karlsruhe/Germany
6.35, 9.7	Spillingwerk GmbH Hamburg/Germany
6.10, 6.71, 6.72, 6.73, 6.80, 6.83, 6.84, 6.100, 6.101, 6.102, 6.121, 6.169, 6.192, 6.193, 6.196, 6.197, 6.198, 6.199, 6.200, 6.203, 6.204, 6.205, 6.206, 6.210, 6.211, 6.212, 6.215, 6.217, 6.218, 6.219, 6.220, 6.221, 6.222,	R. STAHL Schaltgeräte GmbH Waldenburg/Germany

(continued)

Figure	Company, institution, firm
6.223, 6.224, 6.225, 6.227(a)–(c), 6.228, 6.229, 6.230, 6.231, 6.232, 6.239, 6.241, 6.242, 6.245, 6.246, 8.11, 11.5, 11.6	
6.1, 6.2, 6.31, 6.32, 6.34	Oivind Hagen, STATOIL Stavanger/Norway
6.5	Maschinenfabrik SULZER BURCKHARDT AG Winterthur/Switzerland
6.186(a), (b)	Taciak AG Nordkirchen/Germany
6.238, 6.240	Technische Universität Braunschweig Braunschweig/Germany
6.143, 6.144	Victor Products Ltd Newcastle upon Tyne/UK
6.47(a), (b), 6.48	WAGO Kontakttechnik GmbH Minden/Germany
6.45	Weidmüller GmbH & Co. Paderborn/Germany
9.9	WEIER Electric GmbH Eutin/Germany
6.201	WIKA Alexander Wiegand GmbH & Co. KG Klingenberg/Germany

Chapter 1

Basic principles of explosion protection

1.1 Introduction

In general, an explosion is an exothermic chemical reaction between two components. A well-known example is the reaction between the oxygen content of the atmospheric air and a combustible substance like petrol. As an exception, there are very few substances – such as acetylene – which are thermodynamically unstable and tend to exothermic self-decomposition. An explosion can start only with an ignition source and a volume or mass ratio of the two components in such a manner that the reaction zone is sustained by itself. Typical values of the peak explosion pressure – when starting with components at atmospheric pressure in a constant volume – are 1 MPa (10 bar) and a propagation velocity of the reaction zone up to 10^2 m/s (as an order of magnitude).

These basic facts can be summarized as:

Explosions can be avoided, if at least one of the three parts – component 1, component 2 or the ignition source – is absent.

This is the fundamental rule of the primary type of explosion protection: the existence of hazardous fuel–air mixtures shall be prevented by an artificial or natural ventilation of plant installations and/or an equivalent tightness of manufacturing equipment.

In addition, explosion protection by using inert gases can be derived from this rule: in the presence of combustible substances, the oxygen content of the surrounding atmosphere will be reduced to a safe level – some per cent by volume – by dilution with an inert gas such as nitrogen or carbon dioxide.

The rule described above forms a basis for the secondary type of explosion protection, which results in the avoidance of ignition sources, e.g. by using either a flameproof enclosure of a commutator motor or the pressurized enclosure of a gas analyser, or by limiting the electrical values in intrinsically safe circuits.

This secondary type of explosion protection will be the main object throughout all the following chapters. It covers burnable substances like gases, vapours, mists as well as dusts, and usually refers to atmospheric air as the second component forming a hazardous atmosphere. 'Atmospheric conditions' are defined as total pressures from $8 \cdot 10^4$ Pa (0.8 bar) to $1.1 \cdot 10^5$ Pa

(1.1 bar) as absolute pressure values (referring to vacuum) and temperatures from $-20°C$ to $+60°C$. All these values refer to the stationary (non-reacting) mixture of air with a burnable substance.

In the field of explosion protection of electrical apparatus and systems, the upper temperature limit is normally decreased to $+40°C$, and this fact should be noticed throughout all the following chapters.

'Exotic reactions' like the ignition of a H_2–Cl_2 mixture by optical radiation will not be covered in this book.

1.2   Sources of ignition

Combustions of gas–air mixtures as self-sustained chemical reactions occur only within well-defined volumetric ratios. The gas concentration – by volume – is limited by a lower value – Lower Explosive Limit (LEL) – and an upper value – Upper Explosive Limit (UEL) – in order to maintain an expanding reaction zone in the gas–air mixture. Below the LEL and above the UEL an explosive gas–air atmosphere will not be formed. Table 1.1 summarizes LEL and UEL for gases in chemical plants, the oil and gas industry and in coal mines.

Note:

For methane, CH_4, LEL and UEL refer to pure methane.

In coal mines, in the unavoidable presence of coal dust, hybrides (methane–coal dust–air) show decreased values for LEL [2].

Acetylene, C_2H_2, is an important example of a substance with a thermodynamic instability. Without any additional component, it can decompose into hydrogen and soot. Therefore, UEL = 100% (v/v).

The mixing ratio of two gas components (in steady state) can be defined – besides their volume ratio – by their partial pressures.

It is:

The sum of the partial pressures p_i of n components in a multi-component gas mixture equals the total pressure

$$\sum_{i=1}^{n} p_i = p_t$$

Example 1: A mixture with $p_{H2} = 3.2 \cdot 10^4$ Pa for hydrogen, H_2, and $p_{air} = 6.5 \cdot 10^4$ Pa for air forms a total pressure

$p_t = p_{H2} + p_{air}$
$p_t = 9.7 \cdot 10^4$ Pa

Table 1.1   LEL (Lower Explosive Limit) and UEL (Upper Explosive Limit) values for gas–air mixtures, valid for a total pressure of $1.013 \cdot 10^5$ Pa and a temperature of $+20°C$ ($= +68°F$)

Gas	Formula	Gas concentration	
		LEL % (v/v)	UEL % (v/v)
Acetylene	C_2H_2	1.5 (2.3)*	100**
Benzene	C_6H_6	1.2	8.0
Carbon disulphide	CS_2	0.6 (1.0)*	60
Ethane	C_2H_6	3.0 (3.2)*	12.5 (15.5)*
Ethylene	C_2H_4	2.7	28.5 (34)*
Hydrogen	H_2	4.0	75.6 (77.0)*
Methane	CH_4	4.4 (5.0)*	15.0 (16.5)*
Propane	C_3H_8	1.7 (2.1)*	9.5 (10.9)*

* Different values given in the literature
** Thermodynamic instability

And, in addition:

The partial pressures p_i of n components in a multi-component gas mixture show the same ratio as their volumetric ratios (volumes V_i, total volume V_t)

$$p_i/p_j = V_i/V_j, \quad \text{especially } p_i/p_t = V_i/V_t$$

Example 2: The volumetric ratios of the gas–air mixture of example 1 are:

$$\text{for hydrogen:} \quad V_i/V_t = p_i/p_t = 32.99\%$$
$$\text{for air:} \quad V_j/V_t = p_j/p_t = 67.01\%$$

In atmospheric air, the oxygen concentration is 21% (v/v), the nitrogen concentration is 78% (v/v); 1% (v/v) is due to CO_2, argon etc.
   Therefore, the volumetric ratio is:

$$\text{for } O_2: \quad V_{O_2} = 21\% \times 67.01\% = 14.07\%$$
$$\text{for } N_2: \quad V_{N_2} = 78\% \times 67.01\% = 52.27\%$$

Note:

The H_2 concentration is between LEL and UEL, this mixture is burnable!

Example 3: In example 1, the partial pressure of air is $6.5 \cdot 10^4$ Pa.
The oxygen (O_2) content is 21% (v/v), the nitrogen (N_2) content is 78% (v/v).
Calculate the partial pressures of O_2 and N_2!

$$p_i = V_i/V_t \cdot p_t$$

for O_2: $p_{O_2} = 0.21 \cdot 6.5 \cdot 10^4\,\text{Pa} = 1.365 \cdot 10^4\,\text{Pa}$

for N_2: $p_{N_2} = 0.78 \cdot 6.5 \cdot 10^4\,\text{Pa} = 5.07 \cdot 10^4\,\text{Pa}$

The composition of the mixture in example 1 is finally:

for O_2: $V_{O_2}/V_t = p_{O_2}/p_t = 14.07\%$

for N_2: $V_{N_2}/V_t = p_{N_2}/p_t = 52.27\%$

Component	Volumetric concentration %	Partial pressure
H_2	32.99	$3.2 \cdot 10^4\,\text{Pa}$
O_2	14.07	$1.365 \cdot 10^4\,\text{Pa}$
N_2	52.27	$5.07 \cdot 10^4\,\text{Pa}$
Total	99.33	$9.635 \cdot 10^4\,\text{Pa}$
others	0.67	$0.065 \cdot 10^4\,\text{Pa}$

(other constituents of air, like CO_2, Ar, ...)

For the 'others', their partial pressure $0.065 \cdot 10^4\,\text{Pa}$ corresponds to their volumetric ratio:

$$V_{others}/V_t = p_{others}/p_t = 0.67\%$$

If a liquid is in equilibrum with air, a part of this liquid is in the vaporous state and shows a temperature-dependent partial pressure. At the boiling point, the partial pressure equals the pressure of the ambient air. The partial pressure versus temperature diagram is very close to an exponential shape (Fig. 1.1). LEL and UEL, as volumetric values, correspond to partial pressures p_{LEL} and p_{UEL}. And these values define – on the temperature axis – two points, the LEP (Lower Explosive Point) and the UEP (Upper Explosive Point), which say the same as LEL and UEL: below LEP and above UEP, an explosive mixture cannot be formed.

Somewhat higher than the LEP is the flash point (FP), the lowest (experimentally determined) temperature, which allows the vapour–air mixture to be ignited.

It may be expected that FP equals LEP. But the different methods of measuring the FP do not allow an equilibrum in a perfect state.

In general, gas–air or vapour–air mixtures need a certain concentration (or partial pressure) of the burnable substance to form an explosive mixture. This concentration range is limited by LEL and UEL. If there is a liquid in equilibrium with air, the same is valid for its vapour.

Due to the temperature-dependent vapour pressure, the 'range of ignitability' can be defined by temperatures corresponding to the volumetric values LEL and UEL, called LEP and UEP as temperature-based values.

Within the range of ignitability, the ignition of a dangerous mixture is possible if an ignition source is present. This may be as follows.

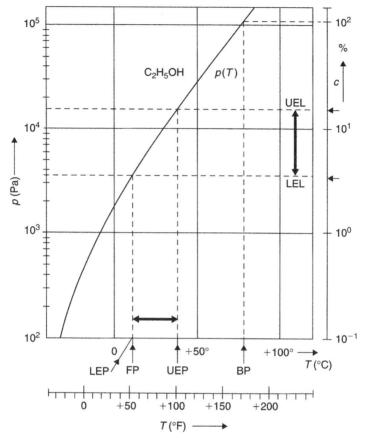

Figure 1.1 *Partial pressure diagram of ethanol, C_2H_5OH, in logarithmic scale. The small deviation of the curve partial pressure p versus temperature T from linearity demonstrates that p(T) is not exactly exponential. The ethanol concentration scale c as a volume ratio refers to an ambient pressure of $1.013 \cdot 10^5$ Pa.*
LEL: Lower Explosive Limit   FP: Flash Point
UEL: Upper Explosive Limit   LEP: Lower Explosive Point
BP: Boiling Point   UEP: Upper Explosive Point
The black crossbars indicate the 'dangerous' concentrations of ethanol in air.

1.2.1   Hot surfaces

Surfaces of electrical apparatus or components exposed to an explosive atmosphere can start a chemical reaction, if the surface temperature exceeds a certain limit, which depends on surface shape, area, material and flow conditions (direction, velocity) of a specified surrounding atmosphere.

It is a fact of experience that the inner surface of an electrically heated glass flask with a delta-shaped longitudinal section shows the lowest temperature limits to start an explosion. This temperature, T_i, experimentally determined as stated in IEC 60079-4, is the lowest value of the total range of experimentally determined ignition temperatures by varying the concentration of the combustible substance in air. Generally, T_i versus concentration is to be found in the 'rich mixture region', at concentrations higher than the stoichiometric point, defining the complete reaction of the combustible substance with the oxygen content of air to their combustion residues such as water or carbon dioxide.

Note:

If a hydrocarbon with the formula C_mH_n reacts with air completely (simplified as a composition of 21% (v/v) O_2 and 79% (v/v) N_2) producing H_2O and CO_2, it is:

$$C_mH_n + \left(m + \frac{n}{4}\right)O_2 + 3.76\left(m + \frac{n}{4}\right)N_2 \rightarrow \frac{n}{2}H_2O + mCO_2 + 3.76\left(m + \frac{n}{4}\right)N_2$$

For this reaction, the volumetric concentrations result as follows:

for the hydrocarbon: $\quad c_{hy} = \dfrac{1}{4.76\left(m + \dfrac{n}{4}\right) + 1}$

for the oxygen: $\quad c_o = \dfrac{m + \dfrac{n}{4}}{4.76\left(m + \dfrac{n}{4}\right) + 1}$

and for the 'nitrogen': $\quad c_N = \dfrac{3.76\left(m + \dfrac{n}{4}\right)}{4.76\left(m + \dfrac{n}{4}\right) + 1}$

Obviously, $c_{hy} + c_o + c_N = 1.0$.

This mixing ratio is called the stoichiometric point.

The gaseous components before and after the chemical reaction show different total volumes.

It is:

$$\text{before reaction:} \quad \Sigma_b = 1 + 4.76\left(m + \frac{n}{4}\right)$$

$$\text{after reaction:} \quad \Sigma_a = \frac{n}{2} + m + 3.76\left(m + \frac{n}{4}\right)$$

The volume ratio $\mu = \dfrac{\Sigma_b^-}{\Sigma_a}$ is

$$\mu = \dfrac{1 + 4.76\left(m + \dfrac{n}{4}\right)}{\dfrac{n}{2} + m + 3.76\left(m + \dfrac{n}{4}\right)}$$

Table 1.2   Stoichiometric mixtures of selected hydrocarbons with air

Gas	Composition	Concentration % (v/v)			Volume ratio
		c_{hy}	c_O	c_N	μ
H_2	$m = 0$ $n = 2$	29.58	14.79	55.62	1.174
CH_4	$m = 1$ $n = 4$	9.51	19.01	71.48	1.000
C_2H_2	$m = 2$ $n = 2$	7.75	19.38	72.87	1.040
C_2H_4	$m = 2$ $n = 4$	6.54	19.63	73.82	1.000
C_2H_6	$m = 2$ $n = 6$	5.66	19.82	74.52	0.974

In Table 1.2, typical values for some stoichiometric hydrocarbon–air mixtures are summarized.

The chemical reactions then read:

$$2H_2 + O_2 + 3.76N_2 \longrightarrow 2H_2O + 3.76N_2$$

$$CH_4 + 2O_2 + 7.52N_2 \longrightarrow CO_2 + 2H_2O + 7.52N_2$$

$$2C_2H_2 + 5O_2 + 18.80N_2 \longrightarrow 4CO_2 + 2H_2O + 18.80N_2$$

$$C_2H_4 + 3O_2 + 11.28N_2 \longrightarrow 2CO_2 + 2H_2O + 11.28N_2$$

$$2C_2H_6 + 7O_2 + 26.32N_2 \longrightarrow 4CO_2 + 6H_2O + 26.32N_2$$

Table 1.3 gives a survey of the T_i values of combustible mixtures (with air) for certain substances often used in chemical plants and in the oil and gas industry.

Regarding combustible dusts, a hot surface can ignite whirled-up dusts (in air). The ignition temperature, T_i, here depends on type and particle size of dust. In addition, stationary dust deposits on hot surfaces start a slowly reacting partial oxidation with the oxygen content of air. With increasing power output of the oxidation process, the temperature within the dust layer climbs up and finally an open fire starts. The lowest temperature of the hot surface which can start such a process is called the glow temperature, T_G. It depends on type and particle size of dust, on the inclination of the surface to the horizontal and on the thickness of the dust layer. The values for T_G given in Table 1.4 refer to a horizontal surface and a dust layer thickness of 5 mm.

1.2.2   Flames

Flames indicate a combustion, or, more generally, an exothermic chemical reaction. Flames act as highly efficient ignition sources.

Table 1.3 Ignition temperatures T_i in °C (Celsius) and °F (Fahrenheit) for mixtures with air, valid for a total pressure of $1.013 \cdot 10^5$ Pa (in alphabetical order)

Substance	Formula composition	T_i	
		°C	°F*
Acetaldehyde	CH_3CHO	140	284
Acetic acid	CH_3COOH	485	905
Acetone	CH_3COCH_3	540	1004
Acetylene	C_2H_2	305	581
Ammonia	NH_3	630	1166
Benzene	C_6H_6	555	1031
Butane	C_4H_{10}	365	689
Butylalcohol	C_4H_9OH	340	644
Carbon disulphide	CS_2	95 (102)**	203 (216)**
Carbon monoxide	CO	605	1121
Ethane	C_2H_6	515	959
Ethanol	C_2H_5OH	425	797
Ethyl acetate	$CH_3COOC_2H_5$	460	860
Ethylene	C_2H_4	425	797
Ethyl ether	$(C_2H_5)_2O$	170	338
Hexane	C_6H_{14}	240	464
Hydrogen	H_2	560	1040
Methane	CH_4	595 (650)**	1103 (1202)**
Methanol	CH_3OH	455	851
Petrol	–	~220	~428
Propane	C_3H_8	470	878
Toluene	$C_6H_5CH_3$	535	995
Town gas	45...55% (v/v) H_2 6...10% (v/v) CO 25...33% (v/v) CH_4 inert residues	~560	~1040
Water gas	50...55% (v/v) H_2 38...42% (v/v) CO inert residues	~600	~1112

* The correlation between temperature in °C, T_C, and temperature in °F, T_F, is described by $T_F = 32 + \left(\frac{9}{5}\right) T_C$
** Different values given in the literature

As shown in Section 1.2, the vapour of a combustible liquid in thermal equilibrium with air can be ignited only at temperatures exceeding the flash point (FP).

Table 1.5 summarizes the flash points (FP) for some combustible liquids.

The propagation of flames in gas–air or vapour–air mixtures can be avoided by using joints with a defined ratio gap, w, w to flame path length,

Table 1.4 Ignition temperatures T_i and glow temperatures T_G for combustible dusts in air with a pressure of $1.013 \cdot 10^5 \, \text{Pa}$. T_G refers to a layer thickness of 5 mm

Dust type	Particle size	T_G		T_i	
	μm	°C	°F	°C	°F
Aluminium	10...15	320	608	590	1094
Anthracite	30...40	380	716	710	1310
Gas coke	10...20	240	464	580	1076
Grain	30...50	300	572	520	968
Hard coal	5...10	250	482	580	1076
Hardwood	70...100	315	599	425	797
Iron	50	350	662	430	806
Lignite	40	230	446	440	824
Milk powder	80	330	626	500	932
Peat	60...90	320	608	480	896
Phosphorus (red)	30...50	340	644	400	752
Soot	10...20	535	995	>690	>1274
Tobacco	50...100	290	554	470	878

or, in other words, joint width, L. If a volume 'a' with an ignition source is connected to another volume 'b' via a joint with dimensions w, L, the flame transmission from the reacting mixture in a to b (containing the same mixture at identical physical parameters of pressure and temperature) is prohibited by suitable ratios w to L. For all combustible substances forming gas–air or vapour–air mixtures, the values w to L can be determined experimentally to prevent flame propagation through the path length L. The apparatus and the method of determination are stated in IEC 60079-1A. Figure 1.2 shows a longitudinal section of the test apparatus. The volume 'a' as a 20 cm^3 sphere is divided by an annular-shaped joint with $L = 25$ mm.

The upper hemisphere can be adjusted by a fine-pitch thread to defined values 'w' for the gap. The exterior volume 'b' equals nearly 2.5 litres. In both volumes, 'a' and 'b', the same gas–air mixture normally with $T = +20°\text{C}$ and a total pressure of $1.0 \cdot 10^5 \, \text{Pa}$ is present. After ignition in 'a', an MESG value (**m**aximum **e**xperimental **s**afe **g**ap) can be found by varying w, so that the flame cannot pass from 'a' to 'b'. These MESG values (ensuring the non-transmission of the flame) show a parabolic function versus the gas concentration c (Fig. 1.3).

This parabola defines the area of flame transmission through the joint. Its vertex is usually nearby or somewhat below the stoichiometric point. All gases show a behaviour similar to that shown in Fig. 1.3. The knowledge of the vertex MESG values of gas–air or vapour–air mixtures is essential for the construction and use of enclosures, type of protection 'flameproof

Table 1.5   Flash points (FP) in °C (Celsius) and °F (Fahrenheit) of combustible liquids

Substance	Chemical formula	FP	
		°C	°F
Acetaldehyde	CH_3CHO	−38...−27	−36...−17
Acetic acid	CH_3COOH	+40	+104
Acetone	CH_3COCH_3	−20	−4
Benzene	C_6H_6	−11	+12
Benzene chloride	C_6H_5Cl	+28...+30	+82...+86
Carbon disulphide	CS_2	−30	−22
Castor oil	−	+229	+444
Ethanol	C_2H_5OH	+11...+13	+52...+55
Formic acid	HCOOH	+42	+108
Lanoline	−	+240	+464
Methanol	CH_3OH	+11	+52
Naphthalene	$C_{10}H_8$	+80	+176
Nitrobenzene	$C_6H_5NO_2$	+88	+190
Nitrotoluene	$CH_3C_6H_4NO_2$	+106	+223
Olive oil	−	+225	+437
Phenol	C_6H_5OH	+82	+180
Phthalic acid	$C_6H_4(COOH)_2$	+168	+334
Phthalic anhydride	$C_6H_4(CO)_2O$	+152	+306
Propyl alcohol	C_3H_7OH	+15	+59
Sulphur	S	+207	+405
Toluene	$C_6H_5CH_3$	+6	+43
Vinyl chloride	CH_2CHCl	−43	−45
Crude oil products			
Benzine	−	+21	+70
Petrol	−	<+81	<+178
Kerosene	−	<−20...+60	<−4...+140
Diesel fuel	−	>+55	>+131
Fuel oil (light)	−	>+55	>+131
Fuel oil	−	>+65	>+149
Motor oil	−	>+185	>+365
Transformer oil	−	>+145	>+293

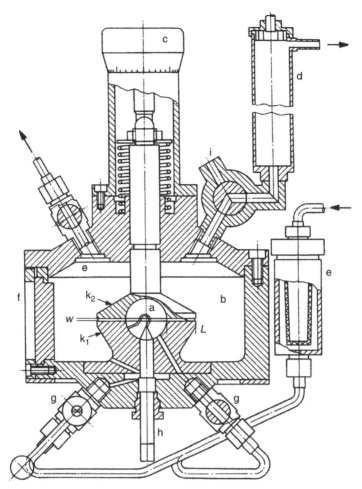

Figure 1.2   *Test apparatus for ascertainment of maximum experimental safe gap according to IEC Standard 60079-1 A.*

a = interior sphere with a volume of 20 cm^3
b = indication chamber, a cylindric enclosure with diameter = 200 mm and
      height = 75 mm
c = micrometer screw with a thread pitch = 0.5 mm
d = pump for adjusting the pressure within the test apparatus
e = flame arrestor
f = observation windows
g = valves
h = high voltage connection to spark gap
i = three way ball valve
k_1, k_2 = parts of the interior sphere
L = joint width (flame path length)
w = gap, adjusted by part 'c'

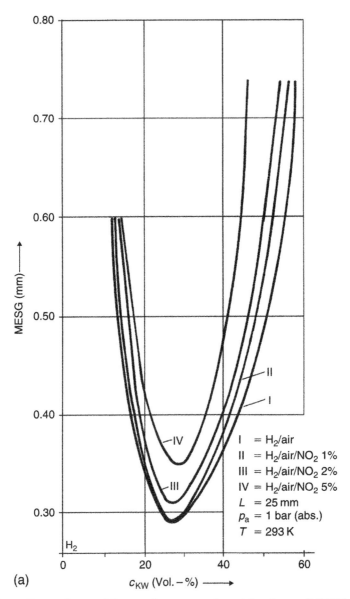

(a)

Figure 1.3 *Dependence of the maximum experimental safe gap (MESG) on the gas concentration c in air.*
$L = 25\,mm$
$p_a = 1.0 \cdot 10^5\,Pa$ (1 bar)
$T = +20°C$ ($=68°F$)
The curve I refers to pure gas–air mixtures, the curves II, III and IV refer to gas–air mixtures enriched with nitrogen dioxide, NO_2, in a concentration of 1%, 2% and 5% (v/v) [56].
1.3(a) MESG versus c for hydrogen–air mixtures. 1.3(b) MESG versus c for acetylene–air mixtures. 1.3(c) MESG versus c for ethylene–air mixtures.
The area above the parabolas represents the region of flame transmission.

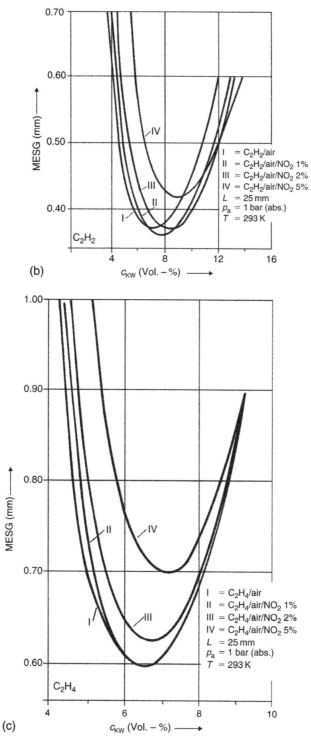

Figure 1.3 (*continued*)

Table 1.6   MESG values for combustible gas–air and vapour–air mixtures (vertex of the parabolic function MESG versus gas concentration) $T = +20°C (= +68°F)$, $L = 25\,mm$, $p = 1.0 \cdot 10^5\,Pa$ (in alphabetical order)

Substance	Chemical formula composition	Gas concentration c (% (v/v))	MESG mm
Acetone	CH_3COCH_3	5.9/4.5*	1.02
Acetylene	C_2H_2	7.15 (8.5)**	0.37
Ammonia	NH_3	24.5/17.0*	3.17
Butadiene-1,3	$CH_2CHCHCH_2$	3.9	0.79
Butane	C_4H_{10}	3.2	0.98
Carbon disulphide	CS_2	8.5	0.34
Carbon monoxide	CO	40.8	0.94
Ethane	C_2H_6	5.65 (5.9)**	0.83 (0.91)**
Ethanol	C_2H_5OH	6.5	0.89
Ethylacetate	$CH_3COOC_2H_5$	4.7	0.99
Ethylene	C_2H_4	6.5	0.60 (0.65)**
Ethylene oxide	$(CH_2)_2O$	8	0.59
Ethyl ether	$(C_2H_5)_2O$	3.47	0.87
Hexane	C_6H_{14}	2.5	0.93
Hydrocyanic acid	HCN	18.4	0.80
Hydrogen	H_2	27	0.29
Methane	CH_4	8.2 (8.9)**	1.14 (1.17)**
Methanol	CH_3OH	11.0	0.92
Pentane	C_5H_{12}	2.55	0.93
Propane	C_3H_8	4.2	0.92
Propylene	C_3H_6	4.8	0.91
Propylene oxide	$H_2COCHCH_3$	4.55	0.70
Town gas	45...55% (v/v) H_2 6...10% (v/v) CO 25...33% (v/v) CH_4 inert residues	~21	0.53

* Values obtained with the UK 8-litre sphere apparatus, not with the IEC-apparatus.
  First value: the most incendive *internal* mixture
  Second value: the most easily ignited *external* mixture
** Different values given in the literature

enclosure', e.g. for motors or switchgear units. Table 1.6 shows the vertex MESG values for some gases or vapours.

The IEC testing apparatus according to Fig. 1.2 has been improved for the application in precompressed gas–air/vapour–air mixtures and for gas–oxygen mixtures [17, 18]. The main changes are as follows (Fig. 1.4):

- rated pressure up to 10 MPa (100 bar), enabling the use of combustible mixtures with a precompression up to 1.1 MPa (11 bar)

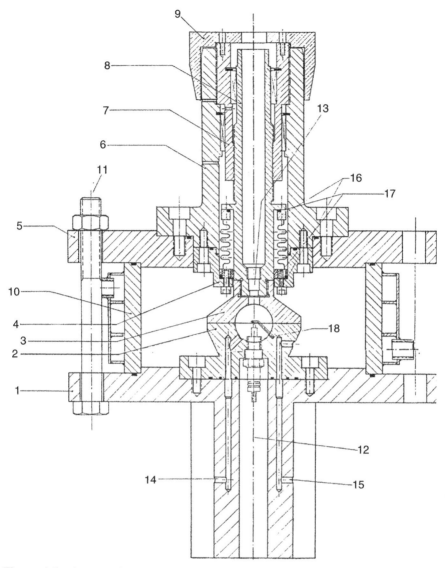

Figure 1.4 *Improved test apparatus for ascertainment of the maximum experimental safe gap [17, 18], longitudinal section drawing.*

 1, 5 = parts of the indication chamber
 2, 3 = parts of the interior sphere with a volume of 20 cm³
6, 7, 8 = adjustment drive, a differential screw with a resulting thread pitch of 50 μm
    10 = water-cooled mantle of the indication chamber
    12 = bore for spark plug
    13 = bore for pressure transducer
    14 = duct to the interior sphere
    15 = duct to the indication chamber
    16 = bolt heads with hexagonal recess
    17 = O-rings
    18 = spark plug

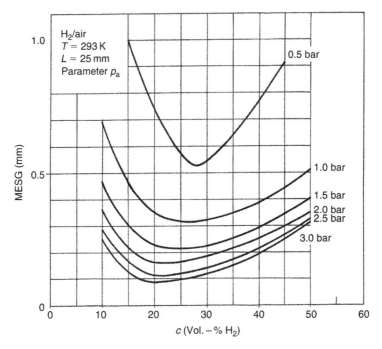

Figure 1.5   *MESG versus gas concentration c for hydrogen–air mixtures [18]*
$L = 25\,mm$
$T = +20°C\ (=68°F)$
The parameter p_a is the static precompression of the H_2–air mixture before ignition, referring to vaccum (1 bar $= 1.0 \cdot 10^5\,Pa$)

- a high precision drive by a differential thread allowing a pitch of 50 μm
- a water-cooled mantle for the exterior volume 'b' to ensure a constant length of this mantle independent of temperature stresses due to ignitions in the exterior volume 'b', and by this way a temperature-independent adjustment of the gap 'w'.

Figs 1.5 and 1.6 show the typical behaviour of the MESG values versus concentration with increasing precompression: the vertex of the parabolas decreases to smaller values for MESG and shifts away from the stoichiometric point to 'poor mixtures'. Fig. 1.7 demonstrates this behaviour in a MESG versus precompression diagram of the mixtures for constant gas concentrations. Fig. 1.8 shows the MESG values versus gas concentration for hydrogen–oxygen mixtures.

As a result of this chapter, it should be stated that the flame propagation can be stopped by suitable dimensioned joints. The MESG values (as given in Table 1.6) are intented for a 'classification' of combustible substances according to their ability to transmit flames through a gap. As for the construction of flameproof enclosures, the MESG values can give a guideline for the constructional gap values for, e.g., shafts and bearings.

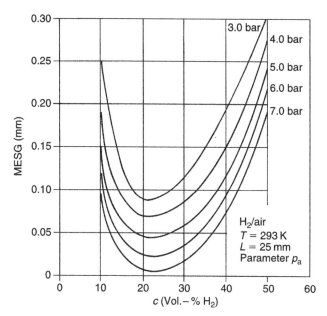

Figure 1.6 *Same as Fig. 1.5.*
p_a in the range 4.0 bar $\leqslant p_a \leqslant$ 7.0 bar (1 bar = 1.0 · 10⁵ Pa)

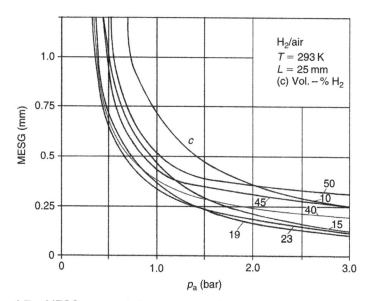

Figure 1.7 *MESG versus static precompression p_a for hydrogen–air mixtures [18]*
Parameter c is the H_2 concentration (v/v)
L = 25 mm
T = +20°C (=68°F)
(1 bar = 1.0 · 10⁵ Pa)

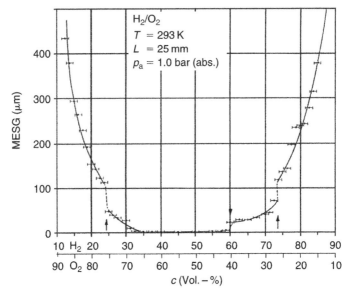

Figure 1.8 *MESG versus H_2 and O_2 concentration for H_2–O_2 mixtures [18] with a static precompression of $1.0 \cdot 10^5$ Pa.*
$L = 25$ mm
$T = +20°C \; (=68°F)$
The 'vertex' of the parabola expands to a rather broad range in H_2 and O_2 concentrations with very low (but not zero!) MESG values. The nature of the indicated 'steps' in this diagram is unknown. Similar effects have been reported in [34].

1.2.3 Sparks and electrical arcs

Switching of currents causes sparks and arcs, e.g. in contactors, circuit breakers, commutator and slipring motors. Generally, these arcs and sparks act as an ignition source for gas–, vapour– or dust–air mixtures. Only when the energy dissipated in a spark or arc does not exceed an experimentally determined level – the minimum ignition energy – can the probability of ignition be neglected. This is correct for most electric circuits in the field of remote controlling, data transmission or communications, but in the field of power circuits, the spark or arc energy exceeds the minimum ignition energy by some orders of magnitude. The minimum ignition energy is determined by a high voltage discharge between two electrodes, powered by a capacitor with capacity C and charge voltage U. The storage energy $W = 0.5\,CU^2$ is considered as the energy content responsible for the ignition. The minimum ignition energy depends on type of substance and concentration, on temperature and pressure and on electrode material and shape. Table 1.7 summarizes the values for gas–air or vapour–air mixtures, and Table 1.8 the values for dust–air mixtures.

Different values for the minimum ignition energy are obtained in low voltage circuits with an ohmic (R) and inductive (L) load. As a switching element, a spark test apparatus with counter-rotating tungsten wires and a cadmium

Table 1.7   Minimum ignition energy (high voltage capacitor discharge) for combustible gas–air and vapour–air mixtures, $T = +20°C$ ($=+68°F$), $p = 1.013 \cdot 10^5$ Pa

Substance	Chemical symbol	Minimum ignition energy μJ
Acetylene	C_2H_2	19
Allylene	CH_3CCH	110
Benzene	C_6H_6	200
Butadiene-1,3	$CH_2CHCHCH_2$	130
Butane	C_4H_{10}	250
Carbon disulphide	CS_2	9
Ethane	C_2H_6	250
Ethylene oxide	CH_2OCH_2	65
Ethyl ether	$(C_2H_5)_2O$	190
Hexane	C_6H_{14}	240
Hydrogen	H_2	19
Methane	CH_4	280
Propane	C_3H_8	250

Table 1.8   Minimum ignition energy (high voltage capacitor discharge) for combustible dust–air mixtures, $T = +20°C$ ($=+68°F$), $p = 1.013 \cdot 10^5$ Pa

Type of dust	Particle size μm	Minimum ignition energy mJ
Aluminium	23	29
Coffee	60	$>500 \cdot 10^3$
Flour	50–60	400–500
Lignite	50–60	232
Methyl cellulose	20–30	12–105
Milk powder	60–80	100–500
Soot	25	>4000
Starch	10–15	250
Sugar	20	90
Sulphur	12	<1
Sulphur	40	3
Sulphur	120	5
Toner (photocopying machines)	<10	<1
Wood	20–30	100

Table 1.9   Minimum ignition energy (low voltage circuits with an ohmic (R) and inductive (L) load), $L \leqslant 1\,\text{H}$, $I < 1\,\text{A}$, for combustible gas–air mixtures

Substance	Chemical symbol	L / R ratio* $L \geqslant 0.01\,H$	Minimum ignition energy µJ
Methane	CH_4	$L/R < 1.5\,\text{ms}$	525
Propane	C_3H_8	$L/R < 1.1\,\text{ms}$	320
Ethylene	C_2H_4	$L/R < 1\,\text{ms}$	160
Hydrogen	H_2	$L/R < 0.5\,\text{ms}$	40

* It is noteworthy for people busy in the field of electrical power engineering that ohmic-inductive circuits with L/R ratios $<1\,\text{ms}$ (even in the range $L/R = 10\,\text{µs}$ or $L/R = 100\,\text{µs}$) are considered as 'inductive circuits' in EN 50020. For comparison, in power circuits, contactors and motor starters for direct current application are rated for L/R ratios within a range $1\,\text{ms} \leqslant L/R \leqslant 15\,\text{ms}$

disc according to IEC 60079-11 and EN 50020 respectively is inserted into the electric circuit. Table 1.9 summarizes the minimum ignition energies for such low voltage circuits.

For comparison, Figs 1.9 and 1.10 give the arcing energy in a contactor (with contacts operating in air) when switching off electrical circuits in a three-phase system with a frequency of 50 cps at 1000 V (phase–phase). In Fig. 1.9 the ratio resistance/impedance has been changed at a constant current of 260 A rms; in Fig. 1.10, a 315 kW rated power 1000 V cage induction motor with different loads (idling/rated load/overload/forced standstill) was simulated. Obviously, the arcing energy shows $10^2 \ldots 10^3\,\text{J}$ (as an order of magnitude) in the switching-off case. This demonstrates that all switching arcs (or sparks) in the field of power engineering are highly efficient ignition sources of combustible gas–air or dust–air mixtures.

Only in the field of data transmission, communications, remote controlling and monitoring, has the very small energy amount, when switching electric circuits, formed a very useful basis for a special type of explosion protection, the so-called 'intrinsic safety' as given in IEC 60079-11 and EN 50020.

1.2.4   Frictional sparks

Two solid surfaces, showing a relative motion at a certain velocity and acting upon one another by a given force, can emit particles with an increased temperature with respect to the temperature of the surfaces. Dependent on materials combination, relative velocity and force, the temperature can exceed 1000°C, the particles now entering the incandescent state. In addition to the temperature rise due to friction, heated particles can react with the oxygen content of the surrounding atmosphere and thus an additional amount of energy caused by the chemical reaction between particle material and oxygen results in increasing the particle temperature. Naturally,

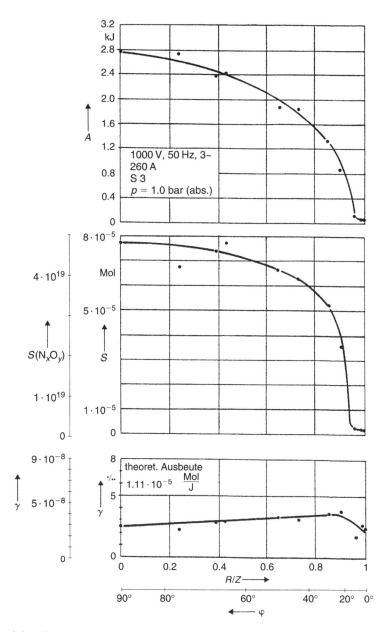

Figure 1.9 *Arc energy A, nitrous oxide production S and yield γ of nitrous oxides in a contactor with contacts operated in air at normal ambient pressure, switching off a three-phase system with 1000 V, 260 A rms at 50 cps, versus the ratio resistance R to impedance Z (or versus the phase angle φ) [16]. The values are arithmetic mean values representing 100 trials each. The yield value γ refers to the theoretical value of $1.11 \cdot 10^{-5}$ mol/joule due to the production of nitrous oxides by electrical arcs in air assuming a complete reaction between N_2 and O_2. The nitrous oxide production marked S (N_xO_y) indicates the quantity of N_xO_y molecules produced in the arcs.*

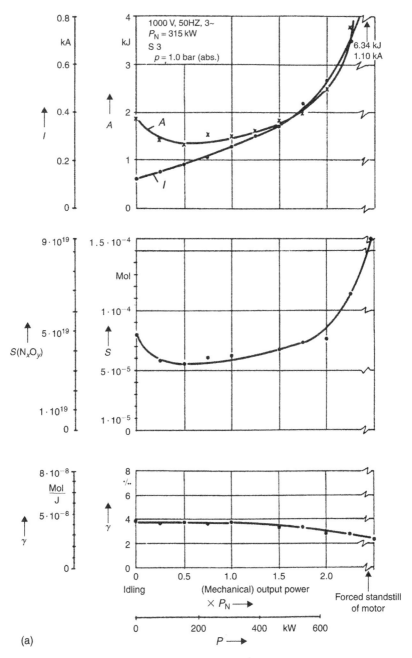

Figure 1.10　*Switching off a simulated 1000 V 315 kW rated power cage induction motor [16]. (a) Arc energy A, motor current I, nitrous oxide production S and yield γ of nitrous oxides versus different load conditions (from idling to forced standstill). Other conditions same as given in Fig. 1.9. The yield value γ refers to the theoretical value of 1.11 × 10⁻⁵ mol/joule due to the production of nitrous oxides by electrical arcs in air assuming a complete reaction between N₂ and O₂. S (NₓOᵧ) indicates the quantity of NₓOᵧ-molecules produced in the arcs.*

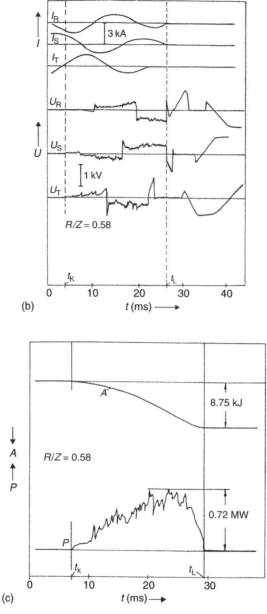

(b)

(c)

Figure 1.10   *(b) Motor currents I and voltages U across the contacts of a contactor operating in air at normal ambient pressure, switching off a simulated 315 kW cage induction motor at forced standstill at I = 1100 A rms. Three-phase system, 1000 V, 50 cps. The non-sinusoidal shape of the voltage after quenching of the arcs is caused by the voltage-limiting function of a thyristor-controlled ohmic voltage divider according to [15, 20]. (c) Power P dissipated in the arcs and its integral, A, summarized over all phases of a three-phase system, same as given in Fig. 1.10(b). The contactor has been stressed by an energy of 8.75 kJ dissipated into heat during a single switching-off cycle [16].*

materials with a high affinity to oxygen show this effect, e.g. aluminium, magnesium or titanium.

Generally, hot particles must be considered as a source of ignition of combustible gas–air or vapour–air mixtures. Even substances with a relatively high minimum ignition energy or a high ignition temperature will be ignited by particles, especially by those with increased temperatures due to an oxidation process.

The influence of oxidation processes on particle temperature is demonstrated by the fact that hot particles work preferably as an ignition source in the 'poor mixture region' of gas–air or vapour–air mixtures, i.e. in a concentration range below the stoichiometric mixture. In the 'rich mixture region' above the stoichiometric mixture, particles are not very efficient as ignition sources, due to the lack of oxygen.

To minimize the risk of ignition by particles generated by frictional effects or impacts, all modern standards for explosion protected electrical equipment restrict the content of aluminium, magnesium and titanium as enclosure materials to a level considered as totally excluding the risk of ignition (e.g. clause 8 in IEG 60079-0 and EN 50014).

A similar situation exists for the particles generated by fault current arcing in flameproof enclosures after passing the flameproof joints (see Section 6.8).

1.2.5   Electrostatics

In separation or frictional processes with at least one electrically non-conductive substance, positive and negative charges (normally forming an electric neutral) may be separated, e.g. in reeling off plastic foils from a drum, pumping or pouring an electric isolating fluid. An electric charge of $Q = 10^{-8}$ As may act as an ignition source of combustible gas–air or vapour–air mixtures if an electric field strength of – roughly estimated – 30 kV/cm as the breakthrough value is exceeded and thus an electric discharge occurs. The two 'layers' of positive and negative charges form a 'capacitor' with a characteristic capacitance C. From $Q = C \cdot U$, there results – even for very small amounts of C, e.g. $C = 10^{-12}$ F as an order of magnitude – a voltage of 10^4 V and a stored energy in this 'capacitor' of $W = 0.5\ CU^2 = 50\ \mu J$. This amount exceeds the minimum ignition energies of some gas–air mixtures, e.g. H_2–air, C_2H_2–air or CS_2–air (see Table 1.7, Section 1.2.3).

In the same way, small metallic parts embedded in electric isolating plastic materials will be pulled up to high voltages if they are blown up by rapidly streaming mixtures of air with combustible components (or without such components, but then the discharge of the stored electric charge will affect nothing).

To avoid dangerous electrostatic effects, the surface resistance of apparatus enclosures and components should not exceed $10^9\ \Omega$ (earthing in an electrostatic sense), or the surface areas of electric isolating parts should be limited (depending on the minimum ignition energy of the combustible substance, e.g. to a maximum area of 100 cm^2) or the velocity of foils should be limited.

1.2.6 High frequency electromagnetic waves (frequency range 10^4 cps to $3 \cdot 10^{12}$ cps, wavelength range $3 \cdot 10^4$ m to 0.1 mm)

The energy respective power transported within the radiation field is absorbed neither by combustible gases or vapours nor by the constituents of the atmospheric air. In this respect, there is no source of ignition. But if metallic structures are considered, e.g. parts of cranes, platforms, catwalks, handrails, pipes or other metallic parts of industrial buildings, they may work as an antenna in the radiation field and even fulfil the resonance condition. If they are not earthed thoroughly and electrically connected with each other, these points of contact may act as 'spark gaps' and produce electric discharges due to the induced voltage. Non-metallic combustible materials, especially with a permanent dipole structure, may absorb the radiated power (like food in a microwave oven) and thus can be heated to their ignition temperature.

1.2.7 Optical radiation (frequency range $3 \cdot 10^{11}$ cps to $3 \cdot 10^{15}$ cps, wavelength range 1 mm to 0.1 μm)

Optical radiation – especially in the infrared range – can be absorbed by combustible gases and vapours by oscillatory transitions of the molecules and cause an increase in temperature. In addition, the radiation may be absorbed by non-metallic solids and effect a temperature rise up to a value sufficient to ignite them or the surrounding gas–air or vapour–air mixtures.

At very high intensities of optical radiation, the electric field strength exceeds the breakthrough value for air – roughly 3 MV/m – and a discharge in the free space (with no electrodes at all) works as an ignition source. In a plane electromagnetic wave, the power density is described by Poynting's vector:

$$S = E \times H$$
(as a vectorial product)
E = electric field strength vector
H = magnetic field strength vector

With respect to Maxwell's equations, it follows:

$$S = (\epsilon\epsilon_o/\mu\mu_o)^{0.5} E^2$$

with

$$\mu_o = 1.256 \cdot 10^{-6} \text{Vs/Am}$$

$$\epsilon_o = 8.854 \cdot 10^{-12} \text{As/Vm}$$

With a good approximation, in air it is:

$$\epsilon = 1 \quad \text{and} \quad \mu = 1$$

In Table 1.10, the peak values for the electric and magnetic field strength are calculated for optical power densities typical for laser systems. Obviously, the breakthrough field strength ($E > 3$ MV/m) occurs in laser beams with a power density >1 MW/cm^2. Figure 1.11 demonstrates this

Table 1.10 Peak values of electric field strength (*E*) and magnetic field strength (*H*) in plane electromagnetic waves with power density *S*

S MW/cm^2	*E* *MV/m*	*H* *A/m*
0.1	0.87	2 320
1	2.75	7 360
10	8.7	23 200

Figure 1.11 *Photo of a laser-induced breakthrough in air. The infrared light pulse had a duration of approx. 0.5 μs and an energy of some joules. The bright spot inside the circle indicates the place of the discharge.*

fact: the radiation of a pulsed high power CO_2 laser with a wavelength at 10.6 μm, pulse duration approx. 0.5 μs, peak power some MW, energy some joules, is focused by a germanium lens.

1.2.8 Other sources of ignition

Electromagnetic radiation in the ultraviolet range or X-rays can form a source of ignition as well as acoustic waves, especially in the ultrasonic range. It is only a question of power density in the radiation field.

Moreover, shockwaves in gas–air or vapour–air mixtures, e.g. generated by adiabatic compression of the filling gas in the mechanical damaged glass tube of a fluorescent lamp, can ignite gas–air mixtures, especially with a low minimum ignition energy such as H_2–air or C_2H_2–air mixtures. But these sources of ignition will not be the main objective of this book.

Chapter 2

The classification of hazardous areas

2.1 Basic principles

Due to the quite different structures of industrial plants equipped with electrical installations, there are different probabilities for the existence of potentially explosive atmospheres formed by gas–air, vapour–air or dust–air mixtures. With respect to economical reasons, the types of explosion protection for electrical apparatus and systems will depend strongly on the explosion risk at the installation point. It is very unusual to construct and build all explosion protected electrical apparatus in such a manner that they can operate permanently in a surrounding hazardous atmosphere with combustible gases, vapours and dusts in air. The common way is to classify different areas in industrial plants according to the probability for the existence of a hazardous atmosphere and to establish adequate types of explosion protection. In other words, industrial plants with potentially explosive atmospheres are divided and classified into 'zones'.

Following the historical development of electrical engineering and explosion protection, zone classification was the objective of national standards and installation rules. Most of the leading industrial countries established an installation practice for chemical plants and the oil and gas industry with two or three zones for areas hazardous due to gas– or vapour–air mixtures and two zones for areas with hazardous dust–air mixtures. Apart from this philosophy, the coal mining industry in most countries tends to avoid an area classification and defines only one category of explosion protection ('firedamp-proof'). More recent standards or directives present a three-zone concept for areas endangered by combustible gas– (vapour–, mist–) air mixtures and dust–air mixtures in industrial plants (other than coal mines).

As time passes, national standards for area classification have been superseded by international standards (IEC, International Electrotechnical Commission, Geneva/Switzerland, and EN, European Standard or European Norm, established by CENELEC, Comité Européen de Normalisation Electrotechnique, Brussels/Belgium), which have been adopted as subsequent national standards. In the member countries of the European Community (EC), a joint area classification is defined by a directive (Directive 1999/92/EC of the European Parliament and of the Council, dated 1999-12-16).

Tables 2.1 and 2.2 survey the different classification systems for hazardous areas.

Table 2.1   Classification systems for hazardous areas

Type of hazardous area	Standard, decree
I Classification system with three zones for areas hazardous due to combustible gas, vapour or mist (Ia) and with two zones for areas hazardous due to combustible dust (Ib).	
	Germany: ElexV/1992 [8]* VDE 0165/1991-02
Ia – areas in which an explosive atmosphere is present permanently or for long periods	zone 0
– areas in which an explosive atmosphere can be expected to occur occasionally (in normal operation)	zone 1
– areas in which an explosive atmosphere is expected to occur only seldom and then only temporarily (not expected in normal operation)	zone 2
Ib – areas in which an explosive atmosphere is present for a long time or often	zone 10
– areas in which an explosive atmosphere is expected to occur occasionally due to dust deposits being stirred up	zone 11

Type of hazardous area	International Standard	Overtaken as German Standard
II Classification system with three zones for areas hazardous due to combustible gas, vapour or mist (IIa) or due to combustible dust (IIb).		
	IEC 60079-10/1995-12 EN 60079-10/1996-01 (identical)	VDE 0165 Teil 101/1996-09
	–	ElexV [8]*
IIa – areas in which an explosive atmosphere is present permanently or for long periods	zone 0	zone 0
– areas in which an explosive atmosphere can be expected to occur occasionally (in normal operation)	zone 1	zone 1

(*continued*)

Table 2.1   (*continued*)

Type of hazardous area	International Standard	Overtaken as German Standard
– areas in which an explosive atmosphere is expected to occur only seldom and then only temporarily (not expected in normal operation)	zone 2	zone 2
IIb	IEC 61241-3/1997-05 –	– ElexV [8]*
– areas in which combustible dust, as a cloud, is present continuously or frequently, during normal operation, in sufficient quantity to be capable of producing an explosive concentration of combustible dust mixed with air, and/or where layers of dust of uncontrollable and excessive thickness can be formed	zone 20	zone 20
– areas not classified as zone 20, in which combustible dust, as a cloud, is likely to occur during normal operation, in sufficient quantities to be capable of producing an explosive concentration of combustible dust mixed with air	zone 21	zone 21
– areas not classified as zone 21 in which combustible dust clouds may occur infrequently, and persist for only a short period, or in which accumulations or layers of combustible dust may be present under abnormal conditions and give rise to combustible mixtures of	zone 22	zone 22

(*continued*)

Table 2.1   (*continued*)

Type of hazardous area	*International Standard*	*Overtaken as German Standard*
dust in air. Where, following an abnormal condition, the removal of dust accumulations or layers cannot be assured then the area is to be classified zone 21		

Type of hazardous area	*Directive 1999/92/EC***
III Classification system with three zones for areas hazardous due to combustible gas, vapour or mist (IIIa) or due to combustible dust (IIIb).	
IIIa – areas in which an explosive atmosphere caused by mixtures of air with combustible gas, vapour or mist is present continuously or for long periods or frequently	zone 0
– areas in which during normal operation an explosive atmosphere caused by mixtures of air with combustible gas, vapour or mist is likely to occur	zone 1
– areas in which during normal operation an explosive atmosphere caused by mixtures of air with combustible gas, vapour or mist is unlikely to occur, but if it does so, it is likely to do so only for a short period	zone 2
IIIb – areas in which an explosive atmosphere formed by a cloud of combustible dust in air is present continuously or for long periods or frequently	zone 20
– areas in which during normal operation an explosive atmosphere formed by a cloud of combustible dust in air is likely to occur	zone 21
– areas in which during normal operation an explosive atmosphere formed by a cloud of combustible dust in air is unlikely to occur, but if it does so, it is likely to do so only for a short period	zone 22

* In the 1996 edition, ElexV [8] defines a three-zone concept for areas hazardous due to combustible dust
** Directive 1999/92/EC shall be transformed into national legislation by all EC member states till 2003-06-30

Table 2.2(a)   Classification system for hazardous areas (USA: National Electrical Code, NEC, and Canada: Canadian Electrical Code, CEC)

Type of hazardous area	Class	Division
Areas hazardous due to:		
– flammable gases, vapours or liquids	I	
• where ignitable concentrations of flammable gases, vapours or liquids can exist all of the time or some of the time during normal operating conditions		1
• where ignitable concentrations of flammable gases, vapours or liquids are not likely to exist under normal operating conditions		2
– combustible dusts	II	
• where ignitable concentrations of combustible dusts can exist all of the time or some of the time under normal operating conditions		1
• where ignitable concentrations of combustible dusts are not likely to exist under normal operating conditions		2
– ignitable fibres and flyings	III	
• where easily ignitable fibres or materials producing combustible flyings are handled, manufactured or used		1
• where easily ignitable fibres are stored or handled		2

Note: The content of Table 2.2(a) describes the traditional classification (IEC standardization not incorporated). Until 1998 the USA and Canada adopted the 'three-zone concept' for areas hazardous due to combustible gases, vapours and mist according to IEC 60079-10, in parallel with the content of Table 2.2(a).

Note:

For areas in medical installations, there exists another classification system, e.g. zones G and M in Germany. Medical rooms are not covered in this book.

Part I of Table 2.1 gives a national (historical) classification system with three zones for areas endangered by gas–air mixtures and two zones for areas endangered by dust–air mixtures.

In addition to the definitions given in standards, national decrees or regulations give examples for the area classification, e.g. in Germany in Ex-VO [11] and ElexV [8], in Ex-RL [45] and in VbF/TRbF [50].

Part II of Table 2.1 shows a present-day classification system with three zones each for areas hazardous due to combustible gas–air and dust–air mixtures. IEC 61241-3 describes a three-zone concept for areas exposed to hazardous dust–air mixtures, accompanied by IEC 61241-1-1 and IEC 61241-1-2, dealing with specifications for apparatus and selection, installation and maintenance of such apparatus. IEC 60079-10 and EN 60079-10 give some guidelines to calculate the

Table 2.2(b)   Protection methods (by courtesy of Underwriters
Laboratories Inc., Northbrook, Illinois/USA)

Area	Protection methods	Applicable certification documents	
		US	Canada
Class I			
Div. 1	Explosionproof	UL 1203	CSA-30
	Intrinsically safe (2 fault)	UL 913	CSA-157
	Purged/pressurized (Type x or y)	NFPA 496	NFPA 496
Div. 2	Nonincendive	UL 1604	CSA-213
	Non-sparking device	UL 1604	CSA-213
	Purged/pressurized (Type z)	NFPA 496	NFPA 496
	Hermetically sealed	UL 1604	CSA-213
	Any Class I, Div. 1 method	–	–
	Any Class I, zone 1 or 2 method	UL 60079 Series	CSA E 60079 Series
Class II			
Div. 1	Dust-ignitionproof	UL 1203	CSA-25 or CSA-E-1241-1-1
	Intrinsically safe	UL 913	CSA 157
	Pressurized	NFPA 496	NFPA 496
Div. 2	Dusttight	UL 1604	CSA-157 or CSA-E-1241-1-1
	Nonincendive	UL 1604	–
	Non-sparking	UL 1604	–
	Pressurized	NFPA 496	NFPA 496
	Any Class II, Div 1 method	–	–
Class III			
Div. 1	Dusttight	UL 1604	CSA-157
	Intrinsically safe	UL 913	CSA-157
Div. 2	Dusttight	UL 1604	CSA-157
	Intrinsically safe	UL 913	CSA-157

Note: In parallel to the 'traditional' protection methods described above, the USA
and Canada have adopted most of the IEC Standards (see Table 3.14).

quantity of gases entering into the surrounding atmosphere from flanges and
other constructional parts of chemical apparatus.

Part III of Table 2.1 shows the area classification as stated in Directive
1999/92/EC of the European Parliament and of the Council (dated 1999-12-16).

This concept incorporates three zones each for hazardous areas endangered by combustible gas– (vapour–, mist–) air or dust–air mixtures.

An example of a 'two-zone concept' for areas hazardous due to combustible substances is given in Table 2.2. It shows the traditional classification (in the USA and Canada) into classes and divisions according to the probability of forming an explosive atmosphere.

Electrical apparatus for coal mines, type 'firedamp-proof', are constructed, built and tested according to the same philosophy as for zone 1 apparatus. Therefore, most of the IEC and European Standards describe constructional requirements and testing procedures for zone 1 apparatus and 'firedamp-proof' apparatus in parallel.

Note:

If a general reference is made to 'explosion protected apparatus' or 'explosion protected systems', then this relies upon zone 1 apparatus. And this will be the guideline throughout all chapters in this book.

Class I Division 1 apparatus comprises three different protection methods:

- explosion-proof enclosure
- purged and pressurized enclosure
- intrinsically safe circuits

which are similar to the corresponding types of protection given in IEC or EN Standards for zone 1 application. Following the European philosophy, explosion-proof and purged and pressurized apparatus are not permitted for zone 0 application.

Generally, all modern standards (IEC, EN based and adopted as national standards) give a 'grouping' of explosion protected electrical apparatus according to their application:

- Group I   for coal mining (underground)
- Group II   for chemical plants and oil and gas industry.

Some years ago, Directive 94/9/EC (the so-called 'ATEX 100a Directive'), dated 1994-03-23 (following the title in French: '**At**mosphères **ex**plosibles ...' and referring to article 100a of the Treaties of the European Community), introduced 'categories' of explosion protected apparatus to give a guideline for their installation with respect to the risks that may be expected in different locations exposed to combustible atmospheres (Table 2.3).

Directive 1999/92/EC (in Annex IIB) gives the correlation between the different zones and the adequate categories of electrical equipment to be installed in these zones.

By overtaking the ATEX 100a Directive into German legislation, ElexV [8] has been revised in 1996, now incorporating the three-zone concept for areas hazardous due to combustible dust.

Table 2.3 The ATEX 100a Directive marking system

Application	Standard or directive					
	IEC 60079-0 EN 50014 Group	IEC 60079-10 EN 60079-10 zone	Directive 1999/92/EC zone	IEC 61241-3 zone	ATEX 100a Directive 94/9/EC Group	Category[1]
Coal mines (underground)	I	–	–	–	I I	M1[2] M2[2]
Chemical plants, oil and gas industry (hazards due to gases, vapours, mist)	II	0 1 2	0 1 2	– – –	II II II	1G[3] 2G[3,4] 3G[3,4]
Chemical plants, industry (hazards due to combustible dusts)	EN 50014 only II	–	20 21 22	20 21 22	II II II	1D[3] 2D[3,5] 3D[3,5]

1 M = Mining
G = Gas
D = Dust
2 Group I – category M1 – apparatus is something like 'zone 0 equipment' for coal mines. For this purpose, some national regulations have already been established, e.g. in Germany clause 30 of ElBergV [7], which allows the continuous operation of specially built and tested electrical apparatus even under methane concentrations exceeding the limit valid for normal Group I apparatus. Recently, CENELEC, Brussels, has drawn up the Standard EN 50303: 2000–07 covering the requirements for Group I – category M1 equipment.
Group I – category M2 – apparatus correspond to the traditional Group I equipment for coal mines ('firedamp-proof')

The ATEX 100a Directive defines:

- Group I – category M1 equipment is intended for use in underground parts of mines as well as those parts of surface installations of such mines endangered by firedamp and/or combustible dusts. Equipment in this category is required to remain functional, even in the event of rare incidents relating to equipment, with an explosive atmosphere present
- Group I – category M2 equipment is intended for use in underground parts of mines as well as those parts of surface installations of such mines likely to be endangered by firedamp and/or combustible dusts. Equipment in this category is intended to be de-energized in the event of an explosive atmosphere

3 The ATEX 100a Directive defines:

- Group II – category 1 equipment is intended for use in areas in which explosive atmospheres caused by mixtures of air and gases, vapours or mists or by dust–air mixtures are present continuously or for long periods or frequently
- Group II – category 2 equipment is intended for use in areas in which explosive atmospheres caused by gases, vapours, mists or dust–air mixtures are likely to occur
- Group II – category 3 equipment is intended for use in areas in which explosive atmospheres caused by gases, vapours, mists or stirred-up dusts are unlikely to occur, but if they do occur, they are likely to do so only infrequently and for a short period only

4 In addition, electrical apparatus of category 1G may be used in zone 1 and zone 2. Electrical apparatus of category 2G may be used in zone 2

5 In addition, electrical apparatus of category 1D may be used in zone 21 and zone 22. Electrical apparatus of category 2D may be used in zone 22

Note:

The new Group II – categories 1D, 2D and 3D cannot be correlated in an easy way to the traditional zone 10 or zone 11 equipment. A guideline may be:

- zone 11 apparatus fulfils the requirements for category 3D
- zone 11 apparatus needs an additional technical quality in order to fulfil the requirements for category 2D, e.g. complete tightness against the ingress of dust
- zone 10 apparatus generally fulfils the requirements for category 2D
- zone 10 apparatus cannot comply with the requirements for category 1D. 1D asks for two independent types of protection, and this contradicts the present practice.

This marking system correlates the adequate zone on the explosion protected apparatus with traditional data, e.g. group, type of protection and temperature class (see Chapter 5).

2.2   Examples of the classification of hazardous areas

This section will demonstrate the area classification into different zones. The examples are based on a German code, the TRbF [50], and cover mainly filling stations. The relevant criteria for the dimension of zone 0, zone 1 and zone 2 are:

- the flash point (FP) of the liquid to be handled and stored. A flash point at low temperatures indicates a highly volatile liquid, in other words, at usual ambient temperatures the partial pressure of the vapour is already high enough so as to exceed the LEL with respect to its volumetric concentration
- the filling rate (volume per time unit) of the liquid entering a tank and displacing the gas–air mixture to the surrounding atmosphere.

In Table 2.4, the linear dimensions of zone 1 and zone 2 are correlated to the filling rate and to the flash point (FP). Figures 2.1 and 2.2 show examples for the area classification according to TRbF [50].

2.3   Explosion protection for zone 0

Until 1999, there exists no international standard for explosion protected electrical apparatus intended for the application in zone 0. Intrinsically safe circuits according to IEG 60079-11 and EN 50020, category 'ia', seemed to be an appropriate type of protection for zone 0 for some time. In the same way, for coal mines the first international standard describing the requirements for electrical apparatus that are intended for operation in the presence of methane/coal dust–air mixtures was prepared in 2000.

In Germany, two attempts have been undertaken in order to establish national standards for this purpose. The first attempt resulted in a draft

Table 2.4 Linear extension R as a function of filling rate (volume per time unit) of combustible liquids with a given flash point

Filling rate of liquid m³/h	Flash point		R m
	°C	°F	
≤60	<0	<32	2
	0–<21	32–<70	1
	21–<35	70–<95	0.5
	35–55	95–131	0.5
≤180	<0	<32	3
	0–<21	32–<70	1.5
	21–<35	70–<95	1
	35–55	95–131	0.5
≤450	<0	<32	5
	0–<21	32–<70	2.5
	21–<35	70–<95	1.5
	35–55	95–131	1
≤900	<0	<32	7
	0–<21	32–<70	3.5
	21–<35	70–<95	2
	35–55	95–131	1
≤1350	<0	<32	8.5
	0–<21	32–<70	4.5
	21–<35	70–<95	2.5
	35–55	95–131	1.5
≤1800	<0	<32	10
	0–<21	32–<70	5
	21–<35	70–<95	2.5
	35–55	95–131	1.5
≤2400	<0	<32	12
	0–<21	32–<70	6
	21–<35	70–<95	3
	35–55	95–131	2
≤3000	<0	<32	14
	0–<21	32–<70	7
	21–<35	70–<95	3.5
	35–55	95–131	2

national standard (VDE 0170/0171 Teil 12/1982-11: Requirements for apparatus for zone 0). The basic philosophy was:

- intrinsically safe circuits, category 'ia' according to EN 50020, with changed diagrams for the minimum ignition current in inductive circuits and for the minimum ignition voltage in capacitive circuits

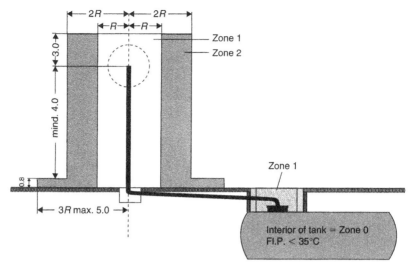

Figure 2.1   *Classification of an area hazardous due to combustible gases/liquids into zones 0, 1 and 2 [50]. The value R is tabulated in Table 2.4 as a function of the filling rate into the tank (liquid with a flash point <35°C). Dimensions given in metres.*

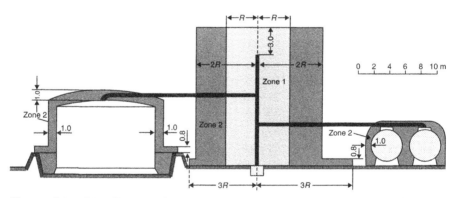

Figure. 2.2   *Classification of an area hazardous due to combustible gases/liquids into zones 0, 1 and 2 [50].*
Flat tanks with 12 m diameter and 7 m height, volume 800 m³ each. Horizontal cylindric tanks with 4 m diameter and 16 m length, volume 200 m³ each. Filling rate 120 m³/h, flash point <0°C. The value R is tabulated in Table 2.4 with $R = 3$ m. The protection area around the exhaust duct extends to a diameter of $6R = 18$ m. The interior of the tanks is classified as zone 0. Dimensions given in metres.

- the combination of two independent types of protection as stated in EN 50014 … EN 50020, EN 50028, e.g. intrinsically safe circuits (and their components) of category 'ib' in a flameproof enclosure, or encapsulated electronic components in a flameproof enclosure, or components in 'increased safety' in a pressurized enclosure
- using separation elements between zone 0 and the electrical apparatus in the surrounding zone 1, which shall have a type of protection according to EN 50014 … EN 50020, EN 50028. As a separation element a 'lantern' or a membrane may be used.

The approval of zone 0 apparatus has been established by PTB, Physikalisch Technische Bundesanstalt, Braunschweig and Berlin.

In recent times, CENELEC (Brussels) started work establishing a zone 0 standard for electrical apparatus which resulted in European Standard EN 50284: 1999-04 with the title:

Special requirements for construction, test and marking of electrical apparatus of equipment group II, category 1 G.

(overtaken as German Standard VDE 0170/0171 Teil 12-1/2000-02).
   The protection measures are summarized as follows:

- intrinsically safe circuits, category 'ia' according to EN 50020
- apparatus according to EN 50028 ('encapsulation') with some modifications
- the combination of two independent types of protection according to EN 50014 ... EN 50020, EN 50028.

Examples:

- a cage induction motor according to EN 50019 'Increased safety – e –' in an enclosure according to EN 50018 'Flameproof enclosure – d –', which in addition complies with the requirements for ingress protection as given in EN 50019
- an inductive transmitter with intrinsically safe electrical circuits category 'ib' according to EN 50020 'Intrinsic Safety – i –' enclosed by 'Encapsulation – m –' as given in EN 50028, or with an enclosure according to EN 50018 'Flameproof enclosure – d –', or protected by 'Powder filling – q –' according to EN 50017
- a tubular fluorescent lamp combined with an alternating current generator, both in 'Increased safety – e –' according to EN 50019 with a common enclosure in 'Pressurization – p –' going along with EN 50016, using pressurized air as combined protective gas and driving power of the generator via a turbine wheel
- zone 1 apparatus behind a separation element (e.g. lantern, membrane).

In the same way as for zone 0, a national 'forerunner' in Germany started work for mining apparatus intended for continuous operation in an explosive atmosphere. The work was done by members of the national committees focused on explosion protection and resulted in a decree stated by the national authority. On the basis of clause 30 of ElBergV [7], the following protection concepts are considered applicable for a continuous operation under a permanently existing explosive atmosphere:

- intrinsically safe circuits, category 'ia' according to EN 50020, with special types of cables
- flameproof enclosures or pressurized enclosures (with leakage compensation by an inert protective gas) according to EN 50018 and EN 50016, with an increased mechanical strength of the enclosure and this enclosure made of corrosion-free materials, containing internal electrical components each protected by a second type of protection, e.g. flameproof enclosure, increased safety, encapsulation or powder filling

- special requirements for accumulators and the sensor cells for catalytic gas sensor.

The approval for this type of mining equipment shall be established by BVS, Bergbau-Versuchsstrecke, Dortmund/Germany

Similar to zone 0, this national 'forerunner' initiates international activities in this field. This leads to a first European Standard 2000-07 EN 50303/1998-04, produced by a joint CEN/TC/305 and CENELEC/TC 31/WG, titled 'Group I, category M1 equipment intended to remain functional in atmospheres endangered by firedamp and/or coal dust.'

The basic technical requirements are as follows:

- electrical equipment shall comply with the European Standards EN 50014– EN 50016, EN 50017, EN 50018, EN 50019, EN 50020 and EN 50028 (see Chapters 3 and 6) in all cases referring to the requirements for Group I apparatus.

Note:

EN 50015 – 'Oil immersion' is excluded.

- restriction on the use of light metals or alloys
- special constructional and test requirements for cables containing more than one intrinsically safe circuit
- limitation of radiated power or peak radiation flux in external pipes and/or optical fibres to:

$$\text{radiated power} < 150\,\text{mW}$$
$$\text{peak radiation flux} < 20\,\text{mW/mm}^2$$

incapable of igniting firedamp or an explosive coal dust cloud in the event of damage to the pipe or fibre

- cells or batteries shall meet the requirements of EN 50020 – intrinsic safety, category ia – in full. Cells or batteries shall, when recharged in an explosive atmosphere, be fed from an intrinsically safe circuit, category ia according to EN 50020, only
- the use of two independent types of protection:

1 Using an outer enclosure with free space inside, the outer enclosure shall be:

a flameproof enclosure – d – /EN 50018

or

a pressurized enclosure – p – /EN 50016,
with leakage compensation by an inert gas or air

Inside these enclosures there shall be used electrical apparatus or components complying with:

EN 50017 'Powder filling – q'
EN 50019 'Increased safety – e'

EN 50020 'Intrinsic safety – i'
EN 50028 'Encapsulation – m'

In the case of a pressurized enclosure outside, a flameproof enclosure according to EN 50018 may by used in addition to those listed above.

2 Using an outer enclosure with no free space inside, the outer enclosure shall be:

with encapsulation – m – /EN 50028, or
with powder filling – q – /EN 50017

Inside these enclosures there shall be used electrical apparatus or components complying with:

EN 50019 – 'Increased safety – e –'
EN 50020 – 'Intrinsic safety – i –'
EN 50028 – 'Encapsulation – m –'

- ingress protection against combustible dust and water assured by a protection degree of (at least) IP 54 according to IEC 60529 (or EN 60529)
- instead of two independent types of protection, equipment may be used which is safe with two faults and complying with the types of protection listed above. At the present time, only intrinsically safe equipment of category ia according to EN 50020 meets this requirement. Thus, intrinsically safe equipment may be used
- generally, the amount of equipment remaining functional in an explosive atmosphere shall be limited to an absolute minimum, e.g. environmental monitoring, communications and remote control systems, gas detectors, detection and prevention systems.

The marking of M1 equipment goes along with the principles laid down in Chapter 5, with the 'special marking' in the case of using two independent types of protection. The 'outer' protection is indicated first, followed by the 'internal' protection, e.g.:

- encapsulated apparatus in a flameproof enclosure EEx d/EEx m
- intrinsically safe circuits 'ib' inside a pressurized enclosure EEx p/EEx ib.

It should be emphasized that underground coal mines are not classified in 'zones'. So EN 50303 is actually not a zone 0 standard.

2.4   Explosion protection for zone 2

As a national predecessor, in Germany the standard VDE 0165 (1991-02) contains installation rules for zone 2 (and, in addition, for zone 0 and 1, zone 10 and 11). The zone 2 concepts are summarized as follows:

- general requirements for all types of explosion protected apparatus, comprising the protection against the ingress of dust and water (Fig. 2.3)

Electrical apparatus and installations for zone 2:

• Zone 0 apparatus
• Zone 1 apparatus
• General requirements for zone 2 apparatus:

IP-degree according to IEC 60529

Outdoor installation	Indoor installation	Built-in-components
≥IP 54	≥IP 40	– with bare live parts
≥IP 44	≥IP 20	– with insulated live parts

• Specific requirements for non-sparking and non-arcing apparatus complying with specified temperature limits

• Protection by specified enclosures for sparking and arcing apparatus, exceeding specified temperature limits

---

Figure 2.3   *The zone 2 concept according to VDE 0165/1991-02. General requirements for all types of electrical apparatus.*

Specific requirements for zone 2 application:

Non-sparking and non-arcing apparatus within specified temperature limits

• Squirrel-cage motors
  ≥IP 44 for outdoor installation
  ≥IP 20 for indoor installation
• Terminal boxes: ≥IP 54
• Oil and dry-type transformers: ≤6 kV
• Stationary luminaires: ≥IP 54
• Movable luminaires: zone 1 and ≥IP 54
• Plug and socket connectors: interlocked
  (no connection/disconnection when energized)
• Fuses with an enclosed fuse link

Figure 2.4   *The zone 2 concept according to VDE 0165/1991-02. Requirements for electrical apparatus without arcs or sparks in normal operation and/or temperatures not exceeding a certain temperature limit, e.g. complying with a given temperature class.*

• apparatus operating without arcs or sparks in normal operation and within a defined temperature range, e.g. transformers, cage induction motors, fuses, sockets and plugs and light fittings have to fulfil specific requirements (Fig. 2.4). Permanently installed light fittings shall have a type of protection against dust or water IP 54 according to IEC 60529, transportable light fittings shall comply with the standards for zone 1

• apparatus with sparking or arcing components in normal operation, e.g. contactors and circuit breakers, or with parts exceeding a defined temperature limit, e.g. heaters, can be protected by an enclosure similar to the 'restricted breathing enclosure' (in German: 'schwadensicheres Gehäuse') or by a pressurized enclosure (Fig. 2.5). 'Schwadensichere Gehäuse' have to pass a type test for tightness: the pressure decrease

> **Protection by specified enclosures in zone 2:**
>
> Sparking and arcing apparatus, exceeding specified temperature limits
>
> - 'schwadensicheres Gehäuse' (similar to a restricted breathing enclosure)
>   - IP-degree $\geqslant 54$
>   - Δp from 400 to 200 Pa within $\Delta t \geqslant 30$ s
> - 'simple' pressurization
>   - no purging procedure required
>   - $p < p_{min}$: alarm

Figure 2.5   *The zone 2 concept according to VDE 0165/1991-02. Requirements for electrical apparatus with arcs or sparks in normal operation and/or temperatures exceeding a defined limit/temperature class. Two types of protection are defined here: the 'schwadensicheres Gehäuse', comparable with a 'restricted breathing enclosure', and 'simple pressurization'.*

from 400 Pa to 200 Pa shall exceed 30 seconds. For pressurized enclosures purging before energizing the internal components is not required. If the internal overpressure falls below the specified minimum overpressure (as stated by the manufacturer), an alarm may be initiated. There is no need for an automatic de-energizing of internal electric parts in such an event.

Installation rules for hazardous areas are given in international standards, IEC 60079-14 (1996-12), subsequent in EN 60079-14 (1997-08), and have been transformed into the German Standard VDE 0165 Teil 1 (1998-08), covering zones 0, 1 and 2. The basic ideas for the installations in zone 2 are:

- the installation of explosion protected electrical apparatus suitable for zone 0 and zone 1
- application of pressurization techniques. Compared with the requirements for zone 1 pressurization (according to IEC 60079-2 or EN 50016), the purging procedure is no longer obligatory
- application of electrical apparatus in type of protection 'n' (deviated from the preceding term 'non-sparking') according to IEC 60079-15. This standard does not define a precisely stated type of protection, as it is the practice in the zone 1 standards, but gives very general requirements to exclude any source of ignition. This is done by a sufficient protection against the ingress of dust and water into the enclosure, by a certain mechanical strength of the enclosure, by a suitable construction of cable entries and by prescribing the clearances and creepages between conducting parts within the enclosure. For some special types of apparatus, such as motors, fuses, light fittings, plugs and sockets, IEC 60079-15 gives additonal constructional requirements to ensure that this group of apparatus will be 'non sparking' in normal operation.

For electrical apparatus outside of this scope, e.g. switchgear, IEC 60079-15 defines a protection technique 'restricted breathing enclosure'.

This concept stands for enclosures tight enough to allow only a very restricted interchange by gas and air respectively between the interior of the enclosure and the surrounding atmosphere. The tightness required may be illustrated by the appropriate type test: an internal overpressure of 400 Pa (with respect to the surrounding atmosphere) shall decrease to 200 Pa in a time interval not shorter than 80 seconds.

A certain disadvantage of this protection concept is that internal sources of flammable gases (e.g. gas analysers) are excluded.

Other protection concepts defined in IEC 60079-15 are:

- enclosed break device, incorporating electrical contacts and which will withstand an internal explosion of the flammable gas (or vapour) without any damage and avoiding the internal explosion to enter the surrounding atmosphere. This type of protection is limited to 660 V and 15 A
- non-incendive components, incorporating electrical contacts (in potentially incendive circuits), where either the contacts or the contacting mechanism or the enclosure in which the contacts are housed are so constructed that this component prevents ignition of the surrounding flammable atmosphere. This type of protection is limited to 250 V and 15 A
- hermetically sealed device, a device sealed by fusion (e.g. application of heat) of metals or metals and glass and not intended to be opened in service. An external atmosphere cannot gain access to the interior
- sealed device, so constructed that they cannot be opened in normal service, with a free internal volume (limited to 100 cm^3). Resilient gasket seals, poured seals and potting compounds are not excluded. The interior is sealed against entry of an external atmosphere
- energy limited apparatus and circuits in which no arc, spark or thermal effect is capable of causing ignition of a flammable atmosphere (under conditions given in this standard). IEC 60079-15 defines 'subdivisions' of non-sparking apparatus:

  nA = non-sparking apparatus
  nC = protection concepts described above, with the exclusion of restricted breathing enclosures
  nR = restricted breathing enclosures

  Examples for marking according to IEC 60079-15:

  Ex nA II T6, Ex nC IIB T3, Ex nR II T4 (see Chapter 4 for the classification of Group II into IIA, IIB and IIC).

The 'n' concept has become a European Standard: EN 50021 (1999-04) and subsequently a German Standard, VDE 0170/0171 Teil 16/2000-02.

EN 50021 shall be considered with EN 50014 in parallel. It should be emphasized that 'n' apparatus are not intended for zone 1 application. EN 50021 applies to Group II apparatus, category 3G, and this indicates zone 2 application.

In a very similar way to IEC 60079-15, EN 50021 contains constructional requirements, e.g. adequate tightness against the ingress of dust and water, creepage distances and clearances, radial gaps (between rotor and stator) of motors and generators, for light fittings and their lamps and sockets, and for cells and batteries. Another main field covered by the content of EN 50021 are type and routine tests (mechanical tests, explosion tests and spark tests with defined gas–air mixtures similar to the tests applied for flameproof apparatus or intrinsically safe circuits).

EN 50021 defines 'subdivisions' of 'n':

- nA for non-sparking apparatus
- nC for sparking apparatus with an adequate protection of contacts, with the exclusion of
  - restricted breathing enclosures
  - limited energy apparatus
  - simple pressurization
- nR for restricted breathing enclosures
- nL for limited energy apparatus
- nP for apparatus with 'simple pressurization'.

These 'subdivisions' and their temperature class shall be marked on the apparatus e.g.:

EEx nA II T3
EEx nP II T1 or
EEx nL IIB T4
EEx nC IIC T5, using the same groupings A, B or C for the different gases as applied for zone 1 equipment (see Chapter 4).

The application of EN 50021 for zone 2 equipment may be considered as a certain move towards zone 1 equipment accompanied by detailed constructional requirements and type testing procedures.

2.5   Explosion protection for electrical apparatus in the presence of combustible dusts

Similar to installation rules, national standards for explosion protection in the presence of combustible dusts were established long before international standards. In Germany, VDE 0165/1991-092 (and this standard referring in parts to VDE 0170/0171 Teil 13/1986-11) defines requirements for electrical equipment operating in areas hazardous due to combustible dusts.

At the present time, IEC 61241-3 combined with IEC 61241-1-1 and with IEC 61241-1-2 as international standards describe construction and testing of electrical apparatus for areas hazardous due to combustible dusts on the basis of a three-zone classification (zones 20, 21 and 22).

Following the IEC activities, CENELEC, Brussels, has drawn up standards covering construction, testing, selection, installation and maintenance of such apparatus based on a three-zone classification:

- EN 50281-1-1 (1998-09) (construction and testing)
  overtaken as German Standard VDE 0170/0171 Teil 15-1-1/1999-10
- EN 50281-1-2 (1998-09) (selection, installation and maintenance)
  overtaken as German Standard VDE 0165 Teil 2/1999-11.

> **Note:**
>
> An additional standard is EN 50281-2-1/1998-09, overtaken as German Standard VDE 0170/0171 Teil 15-2-1/1999-11 (Test methods – Methods for determining the minimum ignition temperatures of dust).

Speaking in the classification scheme of ATEX 100a Directive, these standards cover equipment Group II, categories 1D, 2D and 3D. It may be expected that CENELEC will complete this set of standards in the future.

The following is based on IEC 61241-1-1 and IEC 61241-1-2, which at present form a complete set of standards.

Recalling Chapter 1, explosion protection in the presence of combustible dusts (stirred-up dusts as a cloud or dust layers/deposits on the surface of an electrical apparatus) requires:

- the maximum surface temperature of the apparatus – referred to a maximum ambient temperature of normally +40°C – shall have a certain distance to the ignition temperature T_i of stirred-up dusts or to the glow temperature T_G of dust deposits (for zones 20, 21 or 10 respectively, and for zone 22 with conductive dust)
- the enclosure of the apparatus shall be dust-tight or dust protected (for zones 22 or 11 respectively). Arcing or sparking components are 'encapsulated' in this enclosure as well as parts with temperatures exceeding the margin of safety to the ignition or glow temperatures of the dust
- the surface area of enclosures made of plastics shall be limited to a certain value in order to avoid the generation and accumulation of electrical charges due to rapidly moving dust–air mixtures, or, as an alternative, the resistance of the surface material shall be below 10^9 Ω to ensure an 'earthing in an electrostatic sense'
- the content of magnesium and titanium in light alloys as an enclosure material shall be limited to a certain value (e.g. 6 per cent) to avoid sparking due to frictional effects or impacts.

Before dealing with the different methods to achieve the safety requirements given above, the thermal relations at a hot apparatus surface covered by a dust layer shall be considered. The glow temperatures T_G given in Table 1.4 are valid for a layer thickness of 5 mm. An increasing layer thickness will adversely affect the heat transfer to the surrounding atmosphere, and in consequence, the glow temperature decreases with increasing layer thickness (see Fig. 2.6).

So, two different methods (called 'practice A' and 'practice B' in IEC 61241-1-2) may be used to define the maximum allowable surface temperature of an electric apparatus and to ensure its dust ingress protection:

A The maximum surface temperature shall be measured under dust-free conditions; this temperature T_S shall be:

$$T_S \leqslant 2/3 T_i$$
(T_i = ignition temperature of a dust cloud)

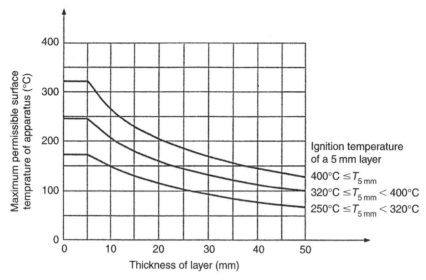

Figure 2.6 *Surface temperature derating diagram for dust layers with a thickness exceeding 5 mm according to practice A/IEC 61241-1-2. For dust layers ≤5 mm, the surface temperature T$_s$ shall be limited to T$_G$ − 75 K (T$_G$ = glow temperature for a 5 mm dust layer). The surface temperature of an electrical apparatus normally refers to an ambient temperature of +40°C.*

and $T_S \leqslant T_G - 75\,\text{K}$
$(T_G = \text{glow temperature of a 5 mm dust layer})$

T_S shall fulfil *both* conditions, i.e., the lowest value is relevant, and normally refers to +40°C ambient temperature.

Where there is a probability that dust layers may be present in excess of 5 mm thickness, T_S shall be reduced. A guideline is given in Fig. 2.6.

B The maximum surface temperature shall be measured with the apparatus covered with the maximum amount of dust that it can retain. This temperature T_S shall be:

$T_S \leqslant 2/3 T_i$
$(T_i = \text{ignition temperature of a dust cloud})$

and $T_S \leqslant T_{G\,12.5\,\text{mm}} - 25\,\text{K}$
$(T_{G\,12.5\,\text{mm}} = \text{glow temperature of a 12.5 mm dust layer}).$

As stated in practice A, T_S shall fulfil both conditions and normally refers to +40°C ambient temperature.

In the same way, two different methods are possible to ensure the dust-tightness or the dust protection of the enclosure of the apparatus:

A A type test according to IEC Standard 60529, category 1 (using artificial depression by a ventilator/vacuum pump) for the types of protection IP 5X or IP 6X

B A type test under a circulating dust–air mixture; the apparatus to be tested is operated at rated load/no load periodically for at least
- six cycles for dust-tight apparatus
- two cycles for dust protected apparatus,

    each period long enough to attain the maximum temperature/the ambient temperature after cooling down. No visible dust shall have entered the enclosure.

The thermal and tightness/protection tests according to practices A and B must not be cross-correlated.

The requirements for explosion protected apparatus suitable for combustible dusts are summarized in Table 2.5.

Table 2.5   Dust explosion protected electrical apparatus

Standard	IEC 61241-1-1 IEC 61241-1-2		VDE 0165/1991-02	
Zone	20, 21 22 with conductive dust	22	10^1	11
1 Thermal aspects				
Practice	A $T_S \leqslant 2/3T_i$ *and* $T_S \leqslant T_G - 75\,\mathrm{K}$	A	A $T_S \leqslant 2/3T_i$ *and* $T_S \leqslant T_G - 75\,\mathrm{K}$	A
Practice	B $T_S \leqslant 2/3T_i$ *and* $T_S \leqslant T_{G\,12.5} - 25\,\mathrm{K}$	B	B not allowed – –	
2 Enclosure tightness/protection				
Practice	A IP 6X	A IP 5X	A IP 65^2	A IP 54^3
Practice	B thermal cycle test $\geqslant$6 cycles Specific requirements for: joints, rods, spindles, shafts, bolts	B $\geqslant$2 cycles –	B not allowed –	

(continued)

Table 2.5   *(continued)*

Standard	IEC 61241-1-1 IEC 61241-1-2	VDE 0165/1991-02
3 Electrostatics	surface area – $\leqslant 100\,cm^2$ *or* thickness $\geqslant 8\,mm$ of the external insulation on metal parts *or* surface resistance $\leqslant 10^9\,\Omega$	surface – resistance $\leqslant 10^9\,\Omega$
4 Marking (dust explosion protection related)	DIP[4]/Practice/Zone DIP A 20 DIP B 20 DIP A 21 DIP A 22 DIP B 21 DIP B 22 maximum surface temperature[5]	St Ex – Zone 10 maximum surface temperature IP degree/Approval/ Certificate number[6]

1 For zone 10 apparatus, VDE 0165 refers to VDE 0170/0171 Teil 13 (1986-11)
2 The type of dust protection according to IEC 60529 may be reduced to IP 20 for parts in intrinsically safe circuits
3 No additional requirements for parts in intrinsically safe circuits
4 DIP = Dust Ignition Protection
   Zone 22 (hazardous due to *conductive* dusts) shall be considered as zone 21 when selecting electrical apparatus
5 The maximum surface temperature may be marked as a temperature value or as a temperature class according to IEC 60079-0 (or both)
6 The certifying body for Zone 10 apparatus in Germany was BVS, Bergbau-Versuchsstrecke, Dortmund

As an example of dust explosion protected apparatus standard IEC 61241-4, 1st edition 2001–3, types of protection 'pD' describes a pressurization technique: the interior of this apparatus is subjected to a continuous pressure from a supply of air (not containing any combustible substances) or other inert gases while electrical components within the enclosure are energized.

The basic safety philosophy is as follows:

• the pressure differential (referring to the pressure of the surrounding atmosphere) is maintained alternatively by static pressurization or by a continuous flow of air or inert gas from a reliable supply source

- pressurization failure automatically initiates an alarm and/or disconnects the apparatus from the electric power supply
- in the case of pressurization with continuous flow, the protective gas may be discharged (after passing the explosion protected apparatus) into a non-hazardous area or into a hazardous area; in this case, a preventing system shall retain hot particles (under normal or fault conditions) in the interior of the enclosure.

A certain disadvantage of this protection technique is the fact that the enclosure (and its components inside) shall be cleaned generally before the apparatus is energized, in order to remove any deposits of combustible dust inside, either following failure of pressurization or following normal shutdown.

Chapter 3

Standards for electrical apparatus and systems in zone 1

3.1 National standardization and approvals

In most industrial countries and those at the threshold of industrialization, the field of explosion protection for electrical apparatus and systems is subject to legislation. The usual procedure, that a manufacturer declares the conformity of a product with a given standard under his own responsibility, is altered in such a way that an administrative body nominated by the government of a state, or, in states with a federal structure, nominated by the authority of a land, issues an approval which declares the compliance with a standard and with legislative acts in parallel. On this basis, the apparatus may be commissioned and put into service.

The administration body often refers to an expert's appraisement and renounces for type testing and inspection. These activities are carried out by certifying bodies, state-owned facilities or privately owned companies.

As an example of such a practice, there may be the situation in Germany before transformating the European Directives 76/117/EEC, concerning explosion protection, and 82/130/EEC, concerning firedamp protection into national legislation. The relevant standard was VDE 0170/0171/1961-02, with amendment VDE 0170/0171 d/1965-02 and with amendment VDE 0170 f/0171 f/1969-01, quoted simply as VDE 0170/0171/1969-01 throughout this book.

Until ElexV [8] and ElZulBergV [9] came into force – ElexV dated 1980-07-01 and ElZulBergV 1984-01-01 – firedamp-proof and explosion protected electrical apparatus had to pass a two-step procedure in order to come into operation:

- certification for compliance with VDE 0170/0171/1969-01 by a certification body. Firedamp-proof apparatus (sometimes in combination with explosion protection) have been certified by BVS, Bergbau-Versuchsstrecke, Dortmund, and explosion protected apparatus by PTB, Physikalisch-Technische Bundesanstalt, Braunschweig and Berlin
- on the basis of such a certificate, an approval was issued by administrative bodies nominated by the different lands in the Federal Republic of Germany.

Here, many apparatus complying with VDE 0170/0171/1969-01 are still in operation. Therefore, this standard shall be considered more in detail. Table 3.1 summarizes the different types of protection, and Table 3.2 demonstrates the

Table 3.1   Types of explosion protection and marking according to VDE 0170/0171/1969-01

Type of protection	Firedamp-proof	Explosion protected
General marking	(Sch)	(Ex)
Flameproof enclosure	d, dz[1]	d_[2]
Increased safety	e	e
Pressurization (with continuous flow only)	f	f
Oil immersion	o	o
Intrinsic safety	i	i
'Plattenschutzkapselung' (protection by flame arrestors)	p[3]	–
Special protection[4]	s	s

1 Type of protection (Sch)dz indicates that this flameproof enclosure has passed successfully a special type test to avoid the propagation of an explosion inside the enclosure initiated by high current fault arcs to the surrounding atmosphere. This test has been prescribed by a decree of the Supreme Authority in Mines.

Test conditions
The ignition source shall be a single-phase 50 cps circuit with 500 V rms no-load voltage and 5 kA short-circuit current. The arc is initiated by a thin copper wire across copper electrodes. The arc duration shall exceed 80 ms.
   The test gas (inside the enclosure to be tested and in the surrounding atmosphere) is a mixture of

8% (v/v) natural gas (mainly methane, CH_4)
12% (v/v) hydrogen, H_2
80% (v/v) air

with a pressure of $1.0 \cdot 10^5$ Pa (absolute)
MESG = 0.62 mm
30 tests shall be carried out

Test acceptance criteria
No transmission of the internal explosion into the surrounding atmosphere shall occur at all 30 tests

2 The dash stands for an additional marking according to Table 3.2(b) as a classification with respect to the MESG values of gas–air mixtures
3 This mark should not be confused with the mark 'p' for 'pressurization' in IEC or European Standards (EN)
4 'Special protection' comprises all protection techniques such as powder filling, static pressurization and encapsulation (by plastics materials). At that time, these techniques started their 'career' in the field of explosion protection and advanced to autonomous types of protection described by IEC or European Standards individually. VDE 0170/0171/1969-01 does not contain any specific requirements for these techniques. It has been the decision of the certifying body that an 's'-apparatus will operate at the same level of safety compared with other well-defined types of protection, e.g. flameproof enclosure 'd'

Table 3.2(a)   VDE 0170/0171/1969-01: Classification of explosion protected electrical apparatus according to the ignition temperature of combustible gas–air or vapour–air mixtures. Maximum surface temperatures of such apparatus

Ignition temperature T_i *in* °C	*Ignition group*	*Maximum surface temperature*[1,2] °C
Firedamp-proof (Sch)	–	200[3]
Explosion protected (Ex)		
$T_i > 450$	G1	360
$300 < T_i \leqslant 450$	G2	240
$200 < T_i \leqslant 300$	G3	160
$135 < T_i \leqslant 200$	G4	110
$100 < T_i \leqslant 135$[4]	G5	80

1 Referring to an ambient temperature of normally +40°C, the maximum temperature rise of the apparatus is given by these values minus 40°C

2 For reasons of safety, the maximum surface temperature is limited generally to 80% referring to the lower limit of the ignition group

3 Despite the high ignition temperature of methane–air mixtures of 595°C/650°C (see Table 1.3), the maximum surface temperature is subjected to a very strong limitation with respect to the unavoidable presence of coal dust with relatively low glow temperatures (see Table 1.4)

4 Lower ignition groups have not been installed with respect to the fact that carbon disulphide, CS_2, shows the lowest known ignition temperature of 95°C/102°C (see Table 1.3)

Table 3.2(b)   VDE 0170/0171/1969-01: Classification of explosion protected apparatus according to the MESG values of combustible gas–air or vapour–air mixtures

Application	*MESG* mm	*Explosion class*
Firedamp-proof	CH_4: 1.14 (1.17)	(Sch) d, (Sch) dz
Explosion protected	MESG > 0.60	(Ex) d1
	$0.40 < MESG \leqslant 0.60$	d2
	$MESG \leqslant 0.40$	d3
	hydrogen, water gas	d3 a
	carbon disulphide	d3 b
	acetylene	d3 c
	all gases/vapours with $MESG \leqslant 0.40$[1]	d3 n

1 Lower explosion classes have not been installed with respect to the fact that hydrogen, H_2, shows the lowest known MESG value of 0.29 mm (see Table 1.6)

classification of such apparatus according to the ignition temperatures of the different gas–air mixtures and their MESG values.

The marking system according to VDE 0170/0171/1969-01 comprises:

first mark:	(Sch)	firedamp-proof
	(Ex)	explosion protected
second mark:		type of protection, explosion class for flameproof enclosures
third mark:		ignition group (for explosion protected apparatus only)

Examples:

1 low voltage switchgear, flameproof enclosure, for coal mines	(Sch)dz
2 series reactor for tubular fluorescent lamp, powder filled, for chemical plants with ethyl ether	(Ex)s G4
3 light fitting with incandescent lamp, increased safety, for coal mines	(Sch)e
4 low voltage switchgear, flameproof enclosure, for chemical plants with ethylene	(Ex)d2 G2
5 high voltage circuit breaker, enclosure and terminal boxes in increased safety, control box as a flameproof enclosure, oil blast circuit breaker, for coal mines	(Sch)eod
6 high power cage induction motor protected by an air flow with overpressure, standstill heaters in flameproof enclosures, for chemical plants with acetylene	(Ex)f d3n G2
7 pressure transducer in an intrinsically safe circuit, for coal mines	(Sch)i
8 pressure transducer in an intrinsically safe circuit, for chemical plants, for all gases (including hydrogen and carbon disulphide)	(Ex)i G5
9 cable connecting box, increased safety, for refinery installation (petrol)	(Ex)e G3

In the former German Democratic Republic (GDR), national standards (TGL 'Technische Güte- und Lieferbedingungen', Technical Regulations for Quality and Delivery) have defined the classification of hazardous areas (TGL 30042) and the requirements for construction and testing of explosion protected electrical equipment (mainly TGL 19491).

TGL 30042 has defined a four-zone concept for areas hazardous due to combustible gases, vapours and mists: EG 1, comparable with zone 0; EG 2 and EG 3, covering zone 1; EG 4, comparable with zone 2. In contradiction to the majority of standards covering the zone 1 requirements with all types of protection such as flameproof enclosure, increased safety, pressurization, oil immersion, intrinsic safety, powder filling, EG 2 excludes the general application of, e.g., increased safety 'e', which has been the domain of EG 3.

The certifying body and testing house has been the IfB, Institut für Bergbausicherheit, Leipzig, Institutsbereich Freiberg/Germany.

3.2 Standardization in Europe – European Standards and Directives

3.2.1 European Standards by CENELEC

In the course of the integration of European states and the removal of trade barriers, national standardization loses importance considerably. The majority of electrical standards comes into existence by international cooperation with IEC, International Electrotechnical Commission, Geneva/Switzerland. On the basis of IEC Standards many European Standards have been developed in the past. This work has been done by CENELEC, Comité Européen de Normalisation Electrotechnique, Brussels/Belgium. European Standards or Norms (EN) often deviate from their IEC origins in their technical content. To avoid double action and to ensure a more effective standardization procedure, IEC and CENELEC are now working in closer cooperation. CENELEC is a private organization and does not act as an institution of the European Union at all. This fact indicates that European Standards need a legal act to come into force. Members of CENELEC are the national committees for standardization of the member states (not identical with the European Union).

Table 3.3 summarizes the members and affiliates of CENELEC.

Table 3.3(a)   Members of CENELEC (dated 1999-04)

State	*Short form*	*Member committee*
Austria	AT	ÖVE
		Österreichischer Verband für Elektrotechnik
		Wien
Belgium	BE	CEB
		Comité Electrotechnique Belge
		BEC
		Belgisch Elektrotechnisch Comité
		Brussels
Czech Republic	CZ	Czech Standards Institute
		CSNI
		Praha
Denmark	DK	DS
		Dansk Standard Electrotechnical Sector
		Charlottenlund
Finland	FI	SESKO
		Finnish Electrotechnical Standards
		Association
		Helsinki
France	FR	UTE
		Union Technique de l'Electricité
		Fontenay-aux-Roses

(continued)

Table 3.3(a)   *(continued)*

State	Short form	Member committee
Germany	DE	DKE Deutsche Elektrotechnische Kommission im DIN und VDE Frankfurt
Greece	GR	ELOT Hellenic Organization for Standardization Athens
Iceland	IS	STRI The Icelandic Council for Standardization Technological Institute of Iceland Reykjavik
Ireland	IE	ETCI Electro-Technical Council of Ireland Dublin
Italy	IT	CEI Comitato Elettrotecnico Italiano Milano
Luxembourg	LU	SEE Service de l'Energie de l'Etat Luxembourg
Netherlands	NL	NEC Nederlands Elektrotechnisch Comité Delft
Norway	NO	NEK Norsk Elektroteknisk Komite Oslo
Portugal	PT	IPQ Instituto Portugues da Qualidade Monte da Caparica
Spain	ES	AENOR Asociación Española de Normalización y Certificación Madrid
Sweden	SE	SEK Svenska Elektriska Kommissionen Kista Stockholm
Switzerland	CH	CES Swiss Electrotechnical Committee Fehraltorf
United Kingdom	GB	BEC British Electrotechnical Committee BSI British Standards Institution London

Table 3.3(b)   Affiliates (dated 1999-04)

State	Committee
Bulgaria	Committee for Standardization and Metrology at the Council of Ministers Sofia
Croatia	DZNM State Office for Standardization and Metrology Zagreb
Cyprus	Cyprus Organization for Standards and Control of Quality Nicosia
Estonia	Estonian Electrotechnical Committee Tallinn
Hungary	MSZH Hungarian Office of Standardization Budapest
Lithuania	Lithuanian Standards Board Vilnius
Poland	PKN Polski Komitet Normalizacyjny Warszawa
Romania	Romanian National Committee for IEC c/o Romanian Institute for Standardization Bucharest
Slovakia	Slovak Office of Standards, Metrology and Testing Bratislava
Slovenia	Standards and Metrology Institute Ljubljana
Turkey	TSE Turkish Standards Institution Bakan Liklar

New affiliates (dated 2000-03)
Bosnia and Herzegovina
Latvia

In CENELEC, different technical committees cover the field of electrotechnology and electronics, such as rotating machinery, communication, signalling and processing systems, high voltage switchgear and controlgear, power electronics, europlug and socket outlets, electrical apparatus for explosive atmospheres insulated bushings, electrical equipment in medical practice, optical radiation safety and laser equipment, EMS products and electronic design automation – only a very short survey of CENELEC activities.

This chapter will focus on the field of explosion protection, covered by Technical Committee TC 31 and its subcommittees (see Table 3.4).

Table 3.4 CENELEC TC 31 and subcommittees (SC) (dated 1999-04)

Committee	Field of activity
TC 31	Electrical apparatus for explosive atmospheres – general requirements
SC 31-1	Installation rules
SC 31-2	Flameproof enclosures 'd'
SC 31-3	Intrinsically safe apparatus and systems 'i'
SC 31-4	Increased safety 'e'
SC 31-5	Apparatus type of protection 'n'
SC 31-6	Encapsulation 'm'
SC 31-7	Pressurization and other techniques
SC 31-8	Electrostatic painting and finishing equipment
SC 31-9*	Electrical apparatus for the detection and measurement of combustible gases to be used in industrial and commercial potentially explosive atmospheres

*TC 216 activities cover industrial and commercial gas detectors not intended for the use in potentially explosive atmospheres

Starting in the 1970s, these committees worked out European Standards for explosion protected electrical apparatus and systems and for electrostatic spraying equipment. Each type of protection (this means a specific technique to ensure that an electrical apparatus cannot ignite a surrounding combustible atmosphere) is covered by a separate standard. This standard refers to nearly all types of equipment, e.g. EN 50018 as the standard for flameproof enclosure covers motors, transformers, switchgear, light fittings, electronic components and analysers in parallel. EN 50014 defines the 'general requirements' for all types of protection, e.g. choice of enclosure materials, classification according to the highest enclosure surface temperature, requirements for cable entries, surface conductivity for plastics and other common details. It is noteworthy that these standards are not specific for a special type of apparatus, e.g. motors, switchgear or electronic components.

Following the progress in electrical engineering, the basic standards have been extended by a certain number of amendments (up to five) all forming the so-called 1st edition of standards. After a general revision and incorporating the technical content of the amendments, the standards were issued as the 2nd edition in the early 1990s. Table 3.5 gives a survey of the European Standards drawn up by CENELEC.

To come into force, one method is for these standards to be transformed into national standards. Table 3.5 lists the German Standards (VDE). Obviously, this procedure causes a certain time delay (1–2 years) due to national acceptance.

Another method – common to all members of the European Union – has been to declare the CENELEC Standards as Harmonized European Norms (HN) on the basis of a directive issued by an authority of the European Union (council or commission). This procedure is described in Section 3.2.2.

Table 3.5 Explosion protection – European Standards drawn up by
CENELEC (draft standards pr EN … not included) (dated 2000-03-31)

Number	Title/Amendment	Edition	Date	German Standard	
				Edition	*Date*
Part A Electrical apparatus for potentially explosive atmospheres					
EN 50014	General requirements	1st	1977-03	1st	VDE 0170/ 0171 Teil 1 1978-05
	Amendment 1		1979-07		A1 1980-09
	Amendment 2		1982-06		A2 1984-05
	Amendment 3		1982-12		A3 1984-05
	Amendment 4		1982-12		A4 1984-05
	Amendment 5		1986-02		A5 1987-01
	General requirements	2nd	1992-12	2nd	1994-03
	General requirements	3rd	1997-06		
	corrigendum		1998-04		
	Amendment 1		1999-02		
	Amendment 2		1999-02		
				3rd	2000-02
EN 50015	Oil immersion 'o'	1st	1977-03	1st	VDE 0170/ 0171 Teil 2 1978-05
	Amendment 1		1979-07		A1 1980-09
	Oil immersion 'o'	2nd	1994-04	2nd	1995-01
	Oil immersion 'o'	3rd	1998-09	3rd	2000-02
EN 50016	Pressurized apparatus 'p'	1st	1977-03	1st	VDE 0170/ 0171 Teil 3 1978-05
	Amendment 1		1979-07		A1 1980-09
	Pressurized apparatus 'p'	2nd	1995-10	2nd	1996-05
EN 50017	Powder filling 'q'	1st	1977-03	1st	VDE 0170/ 0171 Teil 4 1978-05
	Amendment 1		1979-07		A1 1980-09
	Powder filling 'q'	2nd	1994-04	2nd	1995-02
	Powder filling 'q'	3rd	1998-07	3rd	2000-02
EN 50018	Flameproof enclosure 'd'	1st	1977-03	1st	VDE 0170/ 0171 Teil 5 1978-05
	Amendment 1		1979-07		A1 1980-09
	Amendment 2		1982-12		A2 1984-05
	Amendment 3		1985-11		A3 1987-01
	Flameproof enclosure 'd'	2nd	1994-08	2nd	1995-03

(continued)

Table 3.5   *(continued)*

Number	Title/Amendment	Edition	Date	German Standard		
				Edition	Date	
EN 50019	Increased safety 'e'	1st	1977-03	1st	VDE 0170/ 0171 Teil 6 1978-05	
	Amendment 1		1979-07		A1	1980-09
	Amendment 2		1983-09		A2	1984-07
	Amendment 3		1985-12		A3	1987-01
	Amendment 4		1989-10		A4	1990-07
	Amendment 5		1990-08		A5	1992-05
	Increased safety 'e'	2nd	1994-03	2nd		1996-03
EN 50020	Intrinsic safety 'i'	1st	1977-03	1st	VDE 0170/ 0171 Teil 7 1978-05	
	Amendment 1		1979-07		A1	1980-09
	Amendment 2		1985-12		A2	1987-01
	Amendment 3		1990-05		A3	1992-03
	Amendment 4		1990-05		A4	1992-04
	Amendment 5		1990-05		A5	1992-04
	Intrinsic safety 'i'	2nd	1994-08	2nd		1996-04
EN 50028	Encapsulation 'm'	1st	1987-02	1st	VDE 0170/ 0171 Teil 9 1988-07	
EN 50033	Caplamps for mines susceptible to firedamp	1st	1986-06	1st	VDE 0170/ 0171 Teil 14 1987-04	
	Caplamps for mines susceptible to firedamp	2nd	1991-03	2nd	1992-08	
EN 50039	Intrinsically safe electrical systems 'i'	1st	1980-03	1st	VDE 0170/ 0171 Teil 10 1982-04	
Part B Electrostatic spraying equipment						
EN 50050	Electrostatic hand-held spraying equipment	1st	1986-01	1st	VDE 0745 Teil 100 1987-01	
EN 50053 Part 1	Hand-held electrostatic paint spray guns with an energy limit of 0.24 mJ and their associated apparatus	1st	1987-02	1st	VDE 0745 Teil 101 1987-12	
EN 50053 Part 2	Hand-held electrostatic powder	1st	1989-06	1st	VDE 0745 Teil 102	

(continued)

Table 3.5 *(continued)*

Number	Title/Amendment	Edition	Date	German Standard	
				Edition	Date
	spray guns with an energy limit of 5 mJ and their associated apparatus				1990-09
EN 50053 Part 3	Hand-held electrostatic flock spray guns with an energy limit of 0.24 mJ or 5 mJ and their associated apparatus	1st	1989-06	1st	VDE 0745 Teil 103 1990-09

Part C Electrical apparatus for the detection and measurement of combustible gases

Number	Title/Amendment	Edition	Date	German Standard	
EN 50054	Electrical apparatus for the detection and measurement of combustible gases General requirements and test methods	1st	1991-06	1st	VDE 0400 Teil 1 1993-07
	Amendment 1 General requirements and test methods	2nd	1995-05 1998-07	2nd	A1 1996-01 1999-07
EN 50055	Electrical apparatus for the detection and measurement of combustible gases Performance requirements for Group I apparatus indicating up to 5% (v/v) methane in air	1st	1991-06	1st	VDE 0400 Teil 2 1993-07
	Amendment 1 Performance requirements for Group I apparatus indicating up to 5% (v/v) methane in air	2nd	1995-05 1998-07	2nd	A1 1996-01 1999-07
EN 50056	Electrical apparatus for the detection and measurement of combustible gases	1st	1991-06	1st	VDE 0400 Teil 3 1993-07

(continued)

Table 3.5   (*continued*)

Number	Title/Amendment	Edition	Date	German Standard	
				Edition	Date
	Performance requirements for Group I apparatus indicating up to 100% (v/v) methane in air				
	Amendment 1		1995-05		A1 1996-01
	Performance requirements for Group I apparatus indicating up to 100% (v/v) methane in air	2nd	1998-07	2nd	1999-07
EN 50057	Electrical apparatus for the detection and measurement of combustible gases Performance requirements for Group II apparatus indicating up to 100% LEL	1st	1991-06	1st	VDE 0400 Teil 4 1993-07
	Performance requirements for Group II apparatus indicating up to 100% LEL	new	1998-07	new	1999-07
EN 50058	Electrical apparatus for the detection and measurement of combustible gases Performance requirements for Group II apparatus indicating up to 100% (v/v) gas	1st	1991-06	1st	VDE 0400 Teil 5 1993-07
	Performance requirements for Group II apparatus indicating up to 100% (v/v) gas	new	1998-07	new	1999-07

3.2.2   Directives

One of the primary contents of the Treaties of the European Community (EC) is the elimination of trade barriers for technical equipment. These barriers have been based on different technical standards (and their legislation) in the member states of the EC. Referring to article 100 of the Treaties of the EC, some directives have been issued in order to harmonize the standardization for technical products (explosion protected electrical apparatus and systems, elevators). These directives describe precisely the technical requirements and list the relevant standards, in the field of explosion protection CENELEC Standards exclusively. Directive 82/130/EEC refers to mines endangered by firedamp and/or combustible dusts, and Directive 76/117/EEC covers areas hazardous due to potentially explosive atmospheres. The member states of the EC shall transform these directives into their national legislation within 18 months.

This adoption to legislation and the fact that the directives specify all the relevant standards in detail, start a complicated procedure that results in a considerable time delay. The consecutive work of CENELEC in standardization results in an increasing number of amendments to basic standards, and in order to adopt this progress in standardization, the EC authorities have been forced to issue a 'train' of directives covering the new amendments.

Table 3.6 summarizes the explosion protection directives based on article 100. Every new directive forces the EC member states to restart the adoption process to legislation.

Table 3.6   EU Directives referring to article 100 of the Treaties of the European Community, covering the harmonization of standards in the field of explosion protection

Directive	of council = cl of commission = com	Date
Part A For mines endangered by firedamp and/or combustible dusts		
82/130/EEC	cl	1982-02-15
88/35/EEC	com	1987-12-02
91/269/EEC	com	1991-04-30
94/44/EC	com	1994-09-19
98/65/EC	com	1998-09-03
Part B For areas hazardous due to combustible gases, vapours and mists		
76/117/EEC	cl	1975-12-18
79/196/EEC	cl	1979-02-06
84/47/EEC	com	1984-01-16
88/571/EEC	com	1988-11-10
90/487/EEC	cl	1990-09-17
94/26/EC	com	1994-06-15
97/53/EC	com	1997-09-11

This process generates subsequent sets of Harmonized European Norms, forming the basis for a harmonized certification procedure in all EC member states. Table 3.7 gives an overview, condensing the different standards and their amendments to 'generations' of certificates of conformity.

A certificate of conformity (there should be a supplement: 'with Harmonized European Norms') is the usual procedure for type testing and approval

Table 3.7(a)　Part A Harmonized European Norms (HN) – firedamp-proof equipment (Group I apparatus)

Type of protection/ EN Standard/ Amendment	Directive				
	82/130/EEC	88/35/EEC	91/269/EEC	94/44/EC	98/65/EC
General requirements					
1st edition					
EN 50014	$\times^1$	$\times$	$\times$	$\times$	
A1	$\times$	$\times$	$\times$	$\times$	
A2		$\times^6$	$\times$	$\times$	
A3		$\times^6$	$\times$	$\times$	
A4		$\times^6$	$\times$	$\times$	
A5			$\times$	$\times$	
2nd edition					
EN 50014					$\times$
Oil immersion					
1st edition					
EN 50015	$\times$	$\times$	$\times$	$\times$	
A1	$\times$	$\times$	$\times$	$\times$	
2nd edition					
EN 50015					$\times$
Pressurization					
1st edition					
EN 50016	$\times$	$\times$	$\times$	$\times$	
A1	$\times$	$\times$	$\times$	$\times$	
2nd edition					
EN 50016					$\times$
Powder filling					
1st edition					
EN 50017	$\times$	$\times$	$\times$	$\times$	
A1	$\times$	$\times$	$\times$	$\times$	
2nd edition					
EN 50017					$\times$
Flameproof enclosure					
1st edition					
EN 50018	$\times^2$	$\times$	$\times$	$\times$	
A1	$\times$	$\times$	$\times$	$\times$	

(continued)

Table 3.7(a)   (*continued*)

Type of protection/ EN Standard/ Amendment	82/130/EEC	88/35/EEC	91/269/EEC	94/44/EC	98/65/EC
A2		×	×	×	
A3			×	×	
2nd edition					
EN 50018					×
Increased safety					
1st edition					
EN 50019	×	×	×	×	
A1	×	×	×	×	
A2		×	×	×	
A3			×	×	
A4				×	
A5				×	
2nd edition					
EN 50019					×
Intrinsic safety					
1st edition					
EN 50020	×[3]	×[3]	×	×	
A1	×	×	×	×	
A2			×	×	
A3				×	
A4				×	
A5				×	
2nd edition					
EN 50020					×
Encapsulation					
1st edition					
EN 50028			×	×	
Caplamps					
2nd edition					
EN 50033				×	
Type of certificate[4]	CC, S	CC, S	CC, S	CC, S	CC, S
Generation[5]	(A)	B	C	D	E

1 Modified according to supplement B, annex 1 of Directive 82/130/EEC
2 Modified according to supplement B, annex 2 of Directive 82/130/EEC
3 Modified according to supplement B, annex 3 of Directive 82/130/EEC
4 CC = certificate of conformity
  S = control certificate
5 The 1st generation (A generation) is *not* subjected to the certificate marking/ numbering system. Starting with the 2nd generation (B generation) the 'B', 'C', 'D' and 'E' are included in the certificate marking/numbering system, e.g. 89.B … , or 96.D … , of the certifying body
6 Modified according to supplement B, annex 1 of Directive 88/35/EEC

Table 3.7(b)   Part B Harmonized European Norms (HN) – explosion protected equipment (Group II apparatus)

Type of protection/ EN Standard/ Amendment	Directive				
	79/196/EEC	84/47/EEC	88/571/EEC 90/487/EEC	94/26/EC	97/53/EC
General requirements					
1st edition					
EN 50014	×	×	×	×	
A1		×	×	×	
A2		×	×	×	
A3		×	×	×	
A4		×	×	×	
A5			×	×	
2nd edition					
EN 50014					×
Oil immersion					
1st edition					
EN 50015	×	×	×	×	
A1		×	×	×	
2nd edition					
EN 50015					×
Pressurization					
1st edition					
EN 50016	×	×	×	×	
A1		×	×	×	
2nd edition					
EN 50016					×
Powder filling					
1st edition					
EN 50017	×	×	×	×	
A1		×	×	×	
2nd edition					
EN 50017					×
Flameproof enclosure					
1st edition					
EN 50018	×	×	×	×	
A1		×	×	×	
A2		×	×	×	
A3			×	×	
2nd edition					
EN 50018					×
Increased safety					
1st edition					
EN 50019	×	×	×	×	

(continued)

Table 3.7(b) *(continued)*

Type of protection/ EN Standard/ Amendment	Directive				
	79/196/EEC	84/47/EEC	88/571/EEC 90/487/EEC	94/26/EC	97/53/EC
A1		X	X	X	
A2		X	X	X	
A3			X	X	
A4				X	
A5				X	
2nd edition EN 50019					X
Intrinsic safety 1st edition EN 50020	X	X	X	X	
A1		X	X	X	
A2			X	X	
A3				X	
A4				X	
A5				X	
2nd edition EN 50020					X
Encapsulation 1st edition EN 50028			X[1]	X	X
Intrinsically safe electrical systems 1st edition EN 50039			X[1]	X	X
Electrostatic hand-held spraying equipment 1st edition EN 50050			X[1]	X	X
Electrostatic paint spray guns $E < 0.24\,mJ$ 1st edition EN 50053 Part 1			X[1,2]	X[2]	X[2]
Electrostatic powder spray guns $E \leqslant 5\,mJ$ 1st edition EN 50053 Part 2			X[1,2]	X[2]	X[2]
Electrostatic flock spray guns $E \leqslant 0.24\,mJ/5\,mJ$ 1st edition					

(continued)

Table 3.7(b)   *(continued)*

Type of protection/ EN Standard/ Amendment	Directive				
	79/196/EEC	84/47/EEC	88/571/EEC 90/487/EEC	94/26/EC	97/53/EC
EN 50053 Part 3			$\times^{1,2}$	$\times^2$	$\times^2$
Type of certificate[3]	CC, S	CC, S	CC, S	CC, S	CC, S
Generation[4]	(A)	B	C	D	E

1 This is the content of Directive 90/487/EEC
2 Concerns only the clauses dealing with the construction of apparatus
3 CC = certificate of conformity
  S = control certificate
4 The 1st generation (A generation) is *not* subjected to the certificate marking/ numbering system. Starting with the 2nd generation (B generation) the 'B', 'C', 'D' and 'E' are included in the certificate marking/numbering system, e.g. 88.B ... , or 95.D ... , of the certifying body

throughout all EC member states, given in Directives 76/117/EEC and 82/130/EEC. In this procedure, a certifying body, notified by the member state to the authorities of the EC (the so-called 'notified body') carries out the type test according to the relevant standards and (assuming positive results) issues the certificate of conformity. This document ensures free trading and commissioning of the apparatus in all EC member states, without demanding any national approval, and without any reference to the place of business of the notified body.

Besides the certificate of conformity, there is a rather less common way of certification: the 'control certificate'. This relates to apparatus *not* complying with Harmonized European Norms (HN). It is the decision of a notified body to ensure the equivalent level of safety with respect to well-defined types of protection according to the HN. The notified body informs all other notified bodies after an appropriate safety assessment. If there is no contradiction within a 4-month period, the notified body issues the control certificate.

Apparatus approved by a notified body (certificate of conformity, control certificate) shall be marked by the manufacturer with an 'Ex' in a hexagon according to Fig. 3.1(a). This mark ensures the compliance of the individual apparatus with the specimen tested by the notified body and ensures that all routine tests according to the relevant Harmonized European Norms (if a certificate of conformity has been issued) or all routine tests prescribed in the control certificate have been carried out and passed successfully. According to Directive 82/130/EEC, apparatus with a control certificate shall be marked with an additional 'S' in a circle (see Fig. 3.1 (b)).

Table 3.8 gives a list of notified bodies according to Directives 76/117/EEC and 82/130/EEC, referring to article 14 (bodies issuing certificates of conformity or control certificates).

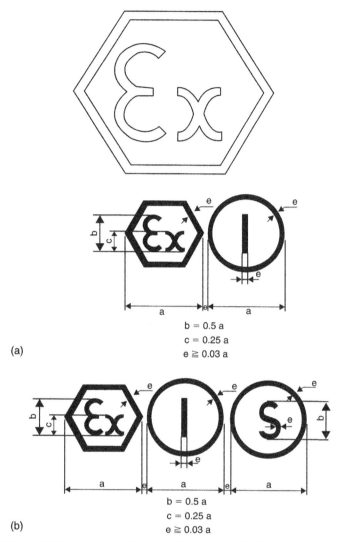

(a)

b = 0.5 a
c = 0.25 a
e ≧ 0.03 a

(b)

b = 0.5 a
c = 0.25 a
e ≧ 0.03 a

Figure 3.1 *(a) 'Ex' mark for explosion protected electrical apparatus complying with Harmonized European Norms (HN). Group I equipment shall be additionally marked with an 'I' in a circle according to Directive 82/130/EEC. (b) Marking of Group I equipment in case of a 'control certificate' according to Directive 82/130/ EEC. For Group II equipment with a 'control certificate', the 'I' in the circle shall be deleted.*

These notified bodies have introduced a type of certificate without any legal status: the 'component certificate', which is marked by a 'U' following the certificate number. This document acts as a working instrument for a simpler handling of components like conductor bushings, terminal blocks or inspection windows. The notified body issuing such a component certificate herewith declares the conformity with the relevant Harmonized European Norm and the relevant type tests. Other notified bodies accept this document without asking for renewed type tests. In this way, components can be handled in a very efficient

Table 3.8 List of notified bodies in the European Union for type testing and certification of electrical equipment for potentially explosive atmospheres

Member state	*Notified body according to Directive 82/130/EEC (article 14)*
Part A Firedamp-proof electrical equipment – Group I	
Austria	–
Belgium	ISSeP Institut Scientifique de Service Public Division de Colfontaine B-7340 Paturages
Denmark	DEMKO Danmarks Elektriske Materielkontrol DK-2730 Herlev
Finland	–

France	• Ineris
	  Institut National de l'Environnement Industriel
	  et des Risques
	  F-60550 Verneuil-en-Halatte
	• LCIE
	  Laboratoire Central des Industries Électriques
	  F-92266 Fontenay-aux-Roses

Germany	BVS Bergbau-Versuchsstrecke Fachstelle für Sicherheit Elektrischer Betriebsmittel der DMT-Gesellschaft für Forschung und Prüfung mbH D-44329 Dortmund
Greece	–
Ireland	–
Italy	CESI Centro Elettrotecnico Sperimentale Italiano I-20134 Milano
Luxembourg	–
Netherlands	–
Portugal	–
Spain	LOM Laboratorio Oficial José Maria Madariaga E-28003 Madrid
Sweden	–

United Kingdom	• HSE (M) (MECS)
	  Health and Safety Executive (MINING)
	  Mining Equipment Certification Service
	  Buxton
	  Derbyshire SK17 9JN GB
	• SIRA Test and Certification Ltd
	  Saighton Lane
	  Saighton
	  Chester CH3 6EG GB

(continued)

Table 3.8   *(continued)*

Member state	*Notified body according to Directive 76/117/EEC (article 14)*

Part B   Explosion protected electrical equipment – Group II

Austria	BVFA Bundesversuchs- und Forschungszentrum Arsenal A-1030 Wien
Belgium	ISSeP Institut Scientifique de Service Public Division de Colfontaine B-7340 Paturages
Denmark	DEMKO Danmarks Elektriske Materielkontrol DK-2730 Herlev
Finland	VTT Technical Research Centre of Finland Espoo FIN-02044 VTT

France	• Ineris
	  Institut National de l'Environnement Industriel
	  et des Risques
	  F-60550 Verneuil-en-Halatte
	• LCIE
	  Laboratoire Central des Industries Électriques
	  F-92266 Fontenay-aux-Roses

Germany	• BVS Bergbau-Versuchsstrecke
	  Fachstelle für Sicherheit Elektrischer Betriebsmittel der
	  DMT-Gesellschaft für Forschung und Prüfung mbH
	  D-44329 Dortmund
	• PTB
	  Physikalisch-Technische Bundesanstalt
	  D-38116 Braunschweig

Greece	–
Ireland	–
Italy	CESI Centro Elettrotecnico Sperimentale Italiano I-20134 Milano
Luxembourg	–
Netherlands	NV KEMA NL-6800 ET Arnhem
Portugal	–
Spain	Laboratorio Oficial José Maria Madariaga E-28003 Madrid
Sweden	–

United Kingdom	• BASEEFA (EECS)
	  British Approval Service for Electrical Equipment
	  in Flammable Atmospheres

(continued)

Table 3.8   *(continued)*

Member state	Notified body according to Directive 76/117/EEC (article 14)
	Electrical Equipment Certification Service
	Buxton
	Derbyshire SK17 9JN GB
	• Sira Test and Certification Ltd
	Saighton Lane
	Saighton
	Chester CH3 6EG GB

way by different notified bodies in an apparatus certification procedure (e.g. notified body A, involved in certifying a high voltage switchgear from manufacturer X according to EN 50019 (Increased safety – e –) accepts the component certificate of a notified body B for a cable entry produced by a manufacturer Y). Generally, parts with a component certificate must not be considered (and operated) as apparatus on their own (e.g. a ballast reactor for a high pressure mercury lamp, with a component certificate for installation in an enclosure complying with EN 50019 Increased safety – e – shall not be used as an individual apparatus under any circumstances). This is based on the fact that a very important aspect related to safety has to be considered: the maximum surface temperature of this component in its environment. This consideration cannot neglect the temperature profile given in a light fitting.

In addition to the Harmonized European Norms, some directives contain – similar to a standard – technical requirements for electrical apparatus and cables. Especially Directive 82/130/EEC modifies the technical content of standards:

- supplement B, annex 1, modifies EN 50014, 1st edition, by incorporating 'earthing in an electrostatic sense' for components with an enclosure made of plastic materials
- supplement B, annex 2, modifies EN 50018, 1st edition, by incorporating
  - restrictions in the volume of insulating materials inside a flameproof enclosure if these materials do not comply with a certain quality against surface leakage currents (comparative tracking index (CTI) test according to IEC 60112)
  - isolating switches for non-intrinsic circuits to ensure opening of the flameproof enclosure with no internal live parts
  - an interlock between the isolating switch and doors or covers with a quick-lock
  - sockets of tubular fluorescent lamps shall be designed as a one-pin cap according to IEC 60061-2
- supplement B, annex 3, modifies EN 50020, 1st edition, by incorporating special constructional requirements and verification tests for cables with intrinsically safe circuits.

Directive 88/35/EEC, supplement B, overtakes the additional requirements to EN 50014, 1st edition, and to EN 50020, 1st edition, from Directive 82/130/EEC, supplement B.

With the progress in standardization the technical content of the directive supplements has been taken into the amendments to the EN basic standards.

In Germany, the transformation of EC directives into national legislation results in two decrees:

- ElexV [8], issued by the Federal Minister of Labour and Social Affairs, covering commercial and industrial equipment, with the exception of railway installations, German Federal Armed Forces, mining industry with the exception of surface installations, shipping, motor vehicles and civil aviation
- ElZulBergV [9], issued by the Federal Minister for Economic Affairs, covering areas in coal mines endangered by firedamp, and areas hazardous due to potentially explosive atmospheres in the field of mining (except coal mines), with the exception of surface installations.

The content common to these two decrees is summarized as follows:

- the two-step procedure (certification and subsequent approval) described in Section 3.1 has been deleted. Generally, a certificate (certificate of conformity, control certificate, type test certificate) issued by the notified body acts as an approval of the electrical apparatus
- type test certificates for explosion protected or firedamp-proof electrical apparatus, based on the national VDE 0170/0171/1969-01, or based on European Standards (EN) with the status of a national VDE Standard (see Table 3.5) issued by PTB or BVS, are German national approvals. The issue of such type test certificates has been limited to 1988-05-01.

3.3   The new approach – Directive 94/9/EC

The procedures for the harmonization of European Standards described in Section 3.2 cause a certain time delay which in many respects is considered as unreasonable to bring technical progress into practice in the field of explosion protection. Similar ideas in other fields of engineering result in a 'new approach' concept for EC directives. Very generally, the directives according to this 'new approach' shall refer to the 'state of the art', laid down in existing standards. The directives shall not contain any listing of standards and in this way the need for subsequent adoption to new standards is deleted. In addition, the commission of the EC has established a new concept for certification and testing. In this concept, there are 'modules' representing different procedures for conformity assessment. The 'new' directives, based upon the new concept and upon article 100a (modified with respect to the original article 100) of the Treaties of the European Community describe a multiple choice for selecting the appropriate modules.

Directive 93/465/EEC, dated 1993-07-22, gives a survey of the different modules (called A … H) for conformity assessment procedures (Fig. 3.2). A short description of these modules is given in Table 3.9. Obviously, module B

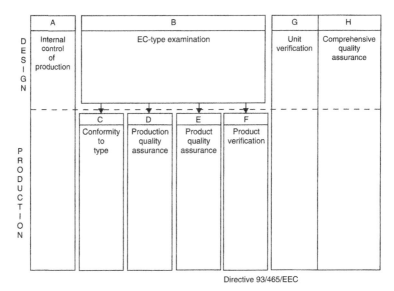

Directive 93/465/EEC

Figure 3.2   *Modules for the assessment of conformity according to Directive 93/465/EEC for the design and production phase.*

is applicable in the design phase of a product, whereas modules C, D, E and F are appropriate in the manufacturing process.

Incorporating these new philosophies, the EC Directive 94/9/EC, dated 1994-03-23 (the so-called ATEX 100a Directive), covers the field of explosion protection.

The content of Directive 94/9/EC is summarized as follows:

- all aspects of explosion protection are covered: areas hazardous due to combustible gases, vapours, mist and dusts, coal mines endangered by firedamp and coal dust
- integration of all zones (see Table 2.3) of hazardous areas
- integration of explosion protection and functional safety aspects for electrical and non-electrical apparatus, safety systems, combustion engines (with internal combustion only)
- the directive defines the objectives of safety, but does not list standards. A general reference is made to the 'state of the art'
- the directive lists the modules A … G (H is *not* in the scope) for conformity assessment procedures.

The 'state of the art' – the second edition (and following editions) of the European Standards (EN) (see Table 3.5) – is considered to form an appropriate basis for construction and testing of explosion protected electrical equipment. Other aspects of safety may require revision of established standards.

Directive 94/9/EC can be applied on a voluntary basis starting on 1996-03-01. The procedures for certification according to the 'old' directives referring

Table 3.9 Modules for the assessment of conformity according to Directive 93/465/EEC (Directive 94/9/EC refers to these modules in its annexes. The number of the corresponding annex is given in brackets [ ])

Module	Title	Description (shortened)
A [VIII]	Internal control of production	• The manufacturer or his EU-resident authorized representative ensures and declares that the products fulfil the requirements given in the directive. They affix the CE mark on each product and provide a written declaration of conformity.
		• The manufacturer provides technical documents enabling the assessment of the product's conformity with the requirements given in the directive. The documents shall cover design, production process and function in such a way as to ensure the assessment.
		• The manufacturer or his EU-resident authorized representative holds these documents at the disposal of the national authorities for at least 10 years after the end of manufacturing this product.
		• The manufacturer ensures, that the production process guarantees the compliance of the products with the technical documents and with the requirements given in the directive.
B [III]	EC-type examination	• A notified body tests and certifies that a specimen provided by the manufacturer or his EU-resident authorized representative and this specimen representing the production complies with the requirements given in the directive.
		• The manufacturer or his EU-resident authorized representative submits technical documents to the notified body. The documents shall cover design, production process and function in such a way as to ensure the assessment of the product's conformity with the requirements given in the directive.
		• The notified body verifies the product's compliance with the technical documents and with standards, carries out the type tests according to these standards and issues the EC-type examination certificate.
		• The notified body involved in the verifying procedure informs all the other EU-notified bodies about this EC-type examination certificate.

(continued)

Table 3.9 (*continued*)

Module	Title	Description (*shortened*)
C [VI]	Conformity to type	• The manufacturer or his EU-resident authorized representative stores a copy of the EC-type examination certificate and the technical documents for at least 10 years after the end of manufacturing this product.
		• The manufacturer or his EU-resident authorized representative – ensures and declares that the products comply with the design according to the EC-type examination certificate and fulfil the requirements given in the directive – affixes the CE mark on each product and provides a written declaration of conformity.
		• The manufacturer carries out the relevant routine tests and affixes the identification number of the notified body during the manufacturing process. This is under the responsibility of this notified body.
		• The manufacturer or his EU-resident authorized representative stores a copy of the declaration of conformity for at least 10 years after the end of manufacturing this product.
D [IV]	Production quality assurance	• The manufacturer ensures and declares that the products comply with the design according to the EC-type examination certificate and fulfil the requirements given in the directive.
		• The manufacturer or his EU-resident authorized representative affixes the CE mark on each product and provides a declaration of conformity.
		• The manufacturer maintains an approved quality assurance system for manufacturing, acceptance control and testing.
		• The manufacturer applies for assessment of this quality assurance system at a notified body. The quality assurance system shall guarantee the product's compliance with the design according to the EC-type examination certificate and with the requirements given in the directive.

- The notified body assesses the quality assurance system.
- The manufacturer is obligated to meet the commitments resulting from the quality assurance system in its approved version and to ensure its efficient operation.
- For the purpose of supervision, the manufacturer grants the notified body access to all facilities for production, acceptance, testing and storage. In addition, he submits technical documents.
- The manufacturer stores the documents for at least 10 years after the end of manufacturing this product.
- The notified body informs all the other EU notified bodies about the approvals of the quality assurance systems.

E [VII]	Product quality assurance

- The manufacturer ensures and declares that the products comply with the design according to the EC-type examination certificate.
- The manufacturer or his EU-resident authorized representative affixes the CE mark on each product and provides a written declaration of conformity.
- The manufacturer maintains an approved quality assurance system for acceptance control and testing.
- The manufacturer applies for assessment of this quality assurance system at a notified body.
- Within the scope of the quality assurance system, each product is subjected to a test to guarantee the compliance with the requirements given in the directive.
- The notified body assesses the quality assurance system.
- The manufacturer is obligated to meet the commitments resulting from the quality assurance system in its approved version and to ensure its efficient operation.
- For the purpose of supervision, the manufacturer grants the notified body access to all facilities for acceptance, testing and storage. In addition, he submits technical documents.

(continued)

Table 3.9 (continued)

Module	Title	Description (shortened)
		• The manufacturer stores the documents for at least 10 years after the end of manufacturing this product.
		• The notified body informs all the other EU notified bodies about the approvals of the quality assurance systems.
F [V]	Product verification	• The manufacturer or his EU-resident authorized representative ensures and declares that the products comply with the design according to the EC-type examination certificate and fulfil the requirements given in the directive.
		• The manufacturer ensures that the production process guarantees the product's compliance with the type described in the EC-type examination certificate and with the requirements given in the directive.
		• The manufacturer or his EU-resident authorized representative affixes the CE mark on each product and provides a declaration of conformity.
		• The notified body carries out the relevant test by controlling and inspections applied to each item to ensure the product's compliance with the relevant requirements given in the directive.
		• The notified body affixes the identification number on each product and provides a written certificate of conformity.
		• The manufacturer or his EU-resident authorized representative stores a copy of the declaration of conformity for at least 10 years after the production date of the item.
G [IX]	Unit verification	• The manufacturer ensures and declares that the product, which has been certified by a notified body, fulfils the relevant requirements given in the directive.
		• The manufacturer or his EU-resident authorized representative affixes the CE mark on each product and provides a declaration of conformity.

| H [none] | Comprehensive quality assurance | <ul><li>A notified body inspects the item and carries out the relevant tests to ensure the compliance with the relevant requirements given in the directive. The notified body affixes the identification number on each product and provides a certification of conformity.</li><li>The manufacturer provides technical documents enabling the assessment of the product's conformity with the requirements given in the directive and the understanding of its conception, production and function.</li><li>The manufacturer ensures and declares that the products fulfil the requirements given in the directive. He affixes the CE mark on each product and provides a declaration of conformity.</li><li>The manufacturer maintains an approved quality assurance system for design, manufacturing, acceptance control and testing.</li><li>The manufacturer applies for assessment of this quality assurance system at a notified body. The quality assurance system shall guarantee the product's compliance with the requirements given in the directive.</li><li>The notified body assesses the quality assurance system.</li><li>The manufacturer is obligated to meet the commitments resulting from the quality assurance system in its approved version and to ensure its efficient operation.</li><li>The manufacturer grants this notified body access to all facilities for development, manufacturing, acceptance control, testing and storage for the purpose of supervision. In addition, he submits technical documents.</li><li>The notified body assesses the quality assurance systems by periodical auditing.</li><li>The CE mark is completed by the identification mark of the notified body.</li><li>The manufacturer stores the documents for at least 10 years after the end of manufacturing this product.</li><li>The notified body informs all the other EU notified bodies about the approvals of the quality assurance systems.</li></ul> |

to article 100 of the Treaties of the European Community, described in Section 3.2, will end on 2003-06-30. In the meantime, the manufacturers have the choice of going along with that well-known practice or applying Directive 94/9/EC. Compliance with this directive will be obligatory for all explosion protected equipment and systems placed on the market/put into use starting on 2003-07-01.

The different combinations of modules for conformity assessment procedures are summarized in Table 3.10.

Explosion protected electrical equipment for zone 1 (Group II, category 2 G) and the 'classical' firedamp-proof apparatus for coal mines (Group I, category M2) are covered by the B module (EC-type examination) and by the C module (conformity to type) or the E module (product quality assurance). Alternatively, with respect to the B module, the notified body issues an EC-type examination certificate (in German: EG-Baumusterprüfbescheinigung), whose marking mainly includes:

- the name of the notified body
- the year of issue
- the abbreviation 'ATEX' to indicate that this certificate refers to Directive 94/9/EC
- the certificate number, e.g. PTB 97.ATEX 2 … , DMT 97.ATEX 1 … .

Table 3.10   Directive 94/9/EC – Combinations of modules – Electrical apparatus and systems (The number of the corresponding annex is given in brackets [ ])

Apparatus		Combinations
Group	*Category*	
I	M1	B [III]: EC-type examination
II	1G, 1D	and
		Production Product
		D [IV]: quality or F[V]: verification
		assurance
		(alternatively)
I	M2	B [III]: EC-type examination
II	2G, 2D	and
		C [VI]: Conformity E [VII]: Product quality
		to type or assurance
		(alternatively)
II	3G, 3D	A [VIII]: Internal control of
		production
I, II	all categories	(alternatively)
		G[IX]: Unit verification

Note: Quality Assurance Notifications according to Directive 94/9/EC normally refer to the number of the annex (and not to the key letter of the module)

Table 3.11 lists the notified bodies according to Directive 94/9/EC. It is note-worthy that competence and responsibility of a notified body may be restricted to certain types of explosion protected equipment (e.g. non-electrical apparatus only), to one group only (e.g. Group II) or to a selection of modules, e.g. to quality assurance related modules. The notified bodies are not obligated to cover the complete field of explosion protection. Nevertheless, some notified bodies cover the total range and can provide a 'single-stop-procedure' to their clients.

Table 3.11   List of notified bodies according to Directive 94/9/EC (dated 1998-10)

State	Notified body	Competence and responsibility	
		Products	*Module*
Austria	TÜV-A TÜV-Österreich A-1230 Wien	Group II, equipment where the source of ignition is primarily electrical	B, C, D, E, F, G
Belgium	–	–	–
Denmark	DEMKO A/S DK-2730 Herlev	Electrical equipment	B, C, D, E, F, G
Finland	–	–	–
France	INERIS Institut National de l'Environnement Industriel et des Risques F-60550 Verneuil-en-Halatte	Apparatus and safety systems all groups/all categories	A, B, C, D, E, F, G
	LCIE Laboratoire Central des Industries Électriques F-92266 Fontenay-aux-Roses	Apparatus and safety systems all groups/all categories	A, B, C, D, E, F, G
Germany	BAM Bundesanstalt für Materialforschung und Prüfung mbH D-12205 Berlin	• Groups I, II categories M1, 1 non-electrical apparatus, safety systems, gas detectors, flame arrestors	B, D, F, G
		• Groups I, II categories M2, 2 and 3 non-electrical apparatus, safety systems, gas detectors	A, B, C, E, G

(continued)

Table 3.11 (*continued*)

State	Notified body	Competence and responsibility	
		Products	Module
	DMT Zertifizierungsstelle der DMT-Gesellschaft für Forschung und Prüfung mbH D-45307 Essen	• Groups I, II categories M1, 1 electrical and non-electrical apparatus, machinery, gas detectors, safety systems	B, D, F, G
		• Groups I, II categories M2, 2 and 3 electrical and non-electrical apparatus, combustion engines (internal combustion), machinery, gas detectors, safety systems	A, B, C, E, G
	DQS Deutsche Gesellschaft zur Zertifizierung von Management-systemen mbH Qualitäts- und Umweltgutachter D-60433 Frankfurt	• Groups I, II categories M1, 1 electrical and non-electrical apparatus, machinery, gas detectors, safety systems	D
		• Groups I, II categories M2, 2 and 3 electrical and non-electrical apparatus, combustion engines (internal combustion), machinery, gas detectors, safety systems	E
	FSA Forschungsge-sellschaft für Angewandte Systemsicherheit und Arbeitsmedizin mbH D-68165 Mannheim	• Group II category 1G non-electrical apparatus, machinery, safety systems	B, D, F, G
		• Group II categories 2 and 3 non-electrical apparatus, combustion engines (internal combustion), machinery, safety systems	A, B, C, E, G
	IBExU Institut für	• Groups I, II categories M1, 1 electrical and	B, F, G

(*continued*)

Table 3.11   (*continued*)

State	Notified body	Competence and responsibility	
		Products	Module
	Sicherheitstechnik GmbH D-09599 Freiberg	non-electrical apparatus, machinery, gas detectors, safety systems	
		• Groups I, II categories M1, M2, 1, 2 and 3 safety systems	B, F, G
		• Groups I, II categories M2, 2 and 3 electrical and non-electrical apparatus, combustion engines (internal combustion), machinery, gas detectors, safety systems	A, B, C, G
	PTB Physikalisch-Technische Bundesanstalt D-38116 Braunschweig	• Group II category 1G electrical and non-electrical apparatus, machinery, electrostatic spraying tools, safety systems	B, D, F, G
		• Group II categories 2G and 3G electrical and non-electrical apparatus, combustion engines (internal combustion), machinery, electrostatic spraying tools, safety systems	A, B, C, E, G
	TÜV Hannover/ Sachsen-Anhalt e.V. TÜV-Cert Zertifizierungsstelle D-30519 Hannover	• Group II category 1G electrical apparatus, safety systems	B, D, F, G
		• Group II categories 2G and 3G electrical apparatus, combustion engines (internal combustion), safety systems	A, B, C, E, G
Greece	–	–	–
Ireland	–	–	–

(*continued*)

Table 3.11   (*continued*)

State	Notified body	Competence and responsibility	
		Products	Module
Italy	CESI Centro Elettrotecnico Sperimentale Italiano I-20134 Milano	all groups, all categories apparatus and safety systems	A, B, C, D, E, F, G
Luxem-bourg	SEE Service de l'Energie de l'Etat L-2010 Luxembourg	all groups, all categories apparatus and safety systems	B, C, D, E, F, G
Nether-lands	KEMA KEMA NV NL-6800 ET Arnhem	• Group II all categories	A, B, C, D, E, F, G
		• Group II category 1 safety systems	B, D, F, G
		• Group II all categories components of apparatus/safety systems	A, B, C, D, E, F, G
Norway	NEMKO AS N-01314 Oslo	Electrical equipment	B, C, D, E, F, G
Portugal	–	–	–
Spain	LOM Laboratorio Oficial José Maria de Madariaga E-28003 Madrid	all groups, all categories apparatus and safety systems	B, C, D, E, F, G
Sweden	Swedish National Testing and Research Institute S-50115 Boras	all groups, all categories apparatus and safety systems (not in the scope: combustion engines)	B, C, D, E, F, G
United Kingdom	EECS (HSE) Electrical Equipment Certification Service, Health and Safety Executive Buxton/Derbyshire SK17 9JN GB	all groups, all categories apparatus and safety systems	A, B, C, D, E, F, G
	ERA Technology Ltd Leatherhead/Surrey HT22 7SA GB	• Groups I, II categories M1, 1 • Groups I, II categories M2, 2 • Group II Category 3	B, C, D, E, F, G

(*continued*)

Table 3.11   *(continued)*

State	Notified body	Competence and responsibility	
		Products	Module
		Protective systems, devices, components	
	SCS Sira Certification Service Sira Test and Certification Ltd Chislehurst/Kent BR7 5EH GB	all groups, all categories apparatus and safety systems	A, B, C, D, E, F, G

Germany has adopted Directive 94/9/EC to national legislation by ExVO [11] and the subsequent revisions of ElexV [8] and VbF [50]. These revisions have reduced the content of ElexV and VbF to aspects of operational safety, excluding all requirements for, e.g., construction and type testing. The revised version of ElexV [8] has adopted the three-zone concept for areas hazardous due to combustible dusts (zones 20/21/22).

3.4   The IEC world

The IEC, International Electrotechnical Commission, Geneva/Switzerland, is a worldwide acting body for the development of consensus electrotechnical standards. IEC global harmonization normally introduces the standards into the member states on a voluntary basis without any adoption by legislation. The IEC members represent 85% of the world's population, generating 95% of the electrical energy worldwide. A great part of European Standards (EN) drawn up by CENELEC, Brussels/Belgium, is based – identical or with modifications – on IEC standards.

Table 3.12 summarizes IEC members and associates. Within the IEC, technical committees and subcommittees are involved in standardization work covering all aspects of electrical engineering, e.g. insulating materials, semiconductors, wires and cables, lamps, motors, generators, transformers, and, of special interest, electrical apparatus for explosive atmospheres. This is the responsibility of Technical Committee TC 31 and eight subcommittees (see Table 3.13). Table 3.14 gives a survey of IEC documents and standards covering the field of explosion protection of electrical apparatus.

As mentioned above, IEC Standards are adopted on a voluntary basis. Thus, compliance with IEC Standards does not guarantee free trading of explosion protected equipment worldwide. Today, mainly two 'blocks' are attainable by compliance only with their own standards:

• the member states of the European Union (see Sections 3.2 and 3.3)

Table 3.12   The IEC (International Electrotechnical Commission) members and associates (as at 1999-12-31)

Part A   Members

Australia	France	Mexico	Slovenia
Austria	Germany	Netherlands	South Africa
Belarus	Greece	New Zealand	Spain
Belgium	Hungary	Norway	Sweden
Brazil	India	Pakistan	Switzerland
Bulgaria	Indonesia	Philippines	Thailand
Canada	Ireland	Poland	Turkey
China	Israel	Portugal	Ukraine
Croatia	Italy	Romania	United
Cyprus	Japan	Russian	Kingdom
Czech Republic	Korea,	Federation	United States of
Denmark	Republic of	Saudi Arabia	America
Egypt	Luxembourg	Singapore	Yugoslavia
Finland	Malaysia	Slovakia	

Part B   Associates

Bosnia and	Iceland
Hercegovina	Latvia
Estonia	Lithuania

Table 3.13   IEC Technical Committee TC 31 – structure and subcommittees – 'Electrical apparatus for explosive atmospheres' (dated 1999-05)

Subcommittee	*Title*	
SC 31-A	Flameproof enclosures	
SC 31-C	Increased safety apparatus	(disbanded)
SC 31-D	Pressurization and associated techniques	(disbanded)
SC 31-G	Intrinsically safe apparatus	
SC 31-H	Apparatus for use in the presence of ignitable dust	
SC 31-J	Classification of hazardous areas and installation requirements	(in standby)
SC 31-K	Encapsulation	(disbanded)
SC 31-L	Electrical apparatus for the detection of flammable gases	

- North America (USA, Canada). Table 2.2 demonstrates the very different philosophy in these states, compared with IEC – or closely related EN – standards.

A main effort to facilitate international trading and commissioning of electrical apparatus for use in (potentially) explosive atmospheres has been undertaken: the so-called 'IECEx-Scheme'; the 'IEC Scheme for Certification to Standards for Safety of Electrical Equipment for Explosive Atmospheres' eliminates the need

Table 3.14 IEC publications covering the field of explosion protection of electrical apparatus and systems (as at 2000-01-01) (not including draft publications)

New number (old number) IEC	Date	Title
60079-0	Ed. 3 1998-04	Electrical apparatus for explosive gas atmospheres Part 0: General requirements
60079-1 (79-1)	Ed. 3 1990-12	Electrical apparatus for explosive gas atmospheres Part 1: Construction and verification test of flameproof enclosures of electrical apparatus
60079-1-am 1 (79-1-am 1)	Ed. 3 1993-08	Amendment No. 1
60079-1-am 2	Ed. 3 1998-05	Amendment No. 2
60079-1	Ed. 3.2 1998-08	Electrical apparatus for explosive gas atmospheres Part 1: Construction and verification test of flameproof enclosures of electrical apparatus
60079-1A (79-1A)	Ed. 3 1975-01	Electrical apparatus for explosive gas atmospheres Part 1: Construction and verification test of flameproof enclosures of electrical apparatus First supplement Appendix D: Method of test for ascertainment of maximum experimental safe gap
60079-2 (79-2)	Ed. 3 1983-01	Electrical apparatus for explosive gas atmospheres Part 2: Electrical apparatus, type of protection 'p'
60079-3 (79-3)	Ed. 3 1990-12	Electrical apparatus for explosive gas atmospheres Part 3: Spark test apparatus for intrinsically safe circuits
60079-4 (79-4)	Ed. 2 1975-01	Electrical apparatus for explosive gas atmospheres Part 4: Method of test for ignition temperature
60079-4-am 1 (79-4-am 1)	Ed. 2 1995-07	Amendment No. 1

(continued)

Table 3.14 (continued)

New number (old number) IEC	Date	Title
60079-4A (79-4A)	Ed. 2 1970-01	Electrical apparatus for explosive gas atmospheres Part 4: Method of test for ignition temperature First supplement
60079-5 (–)	Ed. 2 1997-04	Electrical apparatus for explosive gas atmospheres Part 5: Powder filling 'q'
60079-6 (79-6)	Ed. 2 1995-04	Electrical apparatus for explosive gas atmospheres Part 6: Oil immersion 'o'
60079-7 (79-7)	Ed. 2 1990-08	Electrical apparatus for explosive gas atmospheres Part 7: Increased safety 'e'
60079-7-am 1	Ed. 2 1991-04	Amendment No. 1
60079-7-am 2	Ed. 2 1993-04	Amendment No. 2
60079-10 (79-10)	Ed. 3 1995-12	Electrical apparatus for explosive gas atmospheres Part 10: Classification of hazardous areas
60079-11	Ed. 4 1999-02	Electrical apparatus for explosive gas atmospheres Part 11: Intrinsic safety 'i'
60079-12 (79-12)	Ed. 1 1978-01	Electrical apparatus for explosive gas atmospheres Part 12: Classification of mixtures of gases or vapours with air according to their maximum experimental safe gaps and minimum igniting currents
60079-13 (79-13)	Ed. 1 1982-01	Electrical apparatus for explosive gas atmospheres Part 13: Construction and use of rooms or buildings protected by pressurization
60079-14	Ed. 2 1996-12	Electrical apparatus for explosive gas atmospheres Part 14: Electrical installations in hazardous areas (other than mines)

Number	Ed.	Date	Title
60079-15 (79-15)	Ed. 1	1987-06	Electrical apparatus for explosive gas atmospheres Part 15: Electrical apparatus with type of protection 'n'
60079-16 (79-16)	Ed. 1	1990-05	Electrical apparatus for explosive gas atmospheres Part 16: Artificial ventilation for the protection of analyser(s) houses
60079-17	Ed. 2	1996-12	Electrical apparatus for explosive gas atmospheres Part 17: Inspection and maintenance of electrical installations in hazardous areas (other than mines)
60079-18 (79-18)	Ed. 1	1992-10	Electrical apparatus for explosive gas atmospheres Part 18: Encapsulation 'm'
60079-19 (79-19)	Ed. 1	1993-10	Electrical apparatus for explosive gas atmospheres Part 19: Repair and overhaul for apparatus used in explosive atmospheres (other than mines or explosives)
60079-20 (79-20)	Ed. 1	1996-10	Electrical apparatus for explosive gas atmospheres Part 20: Data for flammable gases and vapours, relating to the use of electrical apparatus
61241-1-1 (1241-1-1)	Ed. 2	1999-06	Electrical apparatus for use in the presence of combustible dust Part 1-1: Electrical apparatus protected by enclosures and surface temperature limitation – Specification for apparatus
61241-1-2 (1241-1-2)	Ed. 2	1999-06	Electrical apparatus for use in the presence of combustible dust Part 1-2: Eletrical apparatus protected by enclosures and surface temperature limitation – Selection, installation and maintenance
61241-2-1 (1241-2-1)	Ed. 1	1994-12	Electrical apparatus for use in the presence of combustible dust Part 2: Test methods Section 1: Methods for determining the minimum ignition temperatures of dust
61241-2-2 (1241-2-2)	Ed. 1	1993-08	Electrical apparatus for use in the presence of combustible dust Part 2: Test methods Section 2: Methods for determining the electrical resistivity of dust in layers
61241-2-3 (1241-2-3)	Ed. 1	1994-09	Electrical apparatus for use in the presence of combustible dust Part 2: Test methods

(continued)

Table 3.14 (continued)

New number (old number) IEC	Date	Title
61241-3	Ed. 1	Section 3: Method of determining the minimum ignition energy of dust–air mixtures
(–)	1997-05	Electrical apparatus for use in the presence of combustible dust
61779-1	Ed. 1	Part 3: Classification of areas where combustible dusts are or may be present
	1998-04	Electrical apparatus for the detection and measurement of flammable gases
61779-2	Ed. 1	Part 1: General requirements and test methods
	1998-04	Electrical apparatus for the detection and measurement of flammable gases
		Part 2: Performance requirements for Group I apparatus indicating a volume fraction up to 5% methane in air
61779-3	Ed. 1	Electrical apparatus for the detection and measurement of flammable gases
	1998-04	Part 3: Performance requirements for Group I apparatus indicating a volume fraction up to 100% methane in air
61779-4	Ed. 1	Electrical apparatus for the detection and measurement of flammable gases
	1998-04	Part 4: Performance requirements for Group II apparatus indicating a volume fraction up to 100% lower explosion limit
61779-5	Ed. 1	Electrical apparatus for the detection and measurement of flammable gases
	1998-04	Part 5: Performance requirements for Group II apparatus indicating a volume fraction up to 100% gas
61779-6	Ed. 1	Electrical apparatus for the detection and measurement of flammable gases
	1999-06	Part 6: Guide for the selection, installation, use and maintenance of apparatus for the detection and measurement of flammable gases

| TR 61 831 | Ed. 1 1999-07 | On-line analyser systems – Guide to design and installation |
| TR 61 832 | Ed. 1 1999-06 | Analyser systems – Guide to technical enquiry and bid evaluation |

Note: USA and Canada have adopted most of the IEC Standards up to 1998. The following survey correlates the IEC Standard with the corresponding national (USA and Canada) standard for Class I apparatus (by courtesy of Underwriters Laboratories Inc., Northbrook, Illinois/USA):

Area	Protection methods	Applicable certification documents		
		US	Canada	IEC
Zone 0	• Intrinsically safe, 'ia' (2 fault)	UL 2279, Pt 11	CSA-E79-11	IEC 60079-11
	• Class I, Div. 1 Intrinsically safe (2 fault) method	UL 913	CSA-157	–
Zone 1	• Encapsulation 'm'	UL 2279, Pt 18	CSA-E79-18	IEC 60079-18
	• Flameproof 'd'	UL 2279, Pt 1	CSA-E79-1	IEC 60079-1
	• Increased safety 'e'	UL 2279, Pt 7	CSA-E79-7	IEC 60079-7
	• Intrinsically safe, 'ib' (1 fault)	UL 2279, Pt 11	CSA-E79-11	IEC 60079-11
	• Oil immersion 'o'	UL 2279, Pt 6	CSA-E79-6	IEC 60079-6
	• Powder filling 'q'	UL 2279, Pt 5	CSA-E79-5	IEC 60079-5
	• Purged/pressurized 'p'	–	CSA-E79-2	IEC 60079-2
	• Any Class I, zone 0 method	–	–	–
	• Any Class I, Div. 1 method (see Table 2.2)	–	–	–
Zone 2	• Nonincendive 'nC'	UL 2279, Pt 15	CSA-E79-15	IEC 60079-15
	• Non-sparking device 'nA'	UL 2279, Pt 15	CSA-E79-15	IEC 60079-15
	• Restricted breathing 'nR'	UL 2279, Pt 15	CSA-E79-15	IEC 60079-15
	• Hermetically sealed 'nC'	UL 2279, Pt 15	CSA-E79-15	IEC 60079-15
	• Any Class I, zone 0 or 1 method	–	–	–
	• Any Class I, Div. 1 or 2 method (see Table 2.2)	–	–	–

for widespread national certification procedures to bring explosion protected equipment into operation. Rules and procedures are described in:

- Document IECEE 04 (1995) 'Rules and Procedures of the Scheme of the IECEE [the IEC System for Conformity Testing to Standards for Safety of Electrical Equipment] for Certification to Standards for Safety of Electrical Equipment for Explosive Atmospheres (IECEx-Scheme)' followed by
- Document IECEx 01 (1999) – Basic Rules and
- Document IECEx 02 (1999) – Rules of Procedures.

The content can be summarized as follows:

- the IECEx-Scheme eliminates the need for multiple national certification
- it provides a route to the ultimate aim of using one international certificate and mark accepted in all participating countries
- the application for a country to participate in the IECEx-Scheme is made on a standard-by-standard basis
- certification bodies and testing laboratories as applicants must reside in a participating country
- IEC Standards and national standards shall be identical; if not, a transitional period (some years) will be necessary to achieve convergence of standards
- certification bodies and testing laboratories are accepted into the IECEx-Scheme following satisfactory assessment of their competence. After assessment (assuming positive results) they are entitled Accepted Certification Body (ACB) and IECEx Testing Laboratory (IECExTL), which is either integral with, or under the complete control of, or belongs to, or works under a written agreement for, an ACB
- a manufacturer of an explosion protected apparatus may apply for an 'IECEx certificate of conformity' to any ACB for the relevant standard, e.g. IEC 60079-1 for a flameproof cage induction motor. The applicant shall supply a technical documentation necessary to specify the explosion protection features of the apparatus. In addition, he shall supply one (or more) items of the apparatus
- the ACB shall arrange for an IECExTL to examine the documentation and the items to verify that the apparatus' design is in conformity with the standard. Assuming positive test results, the IECExTL issues an 'IECEx Assessment and Test Report (ATR)'
- the ACB shall assess the conformity of the manufacturer's quality assurance system with the standard ISO 9002 'Quality systems – Model for quality assurance in production and installation'
- the ACB shall endorse the ATR and issue the IECEx certificate of conformity
- manufacturers holding an IECEx certificate of conformity may affix the 'IECEx mark of conformity' (when writing this book, the mark of conformity was under consideration) to apparatus that they have verified as complying with the certified design.

The final objective of the IECEx-Scheme is worldwide acceptance of a single standard, a single certificate and a single mark.

Table 3.15 lists accepted ACBs and IECExTLs (Part a) and applicant ACBs and IECExTLs (Part b).

Table 3.15(a) IECEx-Scheme – Accepted certification bodies (ACB), test laboratories (TL) and national member bodies (as at 1999–08)

Country	ACB	Date of acceptance	TL	National member body	Standards different to IEC	Transition period
Germany DE	DMT Company for Research and Testing Certification Body 44329 Dortmund *and* DMT Company for Research and Testing Certification Body 45307 Essen	July 1999	Expert Body for Safety of Electrical Equipment (BVS) DMT Company for Research and Testing 44329 Dortmund	Deutsches Komitee der IEC Deutsche Elektrotechnische Kommission im DIN und VDE 60569 Frankfurt	Yes	10 years
	PTB Physikalisch-Technische Bundesanstalt 38116 Braunschweig	July 1999	PTB Physikalisch-Technische Bundesanstalt 38116 Braunschweig			
Sweden SE	Swedish National Testing and Research Institute (SP) 50115 Boras	August 1999	Swedish National Testing and Research Institute (SP) 50115 Boras	SEK Svenska Elektriska Kommissionen 16429 Kista	Yes	10 years
United Kingdom GB	EECS Electrical Equipment Certification Services	August 1999	EECS Electrical Equipment Certification Services	British Electrotechnical Committee British Standards	Yes	10 years

(continued)

Table 3.15(a) (continued)

Country	ACB	Date of acceptance	TL	National member body	Standards different to IEC	Transition period
	Health and Safety Executive Buxton/Derbyshire SK17 9JN		Health and Safety Executive Buxton/Derbyshire SK17 9JN and Explosion and Fire Hazards Laboratory ERA Technology Ltd Leatherhead Surrey KT22 7SA	Institution London W4 4AL		
	SCS SIRA Certification Services Chislehurst Kent BR7 5EH	August 1999	SIRA Test and Certification Ltd Chester CH3 6EG and Explosion and Fire Hazards Laboratory ERA Technology Ltd Leatherhead Surrey KT22 7SA			

Table 3.15(b) IECEx-Scheme – participating countries, applicant accepted certifying bodies (ACB) and applicant IECEx testing laboratories (TL) (as at 1999-08)

Country	National member body	Standards different to IEC	Proposed transition period	Applicant ACB	Applicant TL
Australia AU	Standards Australia Strathfield NSW 2135	Yes	10 years	Quality Assurance Services Strathfield NSW 2135	ITACS International Testing and Certification Services Bowden SA 5007
					TestSafe Australia Londonderry NSW 2753
					SIMTARS Safety in Mines Testing and Research Station Redbank QLD 4075
				SIMTARS Safety in Mines Testing and Research Station Redbank QLD 4075	SIMTARS Safety in Mines Testing and Research Station Redbank QLD 4075
Canada CA	Standards Council of Canada Ottawa, Ontario K1P 6N7	Yes	5 years	Canadian Standards Association Etobicoke, Ontario M9W IR3	Canadian Standards Association Etobicoke, Ontario, M9W IR3 and Edmonton, Alberta T6N IE6
China CN	CSBTS Chinese National	Yes	Group I (mining)	Awaiting further details	Awaiting further details

(continued)

Table 3.15(b) (continued)

Country	National member body	Standards different to IEC	Proposed transition period	Applicant ACB	Applicant TL
	Committee of the IEC Beijing 100088		15 years Group II (industry) 10 years		
Denmark DK	Dansk Standard 2920 Charlottenlund	Yes	10 years	DEMKO 2730 Herlev	DEMKO 2730 Herlev
France FR	Comité Electrotechnique Français UTE 92052 Paris la Defense	Yes	10 years	LCIE Laboratoire Central des Industries Electriques 92260 Fontenay-aux-Roses	LCIE Laboratoire Central des Industries Electriques 92260 Fontenay-aux-Roses
				INERIS Institut National de l'Environnement Industriel et des Risques 60550 Verneuil-en-Halatte	INERIS Institut National de l'Environnement Industriel et des Risques 60550 Verneuil-en-Halatte
Germany DE	Deutsches Komitee der IEC Deutsche Elektrotechnische Kommission im DIN und VDE 60596 Frankfurt	Yes	10 years	Technischer Überwachungsverein Hannover/Sachsen-Anhalt e.V. TÜV-CERT Zertifizierungsstelle 30519 Hannover	TÜV Nord Anlagentechnik GmbH 30519 Hannover

Country					
Hungary HU	Hungarian Approval Service for Ex-proof Electrical Equipment 1037 Budapest	Yes	10 years	Hungarian Approval Service for Ex-proof Electrical Equipment 1037 Budapest	Hungarian Approval Service for Ex-proof Electrical Equipment 1037 Budapest
Italy IT	CEI Comitato Elettrotecnico Italiano 20126 Milano	Awaiting advice	Awaiting advice	CESI Centro Elettrotecnico Sperimentale Italiano 20134 Milano	Awaiting further details
Korea KR	Korea Institute of Industrial Technology IECEE Council of the Rep. of Korea Seoul 152-053	Yes	Awaiting advice	KISKO Korea Industrial Safety Corp. Inchon 403-711	KISKO Korea Industrial Safety Corp. Inchon 403-120
Netherlands NL	Netherlands National Committee of the IEC 2600 GB Delft	Yes	10 years	Awaiting further details	Awaiting further details
Norway NO	Norsk Elektroteknisk Komite (NEK) 0212 Oslo	Yes	Awaiting advice	NEMKO 0314 Oslo	NEMKO 0314 Oslo
Romania RO	Romanian National Committee for the IEC ICPE 74204 Bucharest 3	Yes	10 years	INSEMEX PETROSANI Equipment Ex-Certification Service 2675 Petrosani	INSEMEX PETROSANI Equipment Ex-Certification Service 2675 Petrosani

(continued)

Table 3.15(b) (continued)

Country	National member body	Standards different to IEC	Proposed transition period	Applicant ACB	Applicant TL
Russia RU	GOST The Gosstandard of Russia Moscow 117049	Awaiting advice	Awaiting advice	Awaiting further details	Awaiting further details
Slovenia SL	Standards and Metrology Institute 1000 Ljubljana	Yes	10 years	Slovenian Institute of Quality and Metrology 1000 Ljubljana	Slovenian Institute of Quality and Metrology 1000 Ljubljana
South Africa ZA	South African Bureau of Standards Pretoria 0001	Yes	Awaiting advice	Awaiting further details	Awaiting further details
Switzerland CH	SEV Swiss Electrotechnical Association Dept. Certification and Testing 8320 Fehraltorf	Yes	10 years	SEV Swiss Electrotechnical Association Dept. Certification and Testing 8320 Fehraltorf	SEV Swiss Electrotechnical Association Dept. PCIEX 8320 Fehraltorf
Yugoslavia YU	Federal Institution for Standardization Dept. for Quality and Certification 11000 Beograd	Awaiting advice	Awaiting advice	Awaiting further details	Awaiting further details

Chapter 4

Grouping and classification of explosion protected electrical apparatus

Recalling Chapter 1, there are mainly three specific values for combustible gas– (vapour–, mist–) air mixtures to be considered in order to avoid ignitions of these mixtures:

- the ignition temperatures T_i covering the temperature range between – roughly speaking – 100°C and 650°C (see Table 1.3)
- the MESG values separating the areas for flame transmission and non-transmission, covering the range from 0.3 mm to 3.2 mm (see Table 1.6)
- the minimum ignition energy values for electrical sparks acting as an ignition source, covering the range from (roughly) 10 μJ up to some 10^2 μJ (see Table 1.7).

The ignition temperature as a safety related specific value refers to all surfaces of apparatus enclosures and internal components exposed to a combustible atmosphere, with the exception of components protected by a flameproof enclosure. The MESG values refer to flameproof enclosures only as a guideline for the dimensions of joints (for shafts, doors or covers) to ensure that internal explosions of combustible atmospheres cannot pass to the outer atmosphere. The minimum ignition energy values refer to a special type of protection only, the so-called 'intrinsic safety', restricting currents and voltages in electrical circuits (and in storage elements like capacitors or inductive loads) in such a way that switching (and faults like short-circuits and wire breaks) generates an energy release small enough to avoid the ignition of the combustible mixture.

Very generally, it is considered as quite uneconomical to construct and build explosion protected apparatus in such a way that all specific values mentioned above are met jointly at their lower limits.

So, combustible mixtures are classified in temperature classes according to their ignition temperature (Table 4.1) and in explosion groups according to their MESG values (Table 4.2). It is noteworthy that the maximum surface temperature of apparatus or components (Table 4.1) equals the lower limit of the ignition temperature range, e.g. a T3 apparatus with a maximum surface temperature of 200°C is suitable for combustible mixtures with ignition temperatures exceeding 200°C without any margin of safety.

It must be emphasized that the determination of the ignition temperature according to IEC 60079-4 includes selected conditions (surface shape, area,

Table 4.1   Classification of combustible mixtures of gases, vapours or mist with air according to their ignition temperature T_i

Ignition temperature T_i		Temperature class	Maximum surface temperature	
°C	°F		°C	°F
Part A According to EN 50014 and IEC 60079-0 Electrical apparatus for Group II[1]				
$T_i > 450$	$T_i > 842$	T1	450	842
$300 < T_i \leqslant 450$	$572 < T_i \leqslant 842$	T2	300	572
$200 < T_i \leqslant 300$	$392 < T_i \leqslant 572$	T3	200	392
$135 < T_i \leqslant 200$	$275 < T_i \leqslant 392$	T4	135	275
$100 < T_i \leqslant 135$	$212 < T_i \leqslant 275$	T5	100	212
$85 < T_i \leqslant 100$	$185 < T_i \leqslant 212$	T6[2]	85	185
Part B According to NEC, National Electrical Code, USA, and CEC, Canadian Electrical Code, Canada, for Class I and Class II equipment[3]				
$T_i > 450$	$T_i > 842$	T1	450	842
$300 < T_i \leqslant 450$	$572 < T_i \leqslant 842$	T2	300	572
$280 < T_i$	$536 < T_i$	T2A	280	536
$260 < T_i$	$500 < T_i$	T2B	260	500
$230 < T_i$	$446 < T_i$	T2C	230	446
$215 < T_i$	$419 < T_i$	T2D	215	419
$200 < T_i \leqslant 300$	$392 < T_i \leqslant 572$	T3	200	392
$180 < T_i$	$356 < T_i$	T3A	180	356
$165 < T_i$	$329 < T_i$	T3B	165	329
$160 < T_i$	$320 < T_i$	T3C	160	320
$135 < T_i \leqslant 200$	$275 < T_i \leqslant 392$	T4	135	275
$120 < T_i$	$248 < T_i$	T4A	120	248
$100 < T_i \leqslant 135$	$212 < T_i \leqslant 275$	T5	100	212
$85 < T_i \leqslant 100$	$185 < T_i \leqslant 212$	T6[2]	85	185

1 Group I equipment (for coal mines endangered by firedamp and/or coal dust) is not subjected to any temperature classification. Despite the high ignition temperature of methane–air mixtures (see Table 1.3), the maximum surface temperature is limited to 150°C (or 302°F) in all cases where deposits of coal dust with a low glow temperature (see Table 1.4) may be expected. If such deposits are not likely to occur, the surface temperature is limited to 450°C (or 842°F)

2 Lower temperature classes have not been installed with respect to the fact that carbon disulphide, CS_2, shows the lowest known ignition temperature (see Table 1.3)

3 Article 503 of the NEC (National Electrical Code, USA) limits maximum temperatures for Class III equipment (suitable for ignitable fibres and flyings) to 165°C (equipment *not* subjected to overload) and to 120°C for equipment which may be overloaded. Thus, only temperature classes T3B, T3C, T4 and T4A, T5, T6 are applicable

Table 4.2   Classification of combustible mixtures of gases, vapours or mist with air according to their maximum experimental safe gap (MESG) values

Part A   According to EN50014, IEC 60079-0 and IEC 60079-1 A respectively Electrical apparatus for Group II[1]

Subgroup	*MESG* mm
IIA	MESG > 0.9
IIB	$0.5 \leqslant$ MESG $\leqslant 0.9$
IIC[2]	MESG < 0.5

Part B   According to NEC, National Electrical Code, USA, and CEC, Canadian Electrical Code, Canada, for Class I equipment[3]

Group	*Typical combustible gas*	*MESG* mm
A	Acetylene	0.37
B	Hydrogen	0.29
C	Ethylene	0.60/0.65
D	Propane	0.92

1 Group I equipment (for coal mines endangered by firedamp and/or coal dust) is not subjected to any MESG classification. Methane, CH_4, shows an MESG value of 1.14 mm or 1.17 mm (see Table 1.6)
2 Lower subgroups have not been installed with respect to the fact that hydrogen, H_2, shows the lowest known MESG value of 0.29 mm (see Table 1.6)
3 Class II equipment (intended for areas hazardous to combustible dusts) is subject to a classification as follows:

Group	*Typical combustible dust*
E	metal dust (Division 1 only)
F	coal dust
G	flour, starch, grain

material, flow conditions) in order to achieve the lowest possible values. Normally, conditions are located far away from that specified in IEC 60079-4 and by this way, an undefined but existing 'margin of safety' is incorporated in the philosophy of Table 4.1. In addition, this is a reason for EN 50014 and IEC 60079-0 allowing the increase of the surface temperature of small parts in excess of the values given in Table 4.1 [49].

Firedamp-proof electrical equipment (for coal mines) is not subjected to any classification regarding ignition temperature or MESG values.

It is very uncommon to classify combustible atmospheres according to their minimum ignition energies (see Table 1.7). Figure 4.1 demonstrates that – in double-logarithmic scale – the MESG values are strongly correlated with the minimum ignition energy showing good linearity. The threshold values

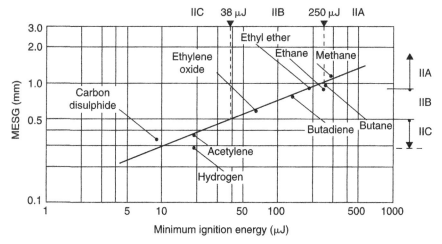

Figure 4.1   *MESG values for selected gas–air mixtures versus their minimum ignition energy. The MESG threshold values, MESG =0.90 mm for IIA/IIB and MESG =0.50 mm for IIB/IIC, correspond to minimum ignition energies of 250 μJ for IIA/IIB and 38 μJ for IIB/IIC.*

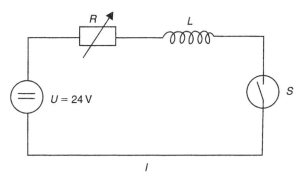

Figure 4.2   *Electric circuit for the determination of the MIC (minimum ignition current) values of gas–air mixtures.*

between Groups IIA/IIB, MESG = 0.90 mm, and IIB/IIC, MESG = 0.5 mm, correspond with minimum ignition energies of roughly 250 μJ (IIA/IIB) and 38 μJ (IIB/IIC).

In practice, combustible mixtures (gases, vapours, mist with air) are classified according to their MIC (minimum ignition current) values. A DC source (Fig. 4.2) with a 24 V voltage is connected via an adjustable resistor R and an air-core inductance $L = 95$ mH to the test apparatus S. In S, a rotating cadmium disc with two grooves and a counter-rotating part with four tungsten wires act as a switching element. This element is surrounded by a pressure resistant enclosure, filled with the mixture to be tested. The test apparatus is described in IEC 60079-11 and in EN 50020. Caused by constructional details, the current

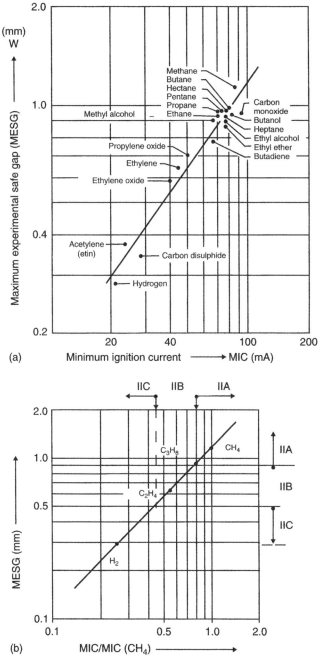

Figure 4.3 *(a) MESG values for selected gas–air mixtures versus their MIC values (minimum ignition current) [35, 44] (b) MESG values for selected gas–air mixtures versus their MIC/MIC (CH₄) ratio, i.e. MIC values normalized to MIC of CH₄–air mixtures. The MESG threshold values MESG = 0.90 mm for IIA/IIB and MESG = 0.50 mm for IIB/IIC correspond to MIC/MIC (CH₄) values of 0.80 for IIA/IIB and of 0.45 for IIB/IIC.*

I is restricted to 3 A, the DC voltage shall not exceed 300 V, *L* is limited to 1 H. Referring to a DC voltage of 24 V and $L = 95\,\text{mH}$, the MIC values are:

102 mA for 8.3% (v/v) methane in air
80 mA for 5.25% (v/v) propane in air
55 mA for 7.8% (v/v) ethylene in air
25 mA for 21.0% (v/v) hydrogen in air

(these gas–air mixtures representing the lowest MIC values with respect to a variation in gas concentration). In [44], linearity between MESG and MIC values in a double-logarithmic scale is reported, with:

$$\text{MESG (mm)} = 0.022 \times \text{MIC}^{0.87}\,\text{(mA)}$$

Often, the MIC values are normalized to the MIC value of methane, CH_4. So, the ratio MIC/MIC (CH_4) shall be considered. Obviously, the normalization to MIC (CH_4) does not change the linearity to MESG as reported in [44]. Figure 4.3 shows MESG versus MIC/MIC (CH_4) in a double-logarithmic scale. MESG = 0.90 mm (IIA/IIB) and MESG = 0.50 mm (IIB/IIC) correspond with MIC/MIC (CH_4) = 0.80 and MIC/MIC (CH_4) = 0.45 respectively.

Thus, these limits give an identical classification of combustible mixtures as the MESG classification and delete the need for a third classification scheme for intrinsically safe circuits (Table 4.3).

Firedamp-proof apparatus, type of protection 'intrinsic safety', according to IEC 60079-11 and/or EN 50020, are not subjected to any classification with respect to MIC or MIC/MIC (CH_4) ratio; it is MIC/MIC (CH_4) = 1 in this case.

'Intrinsic safety' covers two categories of intrinsically safe circuits:

- category 'ia'   ensuring ignition-free operation in normal operation, including electrical faults according to a safety assessment (see Section 6.9)
- category 'ib'   ensuring ignition-free operation in normal operation, including electrical faults according to a safety assessment differing from that for category ia (see Section 6.9).

Categories ia and ib refer to Group I and Group II apparatus.

Table 4.4 gives a survey of the classification of combustible (gas, vapour or mist) mixtures with air according to temperature class and group/subgroup.

Table 4.3   Classification of combustible mixtures of gases, vapours or mist with air into Groups IIA, IIB and IIC with respect to MESG values and MIC/MIC (CH_4) ratio

Group	MESG[1] mm	MIC/MIC(CH₄) ratio[2]
IIA	MESG > 0.9	MIC/MIC(CH_4) > 0.8
IIB	$0.5 \leqslant \text{MESG} \leqslant 0.9$	$0.45 \leqslant \text{MIC/MIC}(CH_4) \leqslant 0.8$
IIC	MESG < 0.5	MIC/MIC(CH_4) < 0.45

1 The MESG classification relates to flameproof enclosures only. Standards: IEC 60079-1 and IEC 60079-1-am 1, EN 50018
2 The MIC/MIC(CH_4) classification relates to intrinsic safety only. Standards: IEC 60079-11, EN 50020

Table 4.4 Examples for grouping and classification of combustible
gas–air mixtures

Part A Cross-correlated

Group/ Temp. class	IIA	IIB	IIC
T1	Acetone Ammonia Carbon monoxide Ethylacetate Methane Methanol Propane	Ethane Town gas	Hydrogen
T2	Butane	Ethanol Ethylene	Acetylene
T3	Hexane		
T4		Ethyl ether	
T5			
T6	(No substance known)	(No substance known)	Carbon disulphide

Part B In alphabetical order

Substance	Ignition temp.°C	T class	Group/ Subgroup
Acetaldehyde	140	T4	IIA
Acetic acid	485	T1	IIA
Acetone	540	T1	IIA
Acetylacetone	340	T2	IIA
Acetyl chloride	390	T2	IIA
Acetylene	305	T2	IIC
Acrylonitrile	480	T1	IIB
Allyl chloride	485	T1	IIA
Ammonia	630	T1	IIA
Amyl acetate	375	T2	IIA
Aniline	617	T1	IIA
Benzene	555	T1	IIA
Benzaldehyde	190	T4	IIA
Benzyl chloride	585	T1	IIA
Bromobutane	265	T3	IIA
Bromoethane	510	T1	IIA
Butadiene	430	T2	IIB
Butane	365	T2	IIA
Butanol	340	T2	IIA
Butene	440	T2	IIB

(*continued*)

Table 4.4   (*continued*)

Substance	Ignition temp. °C	T class	Group/ Subgroup
Butyl acetate	370	T2	IIA
Butyldigol	225	T3	IIA
Butyraldehyde	230	T3	IIA
Carbon disulphide	95	T6	IIC
Carbon monoxide	605	T1	IIA
Chlorobenzene	637	T1	IIA
Chloroethane	510	T1	IIA
Chloroethanol	425	T2	IIA
Chloroethylene	470	T1	IIA
Chloromethane	625	T1	IIA
Chloropropane	520	T1	IIA
Coal tar naphtha	272	T3	IIA
Cresol	555	T1	IIA
Cyclohexane	259	T3	IIA
Cyclohexanol	300	T2	IIA
Cyclohexanone	419	T2	IIA
Cyclohexylamine	290	T3	IIA
Cyclopropane	495	T1	IIB
Decahydronaphthalene	260	T3	IIA
Diacetone alcohol	640	T1	IIA
Diaminoethane	385	T2	IIA
Diamyl ether	170	T4	IIA
Dibutyl ether	185	T4	IIB
Dichloroethane	440	T2	IIA
Dichloropropane	555	T1	IIA
Diethyl ether	170	T4	IIB
Dihexyl ether	185	T4	IIA
Dimethylaniline	370	T2	IIA
Dioxane	379	T2	IIB
Epoxypropane	430	T2	IIB
Ethane	515	T1	IIB
Ethanol	425	T2	IIB
Ethoxyethanol	235	T3	IIB
Ethyl acetate	460	T1	IIA
Ethylbenzene	431	T2	IIA
Ethylene	425	T2	IIB
Ethylene oxide	440	T2	IIB
Ethyl formate	440	T2	IIA
Ethyl mercaptan	295	T3	IIA

(*continued*)

Table 4.4 (*continued*)

Substance	Ignition temp.°C	T class	Group/ Subgroup
Ethyl methyl ether	190	T4	IIB
Ethyl methyl ketone	505	T1	IIA
Formaldehyde	424	T2	IIB
Formdimethylamide	440	T2	IIA
Hexane	240	T3	IIA
Heptane	215	T3	IIA
Hydrogen	560	T1	IIC
Hydrogen sulphide	270	T3	IIB
Isopropylnitrate	175	T4	IIB
Kerosene	210	T3	IIA
Methane (firedamp)*	595	T1	I/IIA
Methanol	455	T1	IIA
Methoxyethanol	285	T3	IIB
Methyl acetate	475	T1	IIA
Methyl acetoacetate	280	T3	IIA
Methylamine	430	T2	IIA
Methylcyclohexane	260	T3	IIA
Methylcyclohexanol	295	T3	IIA
Methyl formate	450	T1	IIA
Naphtha	290	T3	IIA
Naphthalene	528	T1	IIA
Nitrobenzene	480	T1	IIA
Nitroethane	410	T2	IIB
Nitromethane	415	T2	IIA
Nitropropane	420	T2	IIB
Nonane	205	T3	IIA
Paraformaldehyde	300	T2	IIB
Paraldehyde	235	T3	IIA
Pentane	285	T3	IIA
Pentanol	300	T2	IIA
Phenol	605	T1	IIA
Propane	470	T1	IIA
Propanol	405	T2	IIA
Propyl methyl ketone	505	T1	IIA
Pyridine	550	T1	IIA
Styrene	490	T1	IIA

(*continued*)

Table 4.4   *(continued)*

Substance	Ignition temp.°C	T class	Group/ Subgroup
Tetrahydrofurfuryl alcohol	280	T3	IIB
Toluene	535	T1	IIA
Toluidine	480	T1	IIA
Trimethylbenzene	470	T1	IIA
Trioxane	410	T2	IIB
Turpentine	254	T3	IIA
Xylene	464	T1	IIA

Note: Electrical equipment for coal mines is related to Group I, for industrial application in areas hazardous due to methane to Group IIA
* Industrial methane may include other components, e.g. hydrogen (up to some % v/v)

Chapter 5

Marking and selection of explosion protected apparatus

It shall be emphasized that all markings described in this chapter are related to explosion protection only and shall be considered as an *additional* marking to that given in the standards for non-explosion protected apparatus, e.g. the manufacturer's name or his trademark, type code and rated power/voltage/current/speed/torque, serial number etc.

'Selection' in this chapter is strictly focused on safety related data in the field of explosion protection, neglecting all the other aspects like maintenance or financial considerations (this will be the objective of Chapter 9). Here, selection means the appropriate choice of electrical apparatus according to temperature class and MESG or MIC/MIC (CH$_4$) grouping, taking into account the combustible raw materials and products in the relevant plant.

IEC and EN Standards ask for marking the type of protection as an abbreviation completed by the classification scheme (temperature class, explosion group) as described in the previous chapter. Table 5.1 gives a survey of the types of explosion protection and their abbreviations. Table 5.2 summarizes the IEC and EN marking code.

If the apparatus has been type tested and certified (certificate of conformity, control certificate) according to the 'old' article 100 directives, the identification mark of the notified body and the certificate number shall be affixed on the apparatus. In the same way, a component (with a component certificate) shall be marked.

For the certificate number, there is a code common to all notified bodies:

$$\underline{\text{XXXXX}} \quad \underline{\text{XX}} \quad \underline{\text{X}} \quad \underline{\text{XXXX}} \quad \underline{\text{X}}$$

$$\quad 1 \qquad\quad 2 \quad\ \ 3 \qquad 4 \qquad 5$$

1 Abbreviation of the notified body (e.g. BVS, PTB, CESI, LCIE)
  Some notified bodies add a suffix, e.g. PTB-Ex ...
2 Year of issue (abbreviated) of certificate
3 'Generation' of Harmonized European Norms (HN)
  (see Table 3.7). The 'A' generation is *not* marked in this code
4 Number of certificate
5 There may be an additonal mark:
  X = special conditions. This includes instructions for installation, operation and maintenance as well as technical data not complying with or

Table 5.1   Types of protection: electrical apparatus for potentially explosive atmospheres (zone 1)

Type of protection	Abbrevi- ation	Standard		
		IEC	EN	VDE
General requirements	–	IEC 60079-0 IEC 60079-0-am2	EN 50014	VDE 0170/0171 Teil 1
Oil immersion	o	IEC 60079-6	EN 50015	VDE 0170/0171 Teil 2
Pressurized apparatus	p	IEC 60079-2	EN 50016	VDE 0170/0171 Teil 3
Powder filling	q	IEC 60079-5	EN 50017	VDE 0170/0171 Teil 4
Flameproof enclosure	d	IEC 60079-1 IEC 60079-1-am1	EN 50018	VDE 0170/0171 Teil 5
Increased safety	e	IEC 60079-7	EN 50019	VDE 0170/0171 Teil 6
Intrinsic safety	i (ia, ib)	IEC 60079-11	EN 50020	VDE 0170/0171 Teil 7
Encapsulation	m	IEC 60079-18	EN 50028	VDE 0170/0171 Teil 9
Intrinsically safe electrical systems	i	IEC 60079-11	EN 50039	VDE 0170/0171 Teil 10

Note: Compliance is given between EN and VDE Standards
IEC Standards do not fully comply with EN Standards

Table 5.2   Marking of explosion protected apparatus according to IEC and EN Standards

Marking code: $\dfrac{}{1}$ Ex $\dfrac{}{2}\dfrac{}{3}\dfrac{}{4}$

1 Referring to EN: E (='European')
  Referring to IEC: – blank –
2 Type(s) of protection (abbreviations only, see Table 5.1). The dominant/main type of protection, followed by auxiliary types of protection, e.g. a flameproof motor with an 'e'-type terminal box: de, or a pressurized motor with an 'e'-type terminal box and flameproof standstill heaters: ped
3 Group
  I   Firedamp-proof
  II Explosion protected with subgroups IIA/IIB/IIC for flameproof and/ or intrinsically safe apparatus
4 Temperature class (for Group II apparatus only)

deviating from the technical contents of the standards referred to. As an example, there may be an increased/decreased ambient temperature T_{amb} range for the apparatus,

e.g. $-50°C \leqslant T_{amb} \leqslant +60°C$,

or $-5°C \leqslant T_{amb} \leqslant +35°C$

instead of $-20°C \leqslant T_{amb} \leqslant +40°C$

as specified in HNs

U = this mark indicates that the certificate is a component certificate for electrical equipment which cannot operate on its own (e.g. cable entries for enclosures in increased safety – e –).

In addition, the Ex-mark in a hexagon (see Fig. 3.1(a)), in the special case of Group I apparatus complying with the so-called A generation of Harmonized European Norms (HN) completed by an 'I' in a circle, or, for apparatus with a control certificate, completed by an 'S' in a circle (see Fig. 3.1(b)), shall be affixed to all the apparatus tested and certified according to the article 100 directives (see Table 3.6).

If an intrinsically safe apparatus is intended for operating on its own (installed in zone 1, not including non-intrinsically safe circuits), the marking

is Ex ia ..., Ex ib ...

or EEx ia ..., EEx ib ...

Electrical apparatus with combined non-intrinsically safe and intrinsically safe components, e.g. a flameproof enclosure with a power supply for intrinsically safe circuits, with a 690 V input, terminal boxes in increased safety, is considered as an 'associated apparatus' (from the 'intrinsic' point of view). In this case, the abbreviation for intrinsic safety is put in brackets:

Ex de [ia] ..., Ex de [ib] ...

or EEx de [ia] ..., EEx de [ib] ...

If the apparatus with intrinsically safe circuits is installed *outside* a hazardous area, e.g. in a control room in a chemical plant, the marking inserts 'Ex' or 'EEx' and the type of protection in brackets:

[Ex ia] ..., [Ex ib] ...

or [EEx ia] ... , [EEx ib] ...

In this case the apparatus is *not* subjected to the temperature classification T1 ... T6 because of its 'safe' environment. For Group II intrinsically safe circuits, the MESG or MIC grouping IIA/IIB/IIC shall be marked according to the relevant standards (IEC 60079-11, EN 50020).

In a similar way as described above, the brackets indicate an 'associated' apparatus.

Examples:

IEC marking	EN marking
Ex de I	EEx de I
Ex de IIB T3	EEx de IIB T3
Ex ped IIC T1	EEx ped IIC T1
Ex ib I	EEx ib I
Ex ia IIC T5	EEx ia IIC T5
Ex oe II T4	EEx oe II T4
Ex p II T2	EEx p II T2
Ex p [ia] IIB T3	EEx p [ia] IIB T3
Ex d [ib] I	EEx d [ib] I
[Ex ia] IIC	[EEx ia] IIC
[Ex ib] I	[EEx ib] I

In the following, some examples for marking of explosion protected apparatus are given (the manufacturing companies are fictitious). It is assumed that these apparatuses have been manufactured, type tested and certified under the 'old' article 100 directives (Table 3.6), applicable until 2003-06-30.

1 Pressurized power transformer, with terminal boxes in increased safety, for coal mines:

Magnaflux Transformatorenwerke AG, Dortmund	(manufacturer)
HICORE 10/1-1600	(type code)
1.123.88	(serial number)
EEx pe I	(protection code)
BVS 87.1098	(notified body, certificate number)
'Ex' and 'I' mark	(according to Fig. 3.1)

2 Luminaire for high pressure mercury vapour lamp, flameproof enclosure, terminal box in increased safety, containing the series reactor protected by powder filling, for chemical plants:

Eclairages LUX SARL	(manufacturer)
La Trinité-sur-Mer	
EOS 2000	(type code)
Hg 1234	(serial number)
EEx deq IIC T3	(protection code)
T_{amb} $-40°\ldots+40°C$	(special conditions)[1]
PTB Ex-97.D.2998 X[2]	(notified body, certificate number)
'Ex' mark	(according to Fig. 3.1)

3 Voltage transformer, encapsulated, terminals according to increased safety, for coal mines. Component to be inserted in an enclosure in increased safety:

Bellaplast Kunststofftechnik GmbH, Bitterfeld	(manufacturer)
SINUS G 10/100/100-50	(type code)
1001 SIN	(serial number)

EEx me I (protection code)
BVS 93.C.1123 U[3] (notified body, certificate number)

Note: The 'Ex' mark according to Fig. 3.1 shall *not* be used on a component.	

4 Pressure transducer, in an intrinsically safe circuit, category ia, for chemical plants, general application:

MAXWELL SIGNALLING LTD, Aberdeen (manufacturer)
Fluidguard 0-500 (type code)
MS 123 987 (serial number)
EEx ia IIC T6 (protection code)
$U_i = ^4$ (limiting values for
$I_i = ^4$ external circuits)
$L_i = ^4$
$C_i = ^4$
PTB – Ex 97.D.2987 (notified body, certificate number)
'Ex' mark (according to Fig. 3.1)

The following examples show the EN marking for 'associated' apparatus, as explained above:

5 DC power supply for intrinsically safe circuits, Group IIB, to be installed in a non-hazardous zone (e.g. in a pressurized control room):

HUYGENS CONTROLS BV, Den Haag (manufacturer)
Maxipower 12/1 (type code)
C 987/52 (serial number)
[EEx ib] IIB (protection code)
$U_o = ^4$ (limiting values for
$I_o = ^4$ external circuits)
$L_o = ^4$
$C_o = ^4$
PTB Ex-92.C.2345 X[5] (Notified body, certificate number, special condition)
'Ex' mark (according to Fig. 3.1)

6 Process equipment for coal mines (continuous registration of process data), data transmission via intrinsically safe circuits, category ia, to be installed in a non-hazardous place:

PERPETUUM KG, Kulmbach (manufacturer)
Data Logger D1-Sch (type code)
003/98 (serial number)
[EEx ia] I (protection code)
$U_i = ^4$ (limiting values for
$I_i = ^4$ external circuits)

$L_i = {}^4$
$C_i = {}^4$
BVS 97.D.1003 X^5 (Notified body, certificate
                                                            number, special condition)

'Ex' mark   (according to Fig. 3.1)

7 Power supply for intrinsically safe circuits, category ia, for mining application, flameproof enclosure, terminal box in 'increased safety' on the line
   power side:
   Steiermärkische Montangesellschaft,   (manufacturer)
   Mürzzuschlag
   1000-Sch 24/2   (type code)
   1998-02-S 05 8   (serial number)
   EEx de [ia] I   (protection code)
   $U_o = {}^4$ (limiting values for
     $I_o = {}^4$ external circuits)
   $L_o = {}^4$
   $C_o = {}^4$
   BVS 96-C.1123   (Notified body, certificate
                                                             number)

   'Ex' mark   (according to Fig. 3.1)

8 Same transformer as given in 1, additionally equipped with temperature
   sensors in intrinsically safe circuits, category ia:
   EEx pe [ia] I   (protection code)
   $U_i = {}^4$ (limiting values for
     $I_i = {}^4$ external circuits)
   $L_i = {}^4$
   $C_i = {}^4$
   (the other markings remain unchanged)

Notes:

1 The ambient temperature range is extended to $-40°C$ instead of $-20°C$ as
   given in the IEC or EN Standards
2 The 'X' indicates a special condition. In this case, reference is made to the
   extended ambient temperature range
3 The 'U' indicates that this apparatus cannot be used/installed on its own. The
   voltage transformer, as a component, has to be inserted in an enclosure complying with EN 50014 and 50019 'increased safety', ensuring an appropriate
   tightness against the ingress of dust and water. The certificate is a component
   certificate
4 In intrinsically safe circuits, the electric values shall be limited to ensure the
   correct operation of 'intrinsic safety'. This refers obviously to voltages and
   currents, but in addition to energy storage elements, capacitances and
   inductances, e.g. in cables, or as 'internal' values typical for the apparatus
   referred to
5 The 'X' indicates a special condition. In this case, reference is made to the
   installation in a non-hazardous place

For explosion protected electrical equipment manufactured, type tested, certified and commissioned under the 'ATEX 100a' Directive, the marking described in Table 5.2 shall be added by the marking code given in Table 5.3. The 'Ex' in a hexagon has been taken from the 'old' article 100 directives, whereas the 'B', 'C', 'D' or 'E' marks for the 'generation' of Harmonized Norms have been deleted.

After examining the IEC and the 'European' marking scheme, a survey of the NEC (USA) and CEC (Canada) marking scheme in Table 5.4 will close this subject.

Table 5.3 Marking of explosion protected apparatus according to ATEX 100a Directive (94/9/EC) (this is additional to the EN marking as in Table 5.2)

General   CE — Ex — — —
            1 2 3 4 5 6

1 CE mark according to Fig. 5.1
2 Identification number of the notified body (if involved in a module according to Table 3.9)
3 Ex mark according to Fig. 3.1(a)[1]
4 Group of apparatus (see Table 2.3)
5 Category of apparatus (see Table 2.3), M = Mining
6 G = Gas
   D = Dust

Examples	*Apparatus marked …*	*Installed in …*
	CE^{0102} Ex II 1G	zone 0
	CE^{0158} Ex I M1	mining, category 1
	CE^{0158} Ex II 1D	zone 20
	CE^{0102} Ex II 2G	zone 1
	CE^{0158} Ex II 2D	zone 21
	CE^{0158} Ex I M2	mining, category 2
	CE Ex II 3G	zone 2
	CE Ex II 3D	zone 22

1 The Ex mark in a hexagon (see Fig. 3.1(a)) has different denotations:
  • according to the 'old' article 100 directives (see Table 3.6) the Ex mark indicates that the apparatus has passed successfully a type test examination at a notified body
  • according to the ATEX 100a Directive (94/9/EC), supplement II, clause 1.0.5, the Ex mark indicates 'the prevention of explosions' in a very general sense. Thus, apparatus for Group II, category 3 (3G, 3D), intended for installation in zone 2 or zone 22, may be marked with 'Ex' without involving a notified body for type testing

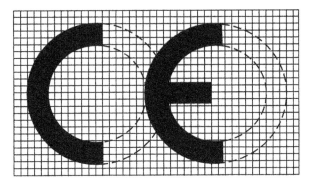

Figure 5.1   *'CE' mark according to Directive 94/9/EC (ATEX 100a). The lattice is not part of the CE mark. It indicates the proportions only.*

Table 5.4.   Marking scheme for Class I apparatus according to National Electrical Code, NEC, USA, and Canadian Electrical Code, CEC, Canada (traditional system)[1]

General	Class I	— — —
		1 2 3

1 Division 1[2]
  Division 2[2]
2 Group A (see Table 4.2 Part B)[3]
      B
      C
      D
3 Temperature classification (see Table 4.1 Part B)

Examples   Class I Div. 1 C T3B
           Class I Div. 2 D T2D

1 Parallel to the traditional protection methods, the USA and Canada have adopted the IEC Standards for explosion protected equipment (see Table 3.14), preferably IEC 60079-0, 60079-1, 60079-2, 60079-5, 60079-6, 60079-7, 60079-11, 60079-15, 60079-18 (at the moment, IEC 60079-2 'Pressurization' is *not* adopted in the USA)
2 Division 1 is comparable with zones 0 and 1 according to IEC classification or the classification according to Directive 1999/92/EC
  Division 2 is comparable with zone 2
3 Very roughly, Groups A, B, C and D may be compared with IEC or EN Groups IIA, IIB, IIC as follows:
  IIA   Group D
  IIB   Group C
  IIC   Group A, Group B

Table 5.5 gives a guideline for the selection of explosion protected electrical apparatus, i.e. the appropriate choice of temperature class and group referring to the hazardous substances in the plant to be considered. Very generally 'lower' temperature classes do comply with the requirements for

Table 5.5 Examples for selection and marking of firedamp-proof and explosion protected electrical apparatus according to IEC and EN Standards[1]

No.	Apparatus, type of protection	Marking
1	Cage induction motor, flameproof enclosure, with terminal box in increased safety	
	• for coal mines	Ex de I
	• driving a hydrogen compressor	Ex de IIC T1
	• operating in an ethylene plant	Ex de IIB T2
2	High voltage circuit breaker, enclosure and terminal boxes in increased safety, control box has a flameproof enclosure, flameproof vacuum circuit breaker	
	• for coal mines	Ex ed I
3	Low voltage switchgear, flameproof enclosure	
	• for coal mines	Ex d I
	• in a refinery	Ex d IIA T3
	• for chemical plants with acetylene	Ex d IIC T2
4	Luminaire for tubular fluorescent lamps, increased safety, encapsulated series reactor, flameproof power factor capacitor	
	• for coal mines	Ex emd I
	• in a refinery	Ex emd IIA T3
	• installed in an ethylene plant	Ex emd IIB T2
	• for chemical plants, general application	Ex emd IIC T6
5	High power cage induction motor, pressurized, terminal box in increased safety, standstill heaters in flameproof enclosures	
	• in a refinery	Ex ped IIA T3
	• in an acetylene plant	Ex ped IIC T2
	• operating a natural gas compressor	Ex ped IIA T1
6	Cable connecting box, increased safety	
	• for coal mines	Ex e I
	• in a refinery	Ex e II T3
	• for chemical plants, general application	Ex e II T6
7	Transformer, oil immersion, with terminal boxes in increased safety	
	• for coal mines[2]	Ex oe I
	• for chemical plants, general application with the exclusion of carbon disulphide	Ex oe II T5
8	Slipring induction motor, increased safety, flameproof slipring enclosure, flameproof standstill heaters, terminal boxes in increased safety	
	• for coal mines	Ex ed I

(continued)

Table 5.5　(*continued*)

No.	Apparatus, type of protection	Marking
	• operating a natural gas compressor	Ex ed IIA T1
	• operating a hydrogen compressor	Ex ed IIC T1
	• in a refinery	Ex ed IIA T3
	• in an ethylene plant	Ex ed IIB T2
9	Pressure transducer with an intrinsically safe circuit	
	• for coal mines	Ex ia I
		Ex ib I
	• chemical plants, general application	Ex ia IIC T6
		Ex ib IIC T6
	• in an ethylene plant	Ex ia IIB T2
		Ex ib IIB T2
	• in a refinery	Ex ia IIA T3
		Ex ib IIA T3

1 For simplification, the IEC marking is given in this table. To comply with EN marking, an 'E' shall be placed in front of the 'Ex'

2 In some countries, e.g. Germany, national installation codes do not allow the installation of oil immersed apparatus in coal mines

'higher' classes, e.g. a T4 apparatus can be used instead of a T3 apparatus (but not vice versa), and a IIC apparatus (or intrinsically safe circuit) can replace a IIA or IIB apparatus or circuit.

Table 5.5 contains the 'minimum' requirements throughout all the examples. Chapter 9 will demonstrate that the design of an apparatus for 'lower' temperature classes (T4, T5, T6) often adversely affects its economic efficiency.

In a similar way, the IIC design, e.g. of a switching unit, asks for enhanced precision in milling and finishing and requires an altered concept of joints for doors and covers in comparison with a IIA or IIB design.

In addition, EN 50015/1994-04 (i.e. 2nd edition) 'oil immersion – o –' excludes the application of mineral oil for Group I apparatus. Other fluids shall be used, e.g. silicon-based fluids.

Chapter 6

The different types of protection – constructional requirements

In the historical development of explosion protection – spanning almost a century – the different constructional details (which 'trim' motors, luminaires, sensors or switchgear to be explosion protected) have been summarized as different 'types of explosion protection'. And these types of protection have been described by individual standards. There are no individual standards for specified electrical apparatus – neither for motors or luminaires nor for transformers, capacitors or sensors. These standards describe technical methods for individual protection concepts, each covering a great variety of different electrical apparatus. To give an example: flameproof enclosures may be built for motors, transformers, light fittings, switchgear and sensors – with an identical physical and technical background. Following this philosophy, this part of the book describes the different types of explosion protection (not the different sorts of apparatus) in individual chapters.

By no means will this chapter claim to fill a standard's place. Therefore, a very careful reading of standards is recommended and in case of any need for clarification the testing houses or notified bodies should be asked for interpretation.

This chapter will seek to answer the question: Why do standards ask for specified temperatures of windings, for limited gap and joint values and for restricted voltages and currents in electric circuits? The following intends to clarify the physical background of the standards' technical content, to explain the methods of explosion protection in electrical engineering, and demonstrate modern explosion protected apparatus to give an appreciation of the 'bandwidth' of the different types of protection.

To start with the latter: Figs 6.1–6.8 give an impression of environmental conditions at the place of duty for modern explosion protected equipment, and at the same time, they may anticipate its efficiency and power as well as its broad range of application.

This chapter deals with protection techniques for Group I and Group II zone 1 equipment, or, with the terminology of the ATEX 100a Directive (see Table 2.3), for Group I, category M2, and for Group II, category 2G, apparatus. In a few sections only, an additional reference will be made to other categories of electrical apparatus.

Figure 6.1   *Gas production platform SLEIPNER in the North Sea, near the western boundary of the Norwegian continental shelf part.*

Figure 6.2   *Pressurized motor, located in zone 1, as a gas compressor drive on SLEIPNER.*
Type of protection: EEx pde[ib] IIC T3; Rated power: 10 800 kW; Rated voltage: 3 kV; Purging volume: 10.8 m^3.

6.1   General requirements

Standards: EN 50014
                      IEC 60079-0
Key letter: – none –

These standards summarize all the constructional requirements, type verifications and tests which refer to all types of protection in common. Thus, each

Figure 6.3   *Shearer-loader at Daw Mill Colliery, Arley, Coventry/United Kingdom.*
Technical data – Machine height: 2051 mm; Length (between cutting drum shafts):
12 570 mm; Length of machine body: 8000 mm; Cutting range: +5100 mm
to −600 mm; Cutting drum diameter: 2400 mm; Cutting depth: 1000 mm; Cutting
drum speed: 29 min^{-1}
Cutter motors (cage induction motors) – Rated power: 2 × 500 kW; Rated voltage:
3.3 kV; Type of protection: EEx dI; Electric haulage unit (with DC motors) – Rated
power: 2 × 54 kW; Rated voltage: 630 V; Type of protection: EEx dI
Hydraulic unit (with cage induction motor) – Rated power: 1 × 60 kW; Rated voltage: 1000 V; Type of protection: EEx dI; Lump breaker (cage induction motor) –
Rated power: 1 × 100 kW; Rated voltage: 3.3 kV; Type of protection: EEx dI
Low voltage casing, high voltage casing, control unit and transformer casing –
Type of protection: EEx dI; General – Tractive force, max.: 643 kN; Speed, max.:
15.4 m/min; Installed motor rated power: 1268 kW; Machine weight: 70 t.

type of protection shall comply with the specific standards, and, in addition,
with the 'General requirements' standard, e.g. a flameproof motor shall comply with EN 50018 *and* with EN 50014 (or with IEC 60079-1 *and* with IEC
60079-0) in full. The frequently used expression 'flameproof motor according
to EN 50018' is somewhat incomplete and should be extended to '... according to EN 50014 and 50018'.

The 'General requirements' cover apparatus grouping and temperature
classification, constructional requirements for enclosures and fasteners,
connection facilities and terminal compartments, cable entries and supplementary requirements for rotating electrical machines, switchgear, plugs and
sockets and luminaires. A third part of the standards cover type verifications
and tests as well as marking requirements.

In the following, some contents of the 'General requirements' will be
explained in more detail.

Figure 6.4   *Battery powered electric locomotive for coal mines, with 3 AC cage induction traction motor.*

Technical data:

Locomotive – Axle assembly code (according to UIC (Union Internationale des Chemins de Fer)): B; Width: 1000 mm; Length over buffers: 6170 mm; Total height: 1800 mm; Wheelbase: 1900 mm; Track gauge: 540 ... 750 mm; Wheel diameter (new/weared down): 600/560 mm; Mass: 15 t; Gear ratio (goods/passenger train): 15.7/9.5:1; Speed, max. (goods/passenger train): 14.4/27 km/h; Starting tractive force (goods/passenger train): 25/15 kN; Electric braking force, max. (goods/passenger train): 22.5/15 kN; Electric braking power: 45 kW

Battery, battery container – Lead–acid accumulators; Rated capacity: 2×600 Ah; Rated voltage: 2×108 V DC (connected in series); Type of protection: EEx el

DC–3 AC convertor – Input voltage: 154–254 V DC; DC link capacitor: 4×6000 μF/350 V; Output voltage: 0–200 V; Output current: 0–300 A; Output frequency: 0–80 cps; Semiconductors: IGBT type; Pulse frequency of IGBT, max.: 1400 cps; Type of protection: EEx dl; Recuperative braking via convertor back to battery

3 AC traction motor – Cage induction motor – Rated continuous power: 45kW; Speed, max.: 2300 min^{-1}; Frequency: 0–78 cps; Voltage: 0–180 V; Torque, max.: 503 Nm; Thermal class: F; Type of protection: EEx dl.

Apparatus grouping and temperature classification are described in Chapter 4, especially in Table 4.1 Part A and Table 4.2 Part A for EN and IEC specifications. As a general rule, the ambient temperature (in service) for explosion protected electrical apparatus is limited to the range from $-20°C$ to $+40°C$. Deviating ambient temperature ranges shall be indicated, e.g. $-30°C \leqslant T_{amb} \leqslant +60°C$, or the apparatus shall be marked with the symbol 'X', which indicates special conditions in general.

Constructional requirements for enclosures cover mechanical aspects to achieve a certain robustness against mechanical damage as well as the avoidance of frictional sparks, and an 'electrostatic safety' referring to enclosures of plastic material only.

Figure 6.5 *Ethylene compressor for ELENAC low density polyethylene plant (France).*
Compressor type: 10 cylinder – 2 stage, reciprocating; Mass flow: 116 t/hour; Inlet pressure: 28 Mpa (280 bar); Outlet pressure: 310 Mpa (3100 bar); Power demand: 23 500 kW; Compressor weight: 270 t.

Figure 6.6 *Synchronous motor for ELENAC ethylene compressor. Explosion protection according to EN 50016 – Pressurization –, Group II.*
Rated power: 23 500 kW; Rated voltage: 11 kV; Purging volume: 1000 m^3; Pressure differential: 380–1500 Pa, depending on location; Certificate: BVS 00.E.2017.

Figure 6.7 *Oxygen analyser for gassy coal mines (Group I). Continuous monitoring of the oxygen content in air (e.g. behind fire dams, in panel entries, in gangways).*
Type of protection: EEx ia I; Certificate: BVS 90.B.1136; Measuring range: 0–25% (v/v) O_2; Power supply: 8 … 16.5 V, max. 65 mA; Outputs: (0–25% (v/v) O_2); Analogue output: 10–100 mV; Frequency shift output: 6–15 cps.

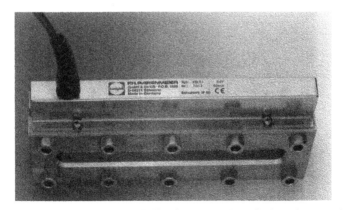

Figure 6.8 *Vessel light fitting with LEDs for inspection windows in vessels, tanks, bunkers and chemical reactors.*
Type of protection: EEx ia IIC T4/T5/T6; Certificate: BVS 98.E.2068 X; The temperature class rating depends on ambient temperature (maximum +60°C) and rated power. Rated power (max.): 2 W; Supply voltage: 24 V AC/DC.

Enclosures (or parts of them) shall guarantee a resistance to impact according to Table 6.1.

An impact with 20 J (to give a rough idea: a mass of 1 kg falling vertically from a height of 2 m) will not damage significantly an enclosure made of steel with a thickness of – say – 2 to 3 mm.

Table 6.1 Resistance to impact for enclosures (according to EN 50014 and IEC 60079-0)

	Impact energy J			
Apparatus group	*I*		*II*	
Risk of mechanical damage	*High*	*Low*	*High*	*Low*
(a) Guards, protective covers fanhoods, cable entries	20	7	7	4
(b) Plastic enclosures	20	7	7	4
(c) Light metal or cast metal enclosures	20	7	7	4
(d) Enclosures of materials other than in (c) with wall thickness				
<3 mm for Group I	20	7	–	–
<1 mm for Group II	–	–	7	4
(e) Light transmitting parts without guard	7	4	4	2
(f) Light transmitting parts with guard	4	2	2	1
Apparatus marking		X		X

In order to avoid frictional sparks (see Section 1.2.4), Group I enclosures shall not contain, by mass

- more than 15% in total of aluminium, magnesium and titanium
- more than 6% in total of magnesium and titanium

Group II enclosures (limiting values for Group II enclosures are currently under consideration in IEC) shall not contain, by mass, more than 6% of magnesium.

Enclosures (or parts of them) made of plastic material shall be constructed so as to prevent the accumulation of electrical charges at the surface when they are rubbed or cleaned *in situ* (see Section 1.2.5). For Group I enclosures, plastic material parts with a surface area projected in any direction of more than 100 cm^2 shall have a surface resistance lower than $1.0 \cdot 10^9$ ohms. For Group II enclosures, a surface resistance lower than $1.0 \cdot 10^9$ ohms or the limitation of surface areas to 100 (400) cm^2 for Group IIA and IIB enclosures and to 20 (100) cm^2 for Group IIC enclosures are assumed to be 'electrostatic safe'. The surface areas given in brackets refer to parts surrounded by conductive earthed frames or to parts additionally protected against the occurrence of dangerous electrostatic charges.

As a general rule, the access to uninsulated live parts within an apparatus shall be possible only with the aid of a tool. Some types of protection, e.g. 'flameproof enclosure – d –' and 'increased safety – e –' require 'special fasteners' complying with the following:

- the thread shall be coarse pitch in accordance with ISO Standard 262, with a tolerance fit of 6 g/6 H in accordance with ISO 965

- the head of the screw or nut shall be in accordance with ISO 4014, 4017, 4032 or 4762, and, in the case of hexagon socket set screws, with ISO 4026, 4027, 4028 or 4029
- for Group I the heads of special fasteners liable to mechanical damage in normal service should be protected (e.g. by the use of shrouds or counter-bored holes).

In many applications, energy storage elements, e.g. capacitors, or 'hot' components which exceed the maximum surface temperature permitted for the temperature class of the apparatus, may be exposed to the environmental hazardous atmosphere when doors or covers are removed in a time shorter than that necessary to allow discharge to a 'safe' value (see Table 6.2) or to cool down to a surface temperature complying with the temperature class rating of the apparatus. In such cases, the apparatus shall be marked with a warning 'AFTER DE-ENERGIZING, DELAY X MINUTES BEFORE OPENING' (X being the value of the delay required), or, alternatively, with the warning 'DO NOT OPEN IN A HAZARDOUS AREA' (IEC Standard 60079-0 gives the following wording: 'DO NOT OPEN WHEN AN EXPLOSIVE GAS ATMOSPHERE MAY BE PRESENT').

The discharge time t_s of a capacitor C with voltage U_o in its operational mode to the 'safe' value W_s via a resistor R can be calculated as follows: at time $t = 0$, C with voltage U_o is disconnected from the voltage source and switched to the resistor R. Then, the capacitor voltage falls according to $U = U_o \cdot e^{-t/RC}$ and the stored energy according to $W = \frac{1}{2}CU_o^2 \cdot e^{-2t/RC}$ or:

$$W = W_o\, e^{-2t/RC}, \quad W_o = \tfrac{1}{2}CU_o^2$$

Obviously, the time constant for the voltage drop down to $1/e$ is RC = twice the energy drop down to $1/e$.

It follows

$$t = -RC/2 \ln W/W_o$$

especially

$$t_s = -RC/2 \ln W_s/W_o$$

To give an example: a Group I frequency convertor contains a capacitor with $C = 2000\,\mu F$ at an operating voltage of $U_o = 1000\,V$ in its DC link. This capacitor shall be discharged to $W_s = 0.2\,mJ$ according to Table 6.2. The discharge resistor has been chosen with $R = 100\,ohm$ (to limit the initial peak current to 10 A!). So, W_o is 1000 J and $RC = 0.2\,s$. The discharge time from W_o to W_s is $t_s = 1.54\,s$, a value which can easily be verified with:

$$U = U_o \cdot e^{-t/RC}$$

Table 6.2 Maximum energy storage in capacitors after removal of doors or covers

Group	Energy storage* Operational charging voltage	
	≥200 V	<200 V
I, IIA	0.2 mJ	0.4 mJ
IIB	0.06 mJ	0.12 mJ
IIC**	0.02 mJ	0.04 mJ

* The maximum energy storage corresponds to the
  minimum ignition energy values for gas–air and
  vapour–air mixtures (see Tables 1.7 and 1.9)
** This includes apparatus generally marked Group II,
  e.g. a pressurized enclosure EEx p II T …

or

$$U_\mathrm{s} = U_\mathrm{o}\, e^{-t_\mathrm{s}/RC}$$

$$U_\mathrm{s} = 0.4528\,\mathrm{V}$$

and

$$W_\mathrm{s} = \tfrac{1}{2}CU_\mathrm{s}^2$$

$$= 0.2\,\mathrm{mJ}$$

For 'small' U and C values, the discharge resistor may be permanently connected to the capacitor. In the example given above, the resulting power loss would be 10 kW. Consequently, a discharge switch shall be fitted which automatically closes the discharge circuit. For pressurized apparatus, a single-pole vacuum tube (contactor type) may be used. Its open position is achieved by a pneumatically actuated drive, the closed position should be maintained by the atmospheric air pressure, and, if necessary, by an additional helical spring.

Back to the standard's content: doors and covers giving access to the interior of enclosures containing remotely operated circuits with switching contacts which can be made or broken by non-manual influences (e.g. electrical, mechanical, magnetic, electromagnetic, pneumatic, hydraulic and others) shall

- either be interlocked with a disconnector which prevents access to the interior, unless it has been operated to disconnect unprotected internal circuits or
- be marked with the warning 'DO NOT OPEN WHEN ENERGIZED'.

In the first case, where it is intended that some internal parts will remain energized after operation of the disconnector, those energized parts shall be protected by either means as below:

- one of the types of protection listed in Sections 6.2–6.9
- protection as follows:
  1 clearances and creepage distances between phases (poles) and to earth in accordance with the requirements of 'increased safety – e – ' (EN 50019, IEC 60079-7) and

2 an internal supplementary enclosure which contains the energized parts and provides a degree of protection of at least IP 30 according to IEC 60529 and

3 marking on the internal supplementary enclosure with the warning 'DO NOT OPEN WHEN ENERGIZED'.

The 'General requirements' standards contain additional requirements for luminaires with the exception of intrinsically safe luminaires. An essential point is: covers giving access to the lampholder and other internal parts of luminaires shall

- either be interlocked with a device which automatically disconnects all poles of the lampholder as soon as the cover opening procedure begins or
- be marked with the warning 'DO NOT OPEN WHEN ENERGIZED'.

In the first case where it is intended that some parts other than the lampholder will remain energized after operation of the disconnecting device, those energized parts shall be protected by either means as below:

- one of the specific types of protection described in Sections 6.2–6.9
- protection as follows:
  1 the disconnecting device shall be so arranged that it cannot be operated manually to inadvertently energize unprotected parts and
  2 clearances and creepage distances between phases (poles) and to earth in accordance with the requirements of 'increased safety – e –' (EN 50019, IEC 60079-7) and
  3 an internal supplementary enclosure (which can be the reflector for the light source) which contains the energized parts and provides a degree of protection of at least IP 30 according to IEC 60529 and
  4 marking on the internal supplementary enclosure with the warning 'DO NOT OPEN WHEN ENERGIZED'.

As a second essential, lamps containing free metallic sodium (e.g. low pressure sodium vapour lamps) are not permitted. This does not confine the use of high pressure sodium vapour lamps.

A very general aspect for the design of an explosion protected apparatus are the requirements for installation and maintenance. In the corresponding standards, e.g. in IEC 60079-14 or EN 60079-14 (overtaken as German Standard VDE 0165 Teil 1/1998-08) for installation, supplementary requirements are given for some types of protection, e.g. 'flameproof enclosure – d –', 'increased safety – e –', 'pressurization – p –' and 'intrinsic safety – i –', and these requirements may influence the design and construction of explosion protected apparatus. To give an example, the pressurization – p – standards ask for 'an automatic safety device to operate' in case of failure of pressurization. The purpose for which this 'device' is used (to disconnect power, to sound an alarm) is the responsibility of the user. The installation standards give more precise statements: in zone 1, the power line to the grid shall be disconnected, and, in addition, an alarm shall be given. In zone 2, the action is restricted to an (acoustic and/or optical) alarm only.

6.2   Oil immersion

Standards: EN 50015
             IEC 60079-6
Key letter: o

An electrical apparatus or component which is not capable of igniting a gas–air mixture in normal operation, e.g. a transformer or a reactor, can be explosion protected by immersion in mineral oil or in another corresponding protective liquid in such a way that an explosive atmosphere inside the apparatus above the level of the protective liquid or outside the enclosure of the o-apparatus cannot be ignited.

This liquid acts as a 'barrier' against any contact between the explosive atmosphere and the apparatus or its component. The o-standards contain some basic requirements:

- the protective liquid shall be mineral oil according to IEC 60296 or shall comply with given limits for fire point ($\geq +300°C$), flash point ($\geq +200°C$), kinematic viscosity ($\leq 100\,cSt$ at $+25°C$), electrical breakdown strength ($\geq 27\,kV$), pour point ($\leq -30°C$) and other specific data
- the protective liquid shall be shielded against the ingress of dust and water. So, IP 66 according to IEC 60529 is required with the exception of pressure relief or breathing openings which shall be protected in accordance with IP 23 at least
- the o-apparatus shall be sealed (with a pressure relief device) or non-sealed (with a breathing device including a siccative)
- the protective liquid shall not adversely affect the properties of materials (especially organic insulating materials) which come into contact with the liquid
- live parts shall be overlayed by the protective liquid not less than 25 mm
- the volume dilation of the liquid caused by temperature variations shall be taken into account
- the temperature at the free surface of the protective liquid shall not exceed the value: flash point $-25\,K$
- o-apparatus shall be fitted with one or more protective liquid level indicator(s).

The o-standards refer to categories 2G and M2 only, i.e. dust explosion protection is not in the scope of EN 50015 or IEC 60079-6. For Group I equipment (category M2), the use of mineral oil as a protective liquid is not permitted.

The main field of application of 'oil immersion – o –' is the explosion protection of Group II transformers in the high power range (up to some 1000 kVA). The terminal compartments often comply with 'increased safety – e –', or plug and socket connectors (in 'e') are used in the high voltage range (see Section 6.7).

In German coal mines high voltage (5 kV or 6 kV 3 AC) oil-blast circuit breakers dominated for many years [17]. Each pole was separated from the others and was filled with some litres of mineral oil. The arcs decompose the oil into gases (mainly hydrogen) and soot, so that the oil shall be refilled after a certain amount of switch-off cycles. The hydrogen (to give a rough idea: 25 litres per pole are generated in a 200 MVA circuit breaker at rated breaking capacity) initiates an instantaneous pressure rise in the pole enclosure and thus facilitates arc extinction. As an essential for the explosion protection of these oil-blast circuit breakers, the hydrogen bubbles ascending in the oil shall be cooled down to a temperature insufficient to ignite the CH_4–air mixture which may be present in the head of the pole. The circuit breakers have been installed in an 'e'-enclosure with some flameproof plug and sockets and 'd'-compartments for relays and protection devices, and so the type of protection has been (Sch) eod according to VDE 0170/0171/1969-01 (see Table 3.1). More recent high voltage switchgear is equipped with vacuum or SF_6 circuit breakers.

6.3   Powder filling

Standards: EN 50017
             IEC 60079-5
Key letter: q

A low energy spark (i.e. with an energy $\int U_{spark} \cdot I \, dt$ lower than 10 J) covered with a layer of appropriate thickness made of granulated material cannot ignite an explosive atmosphere formed by gas, vapour or mist with air above the surface of the layer [2]. The 'free volume' between the granules may be filled with the explosive atmosphere by diffusion, but with an appropriate relation: granule size to layer thickness (see Fig. 6.9), flame propagation from the interior of this layer to the explosive atmosphere above the surface of the layer cannot occur. Obviously, granule sizes smaller than 1 mm and a layer thickness of – as an order of magnitude – 10 mm prevent flame propagation even for hydrogen–air mixtures showing the lowest known maximum experimental safe gap value MESG = 0.29 mm (see Table 1.6, Section 1.2.2), indicating a high 'flame propagation ability' through very narrow channels or joints between surfaces of solids.

So, on this basis, low power electrical components such as semiconductors, capacitors, transformers, relays (which shall be sealed in order to prevent ingress of granules into the contact space and mechanical blocking caused hereby), resistors and fuses can be explosion protected by embedding in a

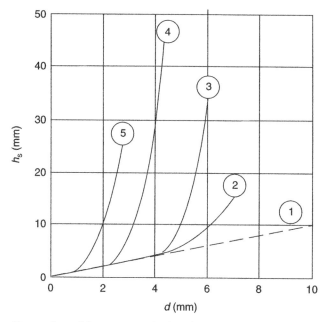

Figure 6.9   *Prevention of flame transmission through a granule layer above a low energy ignition source (≤10 J) [2]. 'Safe granule layer height' h$_s$ versus granule diameter d.*
Diagram 1: Indicates a single layer, i.e. $h_s = d$; Diagram 2: Methane–air mixture; Diagram 3: Propane–air mixture; Diagram 4: Town gas–air mixture; Diagram 5: Hydrogen–air mixture. All combinations h_s/d *above* the diagrams indicate prevention of flame transmission.

filling material such as quartz or solid glass particles (electrical non-conducting, inorganic material).

The q-standards restrict the application of this type of explosion protection to electrical apparatus or components with:

- a rated current less than or equal to 16 A
- a rated voltage not exceeding 1000 V
- a rated power less than or equal to 1000 VA.

The total stored energy of all capacitors in an enclosure of a powder-filled electrical apparatus or component shall not exceed 20 J in normal operation.

The size of granules shall lie within the following sieve limits (according to ISO 565):

- upper limit: metal wire cloth or perforated metal plate with nominal size of opening of 1 mm
- lower limit: metal wire cloth with nominal size of opening of 0.5 mm.

The enclosure of q-apparatus in its normal service condition shall comply at least with IP 54, according to IEC 60529. For higher degrees of protection the enclosure shall be fitted with a breathing device, this complying with

IP 54. To prevent losses of filling material, openings or gaps of the enclosure shall be at least 0.1 mm smaller than the smallest granule size with an upper limit of 0.9 mm.

The minimum distance through the filling material between

- electrically conducting parts of the apparatus or
- insulated components and
- the inner surface of the enclosure

shall comply with the values given in Table 6.3. Conductors used for external connections penetrating the enclosure of the q-apparatus shall be protected with an independent type of protection. They are not within the scope of Table 6.3.

Free volumes (not filled with the filling material) in components, e.g. relays or tubes, are restricted to $30\,cm^3$. For free volumes less than $3\,cm^3$, the minimum distances between the component wall and the inner surface of the enclosure shall comply with the values given in Table 6.3. For free volumes between $3\,cm^3$ and $30\,cm^3$ the values of Table 6.3 apply but with a minimum of 15 mm.

As mentioned above, faults shall not adversely affect the type of protection 'q'. With a certain 'oversizing' of components, the following faults need not be considered:

- resistance values lower than the rated values for:

film type and wire wound resistors and coils with a single layer in helical form, when they are used at no more than two-thirds of their rated voltage and power

Table 6.3   Minimum distances inside the filling material

Maximum voltage AC rms DC^1 V	Minimum distance mm
$U \leqslant 275$	5
$275 < U \leqslant 420$	6
$420 < U \leqslant 550$	8
$550 < U \leqslant 750$	10
$750 < U \leqslant 1000$	14
$1000 < U \leqslant 3000^2$	36
$3000 < U \leqslant 6000^2$	60
$6000 < U \leqslant 10\,000^2$	100

1 Fault conditions such as:
   - short-circuit of any component
   - open-circuit due to any component failure
   - fault in the printed circuitry
   shall not adversely affect the type of protection 'q', i.e. the determination of the maximum voltage in this table shall be based upon a fault analysis
2 The rated supply voltage for q-apparatus shall not exceed 1000 V. The values >1000 V in this table apply to internal circuits only

- short-circuit conditions for:
  plastic foil, ceramic and paper capacitors, when they are used at no more than two-thirds of their rated voltage
- insulation failure of:
  optocouplers and relays designed for segregation of different circuits, when the sum U of the rms values of the maximum voltages of the two circuits is not more than 1000 V and the rated voltage of the relay or optocoupler between the two different circuits is at least 1.5 times U
- transformers, coils and windings complying with type of protection 'increased safety – e –' or with clause 8.1 of EN 50020 or IEC 60079-11 'intrinsic safety – i –' are not subject to fault
- short-circuits, if the creepage distances and distances through filling material between bare live parts or printed tracks are at least equal to the values of Table 6.4.

Additional requirements for q-apparatus, e.g. for means of closing, cable entries and bushings, protective devices for temperature limitation, type and routine verifications and tests, are specified in the q-standards.

Table 6.4 Creepage distances and distances through filling material

Peak voltage V[1]	Creepage distance mm	CTI[2] minimum values	Creepage distance under coating mm	Distance through filling material mm
10	1.5	–[3]	0.6	1.5
30	2	100	0.7	1.5
60	3	100	1	1.5
90	4	100	1.3	2
190	8	175	2.6	3
375	10	175	3.3	3
550	15	175	5	3
750	18	175	6	5
1000	25	175	8.3	5
1300	36	175	12	10
1575	49	175	13.3	10

1 As peak voltage, the maximum peak voltage between the parts shall be considered. If the parts are electrically isolated, the sum of the maximum peak voltages of the two circuits shall be considered as peak voltage. The maximum peak voltage shall be assessed taking into account normal operating conditions (transients being disregarded) and fault conditions as specified in the q-standards.
  Voltages exceeding 1575 V (peak) in internal circuits are always considered subject to fault
2 CTI = comparative tracking index. The method of determination is given in IEC 60112. Examples of insulating materials and their CTI values are summarized in Table 6.17, Section 6.7.1
3 At voltages up to 10 V (peak) the CTI of insulating materials is not required to be specified

Figure 6.10   *Power supply for intrinsically safe electric circuits.*
Type of protection: EEx qd [ia, ib] IIB/IIC T4; Input: 24 V AC/DC via flameproof
plug-and-socket connector; Output: 2 × 5 V DC, 150 mA and 6 × 19 V DC, 40 mA.
The classification according to IIB or IIC depends on external capacitances and
inductances in the 'i'-circuits; Weight: approx. 2 kg.

Figure 6.10 shows a typical example for a q-apparatus. It is a power supply
for intrinsically safe electric circuits to be installed in zone 1 (and zone 2). The
power input (at 24 V AC/DC level) is fed via flameproof plug-and-socket
connectors to the apparatus. So, the power supply can be replaced under
load without conflicting with the requirements of explosion protection.

6.4   Pressurized apparatus

Standards: EN 50016
             IEC 60079-2
Key letter: p

6.4.1   Basic principles of pressurization

An enclosure whose interior is filled with air (without any combustibles) or
inert gas, e.g. nitrogen, argon, carbon dioxide or sulphur hexafluoride showing a positive pressure differential to the environmental potentially explosive atmosphere, indicates the very general principle of pressurization.
The same concept is applied to 'clean rooms' in semiconductor manufacturing or in pharmaceutical production. Gases, vapours, mist or even dust are
prohibited to penetrate into the interior of such a clean room or pressurized
enclosure.

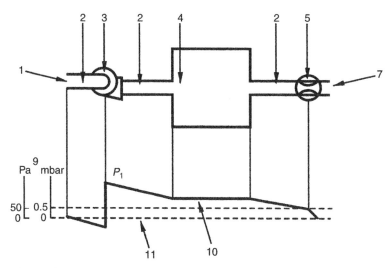

Figure 6.11 *Pressurized apparatus (according to IEC 60079-2). Protective gas outlet without a spark and particle barrier.*
P_1 = pressure of the protective gas, determined by the flow resistance through the ducting, the parts within the enclosure and in certain cases through a choke, and spark and particle barrier, if any; P_2 = pressure of the protective gas, almost constant; P_3 = pressure of the protective gas, determined by the flow resistance of the internal parts, and influenced between A, B and C by the internal cooling fan; P_4 = pressure of the protective gas, determined by the flow resistance of the internal parts and by the uppermost value of pressure of the external air; P_5 = pressure of the external air, caused by the external cooling fan; 1 = protective gas inlet (non-hazardous area required); 2 = ducting; 3 = fan; 4 = enclosure; 5 = choke (where required to maintain the overpressure); 6 = outlet valve; 7 = protective gas outlet; 8 = spark and particle barrier; 9 = overpressure (10^2 Pa = 1 mbar); 10 = internal pressure; 11 = external pressure.

The pressurization (i.e. maintaining a given overpressure inside an enclosure) can be achieved with different methods:

- a specified continuous flow of protective gas (mainly air) supplied by a fan or a pipe with compressed gas passes the enclosure and leaves via an outlet aperture. This method is used preferably to cool the components inside the enclosure, or to supply a specified quantity of breathing air to the personnel inside the pressurized enclosure
- an unspecified flow of protective gas sufficient for leakage compensation to maintain a specified pressure differential enters the enclosure and 'escapes' at all joints not really gas-tight in the enclosure, e.g. windows, covers, doors, cable entries, motor shafts etc.
- an enclosure which can be considered as 'gas-tight' in practice is filled with a certain amount of protective gas ensuring an operational time complying with the requirements of duty. This type of pressurization is known as 'static pressurization'.

The methods described here are illustrated in Figs 6.11 and 6.12 for the continuous flow technique, and in Fig. 6.13 for leakage compensation.

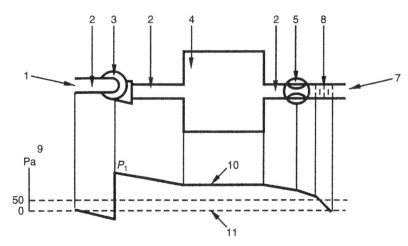

Figure 6.12 *Same as Fig. 6.11, protective gas outlet with a spark and particle barrier.*

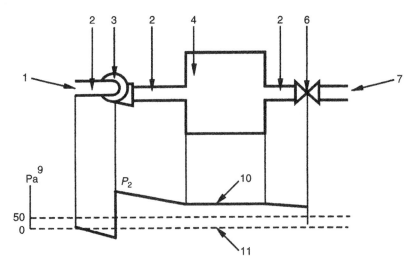

Figure 6.13 *Same as Fig. 6.11, pressurization with leakage compensation, enclosures without moving parts.*

As a very essential requirement, there shall be absolutely no internal source of release of a combustible gas, vapour, mist or liquid. Even small volumes of gases with low LEL values (see Table 1.1 in Section 1.2), e.g. acetylene or benzene, can form an explosive mixture (with air) with a volume exceeding the volume of release by a factor 10^1 to 10^2 (as an order of magnitude). So, one mole of acetylene (=26 g) can form a volume of more than $1\,m^3$ explosive atmosphere. Or, as a second example: a high voltage oil-blast circuit breaker with 200 MVA rated breaking capacity generates 70–80 litres

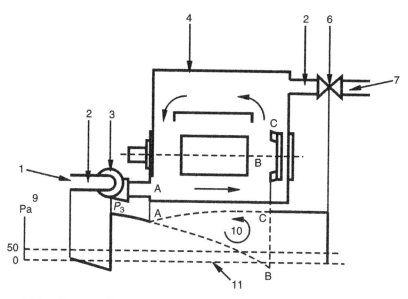

Figure 6.14 *Same as Fig. 6.11, pressurization with leakage compensation, rotating electrical machine with an internal cooling fan.*

of hydrogen in a single switch-off cycle. As a consequence, more than $1\,m^3$ explosive mixture may be present in the p-enclosure. Following the technical development of pressurization, methods have been established to maintain a safe operation in this case, see Chapter 7.

In p-enclosures with fast moving parts (motors, generators, internal cooling fans), the pressure differential (referring to the environmental atmosphere) is strongly influenced by the protective gas movement caused by the rotor or internal fans, which may result in a negative pressure differential at some points of the p-enclosure, e.g. at the suction side of the internal fan. Generally, a positive pressure differential shall be maintained at all points inside the p-enclosure.

So, in this case, the reference point for the determination of the positive pressure differential is the suction side of the internal fan, causing a generally elevated level of internal overpressure at other points inside the p-enclosure. In Fig. 6.14, an example is given for a motor.

Caution needs to be taken with local pressure variations of the environmental atmosphere. A well-known example is the outlet side of an external cooling fan of a surface-cooled motor or generator (see Fig. 6.15). At this point, the environmental potentially explosive atmosphere shows an increased pressure, which shall be 'overcompensated' by an adequate internal overpressure. The p-standards ask for a minimum overpressure of $50\,Pa$ ($=0.5\,mbar$) for Group I p-apparatus or Group II p-apparatus in zone 1.

The protective gas outlet should be located in a non-hazardous area. In such cases, spark and particle barriers at the outlet to guard against the ejection of incandescent particles are not required. For Group I pressurized apparatus and in cases where the protective gas outlet of a Group II apparatus

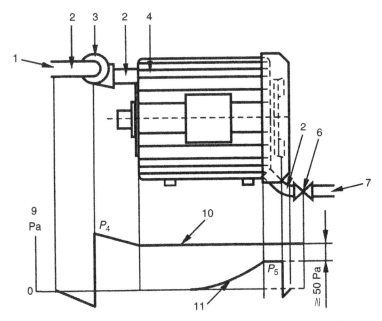

Figure 6.15   *Same as Fig. 6.11, pressurization with leakage compensation, rotating electrical machine with an external cooling fan.*

enters a hazardous area, barriers shall be fitted. Pressurized enclosures with leakage compensation or with static pressurization show outlet apertures in the closed position during operation.

All the facts given above refer to normal operation. How can a p-apparatus start to operate? Due to duty cycles and the thermal cycles caused thereby, a non-operating electrical apparatus located in a hazardous area shall be considered as filled with combustible atmosphere (an example is calculated in Section 6.7.1).

So, a purging process (performed with the protective gas maintaining the positive pressure differential during operation) shall flush the hazardous mixture inside the p enclosure before energizing the apparatus. More precisely: during the purging process, and with air as protective gas, the potentially explosive atmosphere inside the enclosure shall be diluted to a 'safe' value of gas concentration, i.e. 25 per cent of LEL, or, with inert gas as protective gas, the oxygen content of the atmosphere inside the enclosure shall be reduced to 2 per cent (v/v).

To enable an effective purging, the protective gas flow direction during purging should support the 'escape of the unwanted atmosphere' by an appropriate combination of purging flow direction and gas density (the potentially explosive atmosphere, i.e. a gas–air mixture, compared with the protective gas):

For a protective gas with density ρ_p:

- and for p-enclosures for gases and vapours with densities greater than ρ_p, the inlet for the protective gas should be near the top of the enclosure with the outlet near the bottom

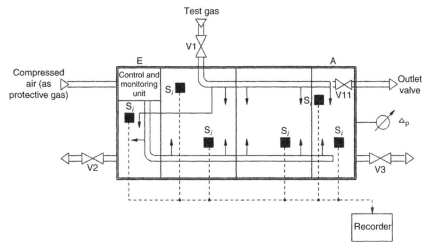

Figure 6.16 *Schematic diagram of the purging procedure of a Group I p apparatus and purging type test set-up.*
Protective gas: air (from a compressed air grid); Mode of protection: leakage compensation.

- and for p-enclosures for gases and vapours with densities below ρ_p, the inlet for the protective gas should be near the bottom of the enclosure with the outlet near the top. For example, Fig. 6.16 shows the purging flow direction (bottom to top) for a Group I enclosure (methane forming the potentially explosive atmosphere) and air as purging and protective gas.

In Fig. 6.16, the protective gas enters via a piping network at the bottom of the enclosure and leaves via outlet valve V11. During normal operation, V11 remains closed. For the purpose of type testing the purging procedure, and as the first step, the test gas (with an identical molecular weight as methane, $=16$) enters via V1 and pushes out the air via V2 and V3. As the second step, with V1–V3 closed, the purging process is started (with V11 opened). Gas sensors S_i ($i = 1$ up to an adequate number of sensors carefully distributed within the enclosure) monitor the flushing of the test gas with the protective gas (air). Valves V1–V3 are fitted for the type test only and will be disassembled after testing.

Table 6.5 summarizes molecular weights and densities of gases and vapours. The gases show extreme variations in density from 0.07 to more than 5, i.e. they cover a range exceeding 70, with respect to air. Obviously, a 'general purpose' p-apparatus is strongly correlated to a very ineffective and uneconomical purging process.

The equivalent values for inert gases commonly used are given in Table 6.6. The inert gases show a similar range in molecular mass and density with reference to air.

The p-standards specify type tests for purging. The combustible gases or vapours forming the environmental atmosphere at the location of duty of the

Table 6.5 Combustible gases and vapours – molecular weight and density (according to [38])

Substance	Chemical formula	Molecular weight g/mole	Density ratio[1, 2] (gaseous state) air = 1
Acetaldehyde	CH_3CHO	44.1	1.52
Acetone	CH_3COCH_3	58.1	2.00
Acetylene	C_2H_2	26.0	0.90
Allyl chloride	CH_2CHCH_2Cl	76.5	2.64
Ammonia	NH_3	17.0	0.59
Amyl acetate	$CH_3COOC_5H_{11}$	130.2	4.49
Aniline	$C_6H_5NH_2$	93.1	3.22
Benzene	C_6H_6	78.1	2.70
Benzaldehyde	C_6H_5CHO	106.1	3.66
Butadiene	$CH_2CHCHCH_2$	54.1	1.87
Butane	C_4H_{10}	58.1	2.05
Butanol	C_4H_9OH	74.1	2.55
Carbon disulphide	CS_2	76.1	2.64
Carbon monoxide	CO	28.0	0.97
Diacetone alcohol	$(CH_3)_2C(OH)CH_2COCH_3$	116.2	4.00
Diesel fuel	–	–	~7.00
Diethyl ether	$(C_2H_5)_2O$	74.1	2.55
Dioxane	$O(CH_2CH_2)_2O$	88.1	3.03
Ethane	C_2H_6	30.1	1.04
Ethanol	C_2H_5OH	46.1	1.59
Ethyl acetate	$CH_3COOC_2H_5$	88.1	3.04
Ethylene	C_2H_4	28.1	0.97
Ethylene oxide	CH_2OCH_2	44.0	1.52
Formaldehyde	$HCHO$	30.0	1.03
Hexane	C_6H_{14}	86.2	2.79
Heptane	C_7H_{16}	100.2	3.46
Hydrogen	H_2	2.0	0.07
Hydrogen cyanide	HCN	27.0	0.93
Hydrogen sulphide	H_2S	34.1	1.19
Kerosene	–	–	4.50–5.50
Methane	CH_4	16.0	0.55
Methanol	CH_3OH	32.0	1.10
Methyl acetate	CH_3COOCH_3	74.1	2.56
Methylamine	CH_3NH_2	31.1	1.07
Methyl formate	$HCOOCH_3$	60.0	2.07
Naphthalene	$C_{10}H_8$	128.2	4.42
Nitrobenzene	$C_6H_5NO_2$	123.1	4.25

(continued)

Table 6.5 (*continued*)

Substance	Chemical formula	Molecular weight g/mole	Density ratio[1,2] (gaseous state) air = 1
Nitroethane	$C_2H_5NO_2$	75.1	2.58
Nitromethane	CH_3NO_2	61.1	2.11
Nitropropane	$C_3H_7NO_2$	89.1	3.06
Pentane	C_5H_{12}	72.2	2.49
Petrol	–	–	~4.00
Phenol	C_6H_5OH	94.1	3.24
Propane	C_3H_8	44.1	1.58
Propanol	C_3H_7OH	60.1	2.07
Styrene	$C_6H_5CHCH_2$	104.2	3.59
Toluene	$C_6H_5CH_3$	92.1	3.18
Town gas	–	–	0.40–0.50

1 The density ratio values refer to air = 1.00 at the same temperature and pressure. For $p = 1.01325 \cdot 10^5$ Pa, $T = 0°C$ (=273 K), the density of air is 1.293 kg/m³, and for $T = +20°C$ (=293 K), it is 1.205 kg/m³

2 The density ratio values are calculated by the molecular weight ratio. For air, the molecular weight is 28.84 (calculated, for simplification, with 79% (v/v) nitrogen, and 21% (v/v) oxygen)

Table 6.6 Molecular weights of inert gases

Gas	Chemical formula	Atomic weight Molecular weight g/mole	Density ratio[1,2] air = 1
Argon	Ar	40	1.39
Carbon dioxide	CO_2	44	1.53
Helium	He	4	0.14
Nitrogen	N_2	28	0.97
Sulphur hexafluoride	SF_6	146	5.06

Notes 1 and 2 are identical with those in Table 6.5

p-apparatus should be replaced by non-flammable and non-toxic gases for the purpose of testing. To ensure the quality of purging:

- p-apparatus for a single specified gas shall be filled with a test gas with a 'similar' density compared with the flammable gas specified (IEC 60079-2 asks for a tolerance of ±10% in density)
- p-apparatus for a specified range of gases shall be filled twice with two different test gases, one for the 'lower end' and the second for the 'higher end' of the gas density range, showing equivalent densities (in IEC 60079-2 specified with a tolerance of ±10% in density).

Looking at Table 6.6, it is obvious that the inert gases commonly available cover a few points only in density range to comply with the requirements for a test gas given in the p-standards. Thus, a mixture of two test gases A, B with molecular weights M_A, M_B can substitute the flammable gas C with molecular weight M_C.

With μ = concentration (by volume) of test gas A and, consequently, $1 - \mu$ = concentration of test gas B, the molecular weight M_C of C can be simulated:

$$\mu \cdot M_A + (1 - \mu) M_B = M_C$$

$$\mu (M_A - M_B) + M_B = M_C$$

$$\mu (M_A - M_B) = M_C - M_B$$

so, for A: $\mu = (M_C - M_B)/(M_A - M_B)$

and for B: concentration $= 1 - \mu$

With the exception of hydrogen, all burnable gases and vapours can be simulated.

1 Example: Test gas for methane ($M_C = 16$) for a Group I apparatus
   Component A shall be helium ($M_A = 4$)
   Component B shall be nitrogen ($M_B = 28$)

For helium $\mu = \dfrac{16 - 28}{4 - 28} = 0.50$

and for nitrogen $1 - \mu = 0.50$.

The test gas is a mixture of 50% (v/v) helium and 50% (v/v) nitrogen
2 Example: Test gas for benzene ($M_C = 78.1$) for a Group II apparatus
   Component A shall be nitrogen ($M_A = 28$)
   Component B shall be sulphur hexafluoride ($M_B = 146$)

For nitrogen $\mu = \dfrac{78.1 - 146}{28 - 146} = 0.5754$

and for sulphur hexafluoride $1 - \mu = 0.4246$

The test gas is a mixture of 57.54% (v/v) nitrogen and 42.46% (v/v) sulphur hexafluoride

According to the p-standards, p-enclosures with air as protective gas and covering all flammable gases shall be tested with helium (to cover lighter-than-air gases) and with either argon or carbon dioxide (to cover heavier-than-air gases). As mentioned above, purging (with air as protective gas) shall be extended to achieve 25% of LEL. In cases covering a wide range of different gases (with different LEL values), the lowest LEL value shall be referred to. The test gas concentration after purging shall not exceed 1 % (v/v) for the helium test and 0.25% (v/v) for the argon or carbon dioxide test.

A 'cross-ventilation' by inlets and outlets at opposite sides of an enclosure supports efficient purging. Enclosures showing a simple geometry without partition walls, manifolds and other obstacles may be purged with a purging volume = 5 × free volume of the enclosure (as a guideline only).

In IEC 60079-2, type of protection 'p' is divided into three categories, px, py and pz, with the following definitions:

px = pressurization that reduces the classification *within the pressurized enclosure* from zone 1 to non-hazardous, or Group I (mining equipment) to non-hazardous

py = pressurization that reduces the classification *within the pressurized enclosure* from zone 1 to zone 2

pz = pressurization that reduces the classification *within the pressurized enclosure* from zone 2 to non-hazardous.

In other words: assuming that there are no internal sources of flammable substances present, type px is the 'general purpose' enclosure which may contain ignition-capable apparatus for application in coal mining and for Group II application in zone 1. With the restriction to Group II application and with no ignition-capable apparatus inside, type py is suitable for zone 1, too. Type pz is a typical Group II zone 2 apparatus which may contain ignition-capable apparatus inside.

Very generally, apparatus which in normal operation constitutes a source of ignition for a specified explosive gas atmosphere is considered as ignition-capable apparatus (including apparatus not protected by a type of protection such as 'd', 'e', 'i', 'm', 'o' or 'q' for type px, and 'd', 'e', 'i', 'm', 'o', 'q', 'n A' or 'n C' for py and pz). Following this philosophy, IEC 60079-2 gives different requirements for px-, py- and pz-enclosures.

6.4.2 Safety devices, control and monitoring units for pressurized enclosures, general safety aspects

It is a fact of experience that pressurized apparatus should be fitted with a control and monitoring unit to ensure an appropriate purging procedure and maintain the protective gas flow and pressure differential which guarantee the safe operation of the apparatus. Purely hand-operated purging procedures are susceptible to errors and slips, and a permanently man-operated flow or pressure control for the protective gas is too remote from all basic principles of economic efficiency. Besides, the standards for pressurization contain requirements for 'safety devices' for zone 1 apparatus (EN 50016) and for type px, py and pz-apparatus (IEC 60079-2, see Table 6.7). So, a 'safety device', or better a control and monitoring unit, is an essential part of a p-apparatus especially for Group I application and for zone 1. It is by no means an imperative that the control unit forms an integral part of the apparatus or has been made by its manufacturer: pressurized apparatus without a control unit shall be marked 'X' and the description documents shall contain all necessary information required by the user to ensure conformity with the requirements of the p-standards.

In many applications, an additional essential for Group I and Group II zone 1 apparatus is an isolating switch which disconnects the p-apparatus from the feeding grid in case of failure of pressurization (the pressure differential or

Table 6.7   Safety devices for p-apparatus (according to IEC 60079-2)*

Design criteria	Type px	Type py	Type pz
Safety device			
• to detect loss of minimum overpressure	Pressure sensor, directly connected to p-apparatus, no valves permitted	Pressure sensor, directly connected to p-apparatus, no valves permitted	Indicator + alarm
• to verify purge period	Timing device, pressure sensor, flow sensor at outlet	Time and flow marked	Time and flow marked
• for a door/ cover that requires a tool to open	Warning	Warning	(No requirement)
• for a door/ cover that does not require a tool to open	Interlock: the electrical supply to the p-apparatus shall be disconnected automatically when they are opened/the supply cannot be restored until they are closed	Warning	Warning

*The original table in IEC 60079-2 contains additional requirements for p-apparatus with a source of internal release of combustible substances. See Chapter 7

the protective gas flow falls below their lower limits specified by the manufacturer). It should be emphasized that the p-standards ask for an 'automatic safety device' to 'operate' in these cases only, the purpose for which the automatic safety device is used (to disconnect power, to sound an alarm, or otherwise ensure the safety of the installation) is the responsibility of the user.

The control and monitoring unit, and, when installed, the isolating switch, shall adhere to the explosion protection requirements, e.g. in zone 1 they shall be explosion protected by one (or more) of the types of protection described in this chapter. The isolating switch may be fitted in a flameproof housing, the control and monitoring unit may contain sensors in 'intrinsic safety – i –', magnetic valves in 'flameproof enclosure – d –' or in 'encapsulation – m –' – or may be non-electric, i.e. the unit is composed of pneumatically operated components only. For complex systems, this way promises a great advantage in installation expenses: the numerous pneumatic/electrical interfaces, such as pressure transducers, flow transmitters and magnetic or electrical operated valves – all explosion protected in their electric part – are dispensable and are replaced by pneumatic/pneumatic counterparts which are not subject to any type of explosion protection.

In the following, a fully pneumatically operated control and monitoring unit as a 'safety device' is described for a Group I pressurized enclosure, e.g. for a frequency convertor.

The protective gas is air, supplied from the compressed air grid in the coal mine. Operational mode is leakage compensation.

The main components are:

1 Air filter with oil separator and air dryer
2 Main air inlet valve
3 Regulated pressure relief valve
4 Purging volume and leakage airflow controller with the following components:
   4a Aperture(s) in the protective gas outlet(s), generating a pressure differential Δp between both sides of the aperture at a given airflow $\dot{Q} =$ dQ/dt with the correlation $\dot{Q}$ proportional to $\sqrt{\Delta p}$
   4b Timer to monitor the purging time t_p. The purging volume is adequate with Δp (corresponding to the specified airflow $\dot{Q}$) being constant or greater than a specified value over the total time period t_p, or, alternatively:
   4c Pneumatic signal convertor which calculates the square root of Δp, resulting in $\dot{Q}$, and a pneumatic integrator for $Q = \int \dot{Q}\, dt$, i.e. determination of purging volume
5a Two limit switches to monitor the lower level of the pressure differential inside the p-enclosure
5b Two limit switches to monitor the highest allowable level of the pressure differential inside the p-enclosure (in case of short-circuit arcing inside the enclosure or of malfunction of the regulated pressure relief valve)
6a Pneumatically operated outlet valve(s) with
6b Position indicators
7 Pressure differential sensors for purging (7a) and operational state of the p-apparatus (7b), acting on the regulated pressure relief valve (3) with (normally) a higher internal overpressure during purging and a somewhat lower internal overpressure in the operational mode
8 Internal air distribution piping to ensure an adequate purging, in this case at the bottom of the p-enclosure forming an upstream airflow to flush the flammable gas (for Group I equipment it is methane, CH_4)
9a An aperture in the protective gas inlet piping, to monitor the leakage air flow, in the operational mode, generating a pressure differential Δp_o, and a pneumatic signal convertor to form the square root of Δp_o and generating $\dot{Q}_o$, the leakage airflow, in order to monitor the correct operation of the p-apparatus. This aperture shall be fitted in a bypass to the main air pipe delivering the purging airflow which is far greater than the airflow required for leakage compensation
9b Two pneumatically operated valves switching the bypass (9a)
10 Logical elements (pneumatic switching elements) as links for connective instructions
11 One (or more) pneumatic emergency off switches
12 (As an option: fire detectors inside the p-enclosure).

The automatic starting and operating sequence of such a control and monitoring unit is described below:

A *Starting conditions*
- pneumatic system   NOT OFF
- fire detectors (12)   NO ALARM*
- internal overpressure (5b)   BELOW UPPER LIMIT
- operational mode   OFF
- doors/covers   LOCKED

B *Start purging*
- outlet valve position indicators (6b)   ACTIVE
- outlet valve(s) (6a)   OPEN
- purge pressure regulation system (3), (4)   ACTIVE
- main air inlet valve (2)   OPEN
- purge air inlet valve (3)   OPEN
- timer (4b)/integrator (4c)   STARTED
- purging indicator (4a)   ACTIVE

Reset-conditions in this period:
- pneumatic system   OFF
- fire detectors (12)   ALARM*
- internal overpressure (5b)   ≥upper limit
- operational mode   ON
- doors/covers   NOT LOCKED
- outlet valve position indicators (6b)   NOT ACTIVE

C *Operational mode* (starting conditions)
- pneumatic system   NOT OFF
- fire detectors (12)   NO ALARM*
- internal overpressure   BELOW THE UPPER
                                            LIMIT (5b)

    and   EXCEEDING THE LOWER
                                            LIMIT (5a)

- PURGING PROCESS TERMINATED
  with a correct air volume (4)
- doors/covers   LOCKED
- outlet valves (6a)   CLOSED
- position indicators (6b)   ACTIVE

D *Start operation*
- leakage compensation (3), (7b), (9)   ACTIVE
- grid isolating switch   ON

Reset conditions in this operational mode:
- pneumatic system   OFF
- fire detectors (12)   ALARM*
- internal overpressure (5b)   >UPPER LIMIT

---

*The fire detectors (fitted as an option) shall have an individual power supply and shall be explosion protected to enable their correct function in the OFF state of the p-apparatus

- internal overpressure (5a)   <LOWER LIMIT
- emergency switch(es) (11)   ACTIVATED
- doors/covers   NOT LOCKED

This short description may give a rough impression of the complexity of a 'safety device', or, speaking more precisely, of an automatic control and monitoring unit. In Table 6.8, the relevant pneumatic data of large volume Group I p-enclosures in German coal mines are summarized.

The air flow values – especially during the purging period – show, that the pneumatic unit asks for adequate pipe diameters and valve sizes. The total volume of such a unit is in the range $1\,m^3$ to $2\,m^3$.

In the same way as for the grid line, output lines to multiple motor drives powered by different p-enclosures shall be fitted with an isolating switch, which disconnects the load from the relevant p-enclosure when pressurization fails, e.g. a conveyor belt with two motor drives 1 and 2 and pressurized frequency convertors 1 and 2. When convertor 1 loses pressurization, motor 1 remains coupled mechanically with drive 2 and is idling. Due to its remanent magnetization, motor 1 feeds back a voltage exceeding all limitations of intrinsic safety – i –, so that this circuit shows no adequate explosion protection. A typical type of protection for an isolating switch is 'flameproof enclosure – d –' (similar to the isolating switch in the grid line).

All electrical components inside a zone 1 pressurized enclosure remaining energized in the non-operational mode of pressurization shall have an independent zone 1 type of explosion protection, e.g. intrinsically safe circuits for data transmission. Batteries remaining in a non-operating p-enclosure may comply with 'encapsulation – m –' or with 'intrinsic safety – i –', in all cases in a separate housing. The reason is that a (secondary) battery may release the gaseous components of decomposed water, i.e. 67% (v/v) hydrogen and 33% (v/v) oxygen, and this fact is considered as contradicting the scope of

Table 6.8   Typical pneumatic data for Group I pressurized enclosures (according to [17] and [31])

Rated overpressure of compressed air grid	$4.0 \cdot 10^5\,Pa$ (4.0 bar)
Minimum overpressure of compressed air grid	$3.0 \cdot 10^5\,Pa$ (3.0 bar)
Rated overpressure of logic elements	$1.4 \cdot 10^5\,Pa$ (1.4 bar)
Minimum pressure differential	$50\,Pa$ (0.5 mbar)
Rated pressure differential	$80\,Pa$ (0.8 mbar)
Maximum pressure differential	$300\,Pa$ (3.0 mbar)
Pressure differential during purging	$200\,Pa$ (2.0 mbar)
Static test pressure differential	$450\,Pa$ (4.5 mbar)
Purging volume	up to $375\,m^3$ (depending on free internal volume)
Purging air flow	up to $900\,m^3/h$
Leakage compensation air flow	up to $50\,m^3/h$

the p-standards according to the content of the 2nd edition of EN 50016 or the 4th edition of IEC 60079-2. A draft amendment opening an extended field of installation of primary and secondary cells and batteries into p-enclosures is under consideration (and may form an integral part of the 3rd edition of EN 50016).

In case of leakage compensation, there is a strong tendency to achieve zero leakage for economical reasons. But 'very low leakage enclosures' (and without a heat exchanger to remove internal thermal losses) tend to exceed the maximum pressure differential, e.g. of 300 Pa (=3.0 mbar) during operation under load.

To give an example: after purging, the internal components and the protective gas start with $T_1 = 283\,K\ (=+10°C)$ and climb to $T_2 = 313\,K\ (=+40°C)$ in steady state. When the protective gas is assumed to be an ideal gas, it will follow the equation

$$p \cdot V = n \cdot R \cdot T$$

p = gas pressure
V = gas volume
n = number of moles
$R = 8.317\ \text{joule} \times \text{mole}^{-1} \times K^{-1}$ ('gas constant')
T = gas temperature

The 'low' temperature state is given by

$$p_1 \cdot V = n \cdot R \cdot T_1 \text{ (constant volume } V)$$

and the 'high' temperature state by

$$p_2 \cdot V = n \cdot R \cdot T_2$$

which will result in

$$p_1/p_2 = T_1/T_2$$

or in $p_2 = T_2/T_1 \cdot p_1$ respectively.

Starting with $p_1 = 1.0 \cdot 10^5\,Pa$ (=1000 mbar) inside (and outside) the p-enclosure, p_2 follows with $1.106 \cdot 10^5\,Pa$ (=1106 mbar) equivalent to a steady state pressure differential of $1.06 \cdot 10^4\,Pa$ (=106 mbar), even in the case of a zero pressure differential when starting to operate. Obviously, the maximum pressure differential of some $10^2\,Pa$ (typical for large volume p-enclosures) will be exceeded by one order of magnitude at least. So, the protective gas flow for leakage compensation should not be pushed down to zero, pressurized enclosures should not be 'too tight'. Within a thermal time constant of some hours, and assuming a time-constant pressure differential, the volume of the protective gas tends to increase from V_1 (initial value) to V_2 in steady state by $V_2 = T_2/T_1 \cdot V_1 = 1.106 \times V_1$, which can be easily calculated with the equation given above (p being constant). Referring to this example, the enclosure shall have a leakage of 0.1 × volume divided by 3 × time constant (after passing 3 × time constant, the thermal conditions inside the enclosure can be considered to be constant, i.e. in steady state). For a $20\,m^3$ pressurized enclosure with a time constant of 3 hours, there shall be a minimum (constructional) leakage of $0.2\,m^3/h$ to avoid any dangerous internal overpressure.

More complex p-apparatus may consist of two (or even more) mechanically separated but pneumatically interconnected enclosures with electrical links. In [31], a pressurized X-ray diagnostic unit has been described for conveyor belt rip detection in German coal mines. The complete apparatus contains three p-enclosures:

- control and analysing unit, high voltage DC supply for X-ray tube and X-ray picture diagnostics
- X-ray tube housing
- X-ray detection unit with convertor into visible light and camera.

These three p-enclosures are equipped with one common 'safety device', i.e. a single control and monitoring unit for purging and pressurization, and they are pneumatically connected 'in series' for the protective gas flow (in this case air is used as the protective gas). The enclosures are connected with flexible metal tubes for the protective gas flow, containing all electric interconnections. Especially for the high voltage DC link to the X-ray tube with a voltage exceeding all Group I installation standards by far, this solution allows the use of this DC link as an internal electric circuit which is out of the scope of these standards. So, pressurization of several enclosures forming one functional unit does not ask for increased expenses for controlling and monitoring the purging and pressurization process.

Hydrocarbon-based insulating materials – commonly used in electrical engineering – form a critical component inside p-enclosures with air as the protective gas: in case of a fire, they generate a considerable amount of energy (see Table 6.9), and, in case of thermal decomposition, they act first of all as a 'hydrogen generator' (see Section 6.4.1). For hydrocarbons as insulating materials, typical contents are 30–75% (by weight) for carbon and 2.5–10% (by weight) for hydrogen! The reactions with the oxygen of air are:

$$C + O_2 \rightarrow CO_2$$

$$2H_2 + O_2 \rightarrow 2H_2O$$

i.e. one mole of carbon forms one mole of carbon dioxide, with the need for one mole of oxygen and two moles of hydrogen forming two moles of water, with the need for one mole of oxygen. For a hydrocarbon with a C content of 50% (by weight) and a H content of 5% (by weight), 1 kg of insulating material, i.e. 41.7 moles of carbon and 25 moles of hydrogen, asks for $41.7 + 25/2 = 54.2$ moles of oxygen for complete combustion. The need for air can be calculated very easily: $1\,m^3$ air with an oxygen content of (roughly)

Table 6.9   Densities and combustion heat values of thermoplastics (according to: Traitzsch, J.: 'Brandverhalten von Kunststoffen; courtesy of Carl Hanser Verlag, Munich/Germany)

Material	Density g/cm^3	Combustion heat MJ/kg
Polyethylene		
• low density	0.91	46.5
• high density	0.96	46.5
Polypropylene	0.91	46.0
Polystyrene	1.05	42.0
Polyvinylchloride	1.40	20.0
Polytetrafluoroethylene	2.20	4.5
Polymethylmethacrylate	1.18	26.0
Polyamide 6	1.13	32.0
Polyethyleneterephthalate	1.34	21.5
Polycarbonate	1.20	31.0
Polyoxymethylene	1.42	17.0

21% (v/v) is equivalent with 210 litres divided by 22.4 litres (molar volume)* resulting in 9.347 moles of oxygen. In other words: 54.2:9.347 = 5.8 m^3 air is needed for a complete combustion of 1 kg hydrocarbon-based insulating material, or 1 m^3 of air (or free volume of a p-enclosure using air as a protective gas) is needed for a complete combustion of 0.172 kg of insulating material. With combustion heat values in the range 4.5–46.5 MJ/kg (see Table 6.9), 1 m^3 of air generates 0.774 MJ up to 8 MJ of combustion heat. In pressurized enclosures for most applications – transformers, switchboards, frequency convertors – the 'insulating material load' exceeds the value of 0.172 kg/m^3, so that the lack of oxygen stops any combustion process. The best way to extinguish a fire inside a p-enclosure is to close all inlet and outlet valves in order to cut off any additional support of oxygen.

To close this Section, Table 6.10 summarizes actions to be taken in case of pressurization faults and emergency.

---

*The molar volume V_0 refers to a temperature of $T_0 = 273$ K (=0°C) and a pressure of $p_0 = 1.013 \cdot 10^5$ Pa (1013 mbar). For different temperatures T and pressures p, the molar volume V shall be calculated according to:

$$p_0 \cdot V_0/T_0 = p \cdot V/T$$

or

$$V = p_0/p \cdot T/T_0 \cdot V_0$$

on the basis of

$$p \cdot V = n \cdot R \cdot T$$
$$p \cdot V/T = n \cdot R = \text{constant}$$

as given above

Table 6.10 Safety-related alarms and actions to be taken in response*

Alarm	Action
1 Internal overpressure too low	• disconnect apparatus from power grid, grid isolating switch OFF • disconnect apparatus from loads with not intrinsically safe inverse voltages (e.g. mutiple drive units), load isolating switches OFF • disconnect all internal electric components which are not explosion protected from their feeding circuits • discharge all internal capacitors (e.g. in the DC link of frequency convertors) to safe values, i.e. the energy $0.5\,CU^2$ shall not exceed: 0.2 mJ for Group I and Group IIA apparatus 0.06 mJ for Group IIB apparatus 0.02 mJ for Group IIC apparatus (assuming a capacitor voltage $\geqslant$200 V)
2 Internal overpressure too high	(This alarm is given due to short-circuit arcing inside the p-enclosure, or fire, or malfunction of the pressure relief valve) Same actions as in (1), in addition: • close all outlet valves immediately • close protective gas inlet valve immediately.
3 Emergency switch ACTIVATED	Same actions as in (1) and (2)
4 Leakage protective gas flow too high	Optical/acoustic alarm for specified time period to give enough time for visual inspection of apparatus. After time period elapsed, same actions as in (1) and (2)
5 Fire detectors ALARM	Same actions as in (1) and (2)

*These actions are not within the scope of the p-standards but they comply with careful safety assessment and installation practice

6.4.3 Techniques for cable entries

In general, there are four methods for cable entries in pressurized apparatus:

• the direct cable entry into the p-enclosure
• the application of terminal compartments complying with 'increased safety – e –' with cable entries into the 'e'-compartment
• the application of terminal compartments complying with 'flameproof enclosure – d –' with d-cable entries (and isolating switches inserted in the d-compartments, if any)
• the use of plug-and-socket connectors complying with 'flameproof enclosure – d –' or 'increased safety – e –'.

Direct cable entries into the pressurized enclosure enable the most economical way to insert external electric circuits into the apparatus. Neither the European Standard EN 50016 nor the International Standard IEC 60079-2 gives special requirements for cable entries. So, the requirements of EN 50014 or IEC 60079-0 'General requirements' shall apply, and, in addition, the type of ingress protection according to IEC 60529 shall be met. EN 50016 asks for IP 40 (at least) with an adjustment upwards to IP 44 for Group I application (in coal mines), whereas IEC 60079-2 asks for IP 4 X (minimum) for zone 1 application. Experience in practice shows that cable entries complying with 'increased safety – e –' fulfil all relevant requirements and guarantee a suitable tightness against leakage of protective gas. The internal connection techniques are identical to those of good industrial practice, e.g. terminal blocks.

Terminal compartments according to 'increased safety – e –' ask for a partition wall between 'e'- and 'p'-compartments. The relevant details for the 'e'-compartment are equivalent with those given in Section 6.7.2. As far as the transition 'e' to 'p' is concerned, bulkhead fittings with rubber packing elements have been shown to save space with the considerable advantage that eddy currents and hysteresis losses in electrical conducting and ferromagnetic partition wall materials can be avoided, thus minimizing the thermal losses in the partition wall.

Due to unavoidable leakage losses at the p-side caused by the bulkhead fittings or different transition techniques and in the 'e'-compartment caused by cable entries and doors or covers, the 'e'-compartment will show a positive internal pressure differential (referred to the environmental atmosphere) somewhat lower than that in the pressurized room. Protective gases other than air shall not be used in this case in order to avoid an atmosphere different from air in the 'e'-compartment. The reason is that the constituents of air (mainly nitrogen and oxygen) show ionization energies different to those of gases often used as protective gases which may result in a stronger ionization in the electric field, so that partial discharges with their negative effects on operational lifetime and reliability of organic insulating materials [20], [30] may be started. In addition, minimum clearances in the e-standards EN 50019 and IEC 60079-7 for explosion protected components and apparatus and in IEC 60664-1 for electric components and apparatus in general refer to air as a dielectric.

In the case of air as a protective gas, the somewhat increased air pressure inside the e-terminal compartment causes a 'shift to the safe side': increased gas pressure (p) reduces partial discharge strength S_{PD} (if any) in a given electric field strength according to:

$$S_{PD} = S_{PDO} \times \exp[-\beta \times (p - 1.0)]$$

S_{PDO} = partial discharge strength at $1.0 \cdot 10^5$ Pa (=1.0 bar absolute pressure)

β = a constant which refers to gas as dielectric, conductor (or electric field) geometry and voltage

p = (absolute) pressure of gas, in bar

In many cases, β has been found with some $10°$ bar^{-1} [20]. A typical example for partial discharges in air and their pressure dependence is given in Fig. 6.17.

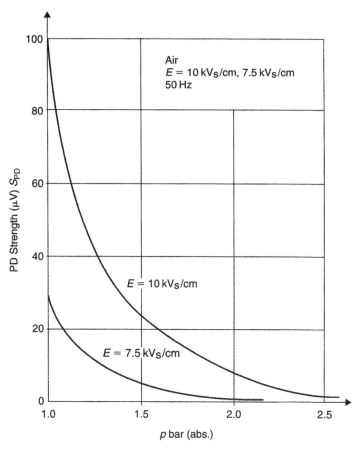

Figure 6.17   *Partial discharge strength S_{PD} in air versus air pressure p [20]. Sinusoidal voltage with 50 cps and electrical field strengths E 10 kV$_{peak}$ · cm^{-1} and 7.5 kV$_{peak}$ · cm^{-1} in a quasi-homogeneous field between plane aluminium lattices.*

Flameproof – d – terminal compartments according to EN 50018 or IEC 60079-1 may be fitted with isolating switches in the grid line and in the outgoing lines to loads with a reverse voltage due to remanent magnetization. The electrical connections between 'd' and 'p' compartments through a partition wall can use the same principles as for 'e' to 'd' or 'd' to 'd' transitions, i.e. bulkhead fittings (flameproof!) or cable bushings.

Considering the unavoidable leakages from 'p' to 'd', they will increase the pressure in the flameproof compartment slightly. In case of air as a protective gas, the explosion pressure will be increased proportionally to the absolute pressure (see Section 6.8.1), and the MESG (maximum experimental safe gap) values of combustible gas–air mixtures are reduced (see Table 1.6, see Figs 1.5–1.7). To remain in the scope of the standards for explosion protected apparatus, the pressure shall be in the range from 8 · 10^4 Pa (=0.8 bar) to 1.1 · 10^5 Pa (=1.1 bar).

This may indicate that 'small' overpressures inside the d compartment, in line with these limitations, will not adversely affect this type of protection.

Figure 6.18   *Flameproof (EEx dI) plug-and-socket connectors for outgoing power lines (1140 V 3 AC, up to 450 A) fitted at the sidewall of an EEx p I switchboard (this is shown in Fig. 6.19).*

When inert gases are used as protective gas, they cause a sort of 'inertization' in the flameproof enclosure. Reducing the oxygen concentration in a gas–oxygen–inert gas mixture down to values of 5 ... 12% (v/v) prevents ignition of this mixture [2]. So, things move to 'the safe side'.

Plug-and-socket connectors complying with 'increased safety – e –' (see Section 6.7.2) or with 'flameproof enclosure – d –' (see Section 6.8.10) provide a simple installation and maintenance of pressurized apparatus, especially in coal mines where frequent and rapid changes of place are required. Figure 6.18 shows flameproof plug-and-socket connectors fitted at the sidewall of a Group I pressurized switchboard.

6.4.4   Energy distribution systems

Pressurization opens the way to a large-scale integration of components such as transformers, circuit breakers and contactors as well as frequency convertors into a single enclosure, thus replacing the multiple-housing switchboards in

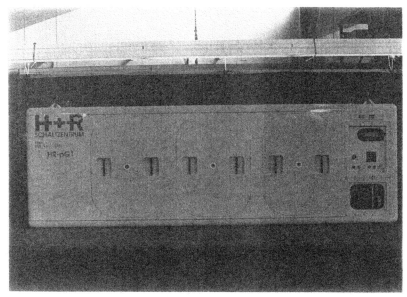

Figure 6.19 *Total view of a pressurized switchboard for coal mines.*
Type of protection: EEx p I; Certificate: BVS 93.C.1134 X; Rated voltage: 1140 V
3 AC; Rated frequency: 50 ... 60 cps; Rated current: 1000 A; Rated short-circuit
current: 25 kA rms; Dimensions: Height × width × depth: 1050 × 3300 × 820 mm³;
Weight (total): 2750 kg; Internal volume: 2.8 m³; Protective gas: air; Mode of
operation: leakage compensation; Purging volume: 14 m³; Purging time: 5 ... 9 min;
Minimum overpressure: 200 Pa (=2 mbar); Rated overpressure: 1500 Pa
(=15 mbar); Maximum overpressure: 7000 Pa (=70 mbar); Power line and protective
gas switch-off at an overpressure of: 12 000 Pa (=120 mbar); Static test
overpressure: 18 000 Pa (=180 mbar); Minimum pressure of compressed air grid:
3.5 bar; Air consumption for leakage compensation losses and control and
monitoring unit: <0.25 m³/h
The switchboard can be fitted with six power modules (maximum), each equipped
(alternatively) with:
• 6 vacuum contactors 150 A
• 3 vacuum contactors 300 A
• 1 vacuum circuit breaker 1000 A
• a 10 kVA power supply for light fittings.

'increased safety – e –' or 'flameproof enclosure – d –'. Examples for Group I
pressurized enclosures are given in Fig. 6.19 – a low voltage switchgear with
flameproof plug-and-socket connectors for external power lines – and in
Fig. 6.20 – a frequency convertor with terminal compartments according to
'increased safety – e –'.

   A certain disadvantage may be seen in the concentration of components
for several load circuits in one single housing and the need to switch off
jointly all these outgoing lines in case of maintenance or repair of one load
circuit only. A solution has been found [31] with the 'multiple compartment

(a)

(b)

Figure 6.20 *Pressurized frequency convertor during assembly at the manufacturer's site.*
Type of protection: EEx pe I (terminal compartments for grid line and outgoing power lines according to 'Increased safety – e –'); Rated grid voltage: 500/1000 V 3 AC; Rated grid frequency: 50 cps; Rated convertor power: 170 kVA; Outgoing line voltage: 0–500 V 3 AC.

technique', see Fig. 6.21. In a single housing, four pressurized compartments, which can be purged and pressurized independently, are fitted:

- a compartment common to the remaining three switchgear compartments, containing the grid line terminal, the busbar, the isolating switches for the switchgear compartments and the terminals for the outgoing circuits
- three compartments containing all components (e.g. circuit breakers, fuses, contactors, current and voltage transformers, relays, overcurrent protection devices) for the outgoing circuits.

(c)

(d)

Figure 6.20 (*continued*)

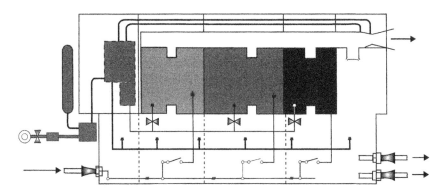

Figure 6.21 *Pressurized apparatus, Group I, 'multiple compartment technique' [31].*

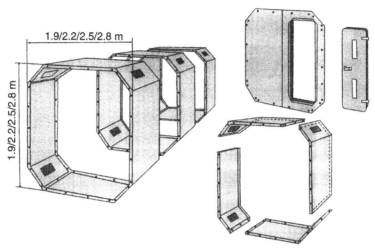

Figure 6.22   *Schematic assembly of a pressurized enclosure for a large-volume Group I energy distribution unit [31].*

In case of repair or maintenance, a single switchgear compartment will be 'pneumatically isolated', i.e. the individual protective gas inlet and outlet valves are closed, the isolating switch is set in its OFF position and an (electrical or pneumatical) interlock with the door of the compartment is set off. Thus, the door can be opened to enable access to the internal components.

After maintenance or repair, the door will be closed and subsequently locked, and inlet and outlet valves are opened to start the purging procedure and subsequent normal operation. The mode of operation is leakage compensation, i.e. the protective gas outlet pipe and the common purge flow or volume measuring device are active during purging only.

The next 'step of component integration' has been achieved with walk-in pressurized enclosures [17], [31] which enable the personnel admittance via an airlock.

The housing of such a p enclosure is made from a certain number of octagon-shaped steel elements (see Fig. 6.22), forming a self-supporting tube after assembly. The sectional area is in the range from $1.9 \times 1.9\,\text{m}^2$ to $2.8 \times 2.8\,\text{m}^2$, and with a length of 0.85 m for a sectional element, a variable length in 0.85 m steps can be achieved according to the customer's requirements, with a maximum of 20 elements (=17 m length) for Group I p-apparatus. Here, high voltage level (10 kV, 6 kV) installations with circuit breakers, contactors and isolating switches are combined with the corresponding transformers (cast resin type) to the low voltage level (1000 V), followed by the complete installation for numerous motor drives, mainly fed by frequency convertors (Figs 6.23 and 6.24).

6.4.5   Gas analysers

Pressurization, with its great diversity of apparatus to be explosion protected for zone 1 (and zone 2) applications, is the favoured protection technique for

Figure 6.23 *Pressurized Group I energy distribution unit at the manufacturer's site.* Type of protection: EEx p I; Doors of the control and monitoring unit (left side) and airlock (right side) in closed position.

Figure 6.24 *Same as Fig. 6.23, doors of control and monitoring unit and airlock opened. The closed internal door of the airlock is shown at the right.*

complex gas analysers such as gas chromatographs or mass spectrometers. Pressurization enables normal laboratory equipment to be operational in hazardous areas with or without modest modifications only. All components carrying the process gas to be analysed which may contain combustible components or oxygen or carrying a combustible sleeve gas, e.g. the piping to the process apparatus, valves, flow or pressure control devices and the analyser head, shall be assessed very carefully against any possibility of leakage. Summarizing all these components in their entirety as a 'containment system', only in the case of

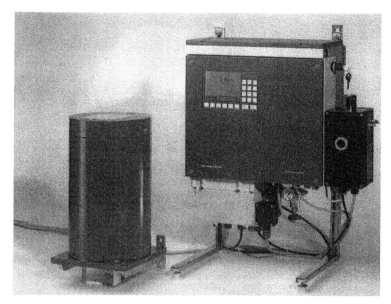

Figure 6.25 *Analyser head (left) and pressurized control unit (right) of an analyser system to be installed in zone 1.*
Data for the central control unit – Type of protection: EEx ped [ib] IIC T4; Certificate: BVS 97.D.2020; Protective gas: air; Mode of operation: leakage compensation; Internal volume: 36 litres; Purging flow rate: $7.5\,m^3/h$; Protective flow rate, operational mode: $0.2\,m^3/h$; Purge duration: typical 4 ... 10 min, depending on internal components; Minimum pressure differential (operational mode): $\geqslant 50\,Pa$ ($=0.5$ mbar); Maximum pressure differential (operational mode): $\leqslant 1500\,Pa$ ($\leqslant 15$ mbar); IP code of ingress protection: IP 54; Overall dimensions – Width: approx. 564 mm; Height: approx. 598 mm; Depth: approx. 199 mm; Weight: approx. 28 kg.

an 'infallible containment system' can zero leakage be assumed. EN 50016 and IEC 60079-2/4th edition give definitions for 'infallibility', e.g. metallic or ceramic components with welded or brazed joints or glass to metal sealing, and contain adequate test procedures. In this case only (zero leakage) a pressurization technique with air or inert gases as protective gases without any additional requirements can be applied. In case of any possibility of leakage, additional precautions shall be made, e.g. dilution (this will be the content of Chapter 7).

Apart from inserting the complete gas analyser into a p-enclosure, the analyser head may be explosion protected for zone 1 application (e.g. according to 'flameproof enclosure – d –') on its own, whereas the central control unit may be installed in a non-hazardous area, or alternatively, in a pressurized enclosure (Fig. 6.25) when zone 1 installation is required.

6.4.6 Motors and generators

Pressurized motors and generators are not subject to special requirements for temperature limitations concerning windings and rotor, there are no stringent

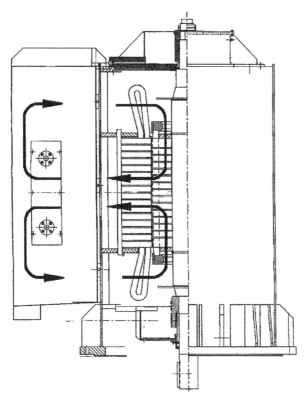

Figure 6.26 *Longtitudinal section of a Group II pressurized squirrel cage motor with vertical axis, indicating the internal protective gas circulation for cooling via gas/water heat exchangers (one shown at the left side).*

conditions for the radial air gap in comparison with rotating machines complying with 'increased safety – e –' (see Section 6.7.3), nor do restrictions exist for the radial clearance of bearings as given for flameproof motors or generators (see Section 6.8.11). So, p-machines show a good economy for constructional parts and materials, combined with an efficiency as high as that for non-explosion protected industrial equipment.

For large power drives, e.g. for compressors, two development areas determine the design of rotating machines:

- cooling by internal protective gas circulation and the application of integrated heat exchangers with an external coolant (mainly water, in special cases air). Figure 6.26 shows a longtitudinal section drawing of a p-motor with vertical axis. The gas circulation for cooling is indicated. Figure 6.27 gives an impression of the completed squirrel cage motor
- replacement of motors with a speed-increasing gear (for turbomachines) often combined with a hydraulic speed convertor by gearless high speed drives, i.e. a frequency convertor powered motor. The 'breakthrough' of this technique has been supported mainly by the introduction of active magnetic bearings for both compressor and motor, allowing for 'dry'

Figure 6.27   *Pressurized squirrel cage motor as a water injection pump drive.*
Type of protection: EEx p II T3; Rated power: 2500 kW; Rated voltage: 11 kV; Rated
frequency: 50 cps; Rated speed: 2986 min⁻¹.

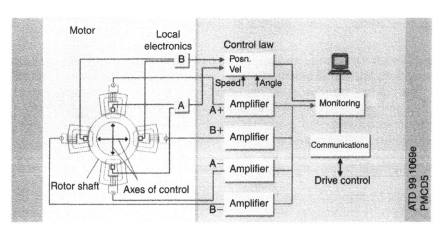

Figure 6.28   *Active magnetic bearing for high speed compressor drives. The rotor
shaft is positioned in both the vertical and horizontal axes by actively controlled
magnetic fields, generated by four levitation coils.*

(i.e. oil-less) systems. The implementation of active magnetic bearings
(Figs 6.28 and 6.29) brings operational advantages, e.g. they are maintenance
free, the absence of a lube and seal oil system simplifies the auxiliary equipment and significantly increases the system availability. In Fig. 6.30,

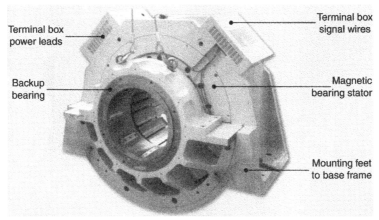

Terminal box
power leads

Terminal box
signal wires

Backup
bearing

Magnetic
bearing stator

Mounting feet
to base frame

Figure 6.29 *Active magnetic bearing for a 23 000 kW 6300 min^{-1} natural gas compressor drive. Mechanical backup bearings 'catch' the rotor and brake it to standstill in case of active control malfunction.*
Power demand of levitation coils: 9 kW.

Figure 6.30 *Natural gas compressor at Groningen gas field/The Netherlands.*
Motor – Pressurized non-salient pole (synchronous) motor, frequency convertor fed.
Type of protection: EEx pe [ib] IIC T3; Rated power: 23 000 kW at 5400 min^{-1}; Speed range: 600 ... 6300 min^{-1}; Voltage: up to 2 × 3600 V; Current: up to 2 × 2030 A;
Rated frequency: 90 cps at 5400 min^{-1}; Total weight: 70 t; Weight of rotor: 9.3 t
Compressor – Natural gas depletion compressor, Output pressure: 65.5 bar; Motor and compressor equipped with active magnetic bearings.

a high speed pressurized 23 000 kW motor combined with a natural gas compressor is shown in its entirety.

Additional examples for pressurized motors are given in Figs 6.31–6.33. These examples demonstrate the great bandwidth of application of 'pressurization' for large drives.

Figure 6.31   *Pressurized motor as a compressor drive in zone 1.*
Type of protection: EEx ped [ib] IIC T3;
Rated power: 10 800 kW; Rated
voltage: 3 kV; Purging volume: 10.8 m^3.

Figure 6.32   *Pressurized motor, vertical axis, in zone 1.*
Type of protection: EEx p [ib] IIC T3; Rated power: 4300 kW; Rated voltage: 13.8 kV.

The installation of generators in hazardous areas is focused on special applications. The specific advantages of 'pressurization' as demonstrated for motors apply in full for generators. Two examples shall be given:

- some natural gas sources (mainly methane, CH$_4$) contain more CO$_2$ than is acceptable according to the delivery specifications. In a CO$_2$ removal process, this gas expands to a lower pressure level. At the SLEIPNER platform two pressurized generators powered by Pelton turbines convert the

Figure 6.33   *Squirrel cage motor as an extruder drive in zone 1.*
Type of protection: EEx p II T4; Rated power: 2000 kW; Rated voltage: 3.3 kV; Rated frequency: 60 cps; Synchronous speed: 1200 min^{-1}.

Figure 6.34   *Pressurized synchronous generator in zone 1, powered by a Pelton turbine in a CO_2 removal process.*
Type of protection: EEx pde [ib] IIC T3; Rated power: 3200 kW; Rated voltage: 13.8 kV.

energy attained during the expansion process into electric energy (Fig. 6.34 shows one of these pressurized generators)
• natural gas passes several pressure levels in the pipeline network on its way to the customer. The 'classical' solution to insert pressure relief valves has been replaced by expansion engines to drive a generator (more detailed information can be found in Section 9.2).

Figure 6.35   *Pressurized synchronous generator powered by a four-cylinder reciprocating natural gas expansion engine.*
Generator – Type of protection: EEx p II T2; Protective gas: air; Operational mode: leakage compensation; Rated power: 1050 kVA, 840 kW; Rated voltage: 400 V 3 AC; Rated frequency: 50 cps; Rated speed: 1000 min^{-1}; Cooling: air/water heat exchanger
Expansion engine – Gas: Natural gas, CH_4 content 96% (v/v); Gas inlet pressure: 51 bar (abs.); Gas preheating: from approx. $+10°C$ to $+60 \ldots +95°C$; Gas outlet pressure: 3.5 bar (abs.); Gas outlet temperature: approx. $+12°C$; Gas flow: 15 500 Nm^3/h; Number of cylinders: 4; Expansion stages: 2; Bore: 360/180 mm; Stroke: 126 mm; Rated speed: 1000 min^{-1}; Rated power: 895 kW.

In Fig. 6.35, a pressurized synchronous generator is coupled with a four-cylinder reciprocating expansion engine.

6.5   Encapsulation

Standards: EN 50028
              IEC 60079-18
Key letter: m

Electrical components which could ignite a hazardous atmosphere either by sparking, arcing or by thermal effects, e.g. the contacts of relays, semiconductors, the windings of transformers or solenoids, can be explosion protected by enclosing them in a compound to avoid immediate contact with the environmental atmosphere.

A compound, in the understanding of this chapter, is a thermosetting, thermoplastic, epoxy resin (cold curing) or elastomeric material with or without fillers and/or additives after its solidification. The compound itself as well as the production method during encapsulation shall be described in detail:

- name and address of the manufacturer of the material
- the exact and complete reference of the material, including colour, type and percentage of fillers and additives (if any)
- surface treatments (e.g. varnishes)
- temperature range of the compound
- continuous operating temperature of the compound

in order to identify material and encapsulation process.

Two different methods of 'encapsulation' have been established:

- embedding the component, i.e. a process of completely encasing this component by pouring the compound over it in a mould, and removing the enclosed component from the mould after solidification of the compound
- potting, i.e. an embedding process in which the mould remains attached to the encased electrical component.

The type of protection 'm' is restricted to rated voltages not exceeding 11 kV due to apprehensions that partial discharges in strong electric fields may adversely affect the dielectric and mechanical properties of insulating materials [20], [30].

If there is no specific marking on the apparatus, m-apparatus connected to an external power line shall be suitable for a short-circuit current of 4 kA (this is the EN 50028 wording), or, more stringently, for a prospective short-circuit current of 4 kA* according to the IEC 60079-18 wording.

Generally, the encapsulation shall be made without voids. But it is permitted to encapsulate components (relays, Reed contacts) each designed with an internal free volume up to $100 \, cm^3$. The thickness of compounds between such components shall be at least 3 mm (for very small internal free volumes $<1 \, cm^3$, the thickness may be reduced to 1 mm).

---

* The 'prospective short-circuit current' defines the short-circuit current which can be expected in case of replacing the short-circuit point by an (ideal) electrical connection with zero impedance. Due to a short-circuit point impedance greater than zero in practice, the 'prospective short-circuit current' indicates the upper limit of a short-circuit current in a defined grid at a given point.

In AC grids, the 'initial symmetrical short-circuit current', I_K'', defined as the rms value of the symmetrical AC component at the occurrence of the short-circuit, the 'peak short-circuit current, I_P, the maximum instantaneous value of the short-circuit current, and the 'steady state short-circuit current', I_K, defined as the rms value of the short-circuit current after fading of all transients, shall be differentiated in general. For 'far-from-generator short circuits', $I_K'' = I_K$.

As far as thermal properties are concerned, I_K shall be taken into account. For mechanical properties (withstanding the mechanical forces of currents and their magnetic fields), I_P shall be referred to. As a 'worst case condition', in a grid with $R/X = 0$, it is $I_P = 2 \cdot \sqrt{2} \cdot I_K''$.

Table 6.11   Clauses of 'General requirements' standards not applicable for 'm' apparatus

Clauses of		Content
EN 50014	IEC 60079-0	
4.1	4.2	Maximum surface temperature (for Ex components)
5.2	5.2	Time delay before opening an enclosure
6.1		Enclosure materials
6.2		Temperature index
6.4	6.2	Threaded holes in enclosures of plastic material
8	8	Fasteners
9	9	Interlocking devices
15	15	Cable and conduit entries
17	17	Switchgear
18	18	Fuses
20.2	20.2	Warning label for luminaires
22.4.3.1	22.4.3.1	Test for resistance to impact (for Ex components)
22.4.4	22.4.4	Test for degree of protection for enclosures
22.4.8	22.4.8	Test in explosive mixtures
22.4.9	22.4.9	Test of clamping of non-armoured cables in cable entries
22.4.10	22.4.10	Test of clamping of armoured cables in cable entries

Switching contacts shall have an additional housing before encapsulation. If the rated contact current exceeds 6 A, the additional housing shall be inorganic (glass, ceramics).

Some clauses of the 'General requirements' standards (EN 50014, IEC 60079-0) are not applicable for 'm'-apparatus, see Table 6.11.

The type of protection 'encapsulation – m –' shall be maintained in the case of overloads and of any single internal electrical fault which could cause either an overvoltage or an overcurrent, e.g.

- short-circuit of any component
- interruption caused by any component fault
- fault in printed circuitry

including a change in the component's characteristics.

Components as specified below and distances through the compound as given in Table 6.12 are considered as not subject to fault:

1 The following components are considered as not subject to a short-circuit fault or to a lower resistance than the rated value (when encapsulated):
   - film-type resistors
   - wire resistors with a single layer in helical form
   - coils with a single layer in helical form

Table 6.12   Minimum distances through the compound

Rated voltage for insulation V rms*	*Minimum distance*** mm
380	1
500	1.5
660	2
1 000	2.5
1 500	4
3 000	7
6 000	12
10 000	20

* The rated voltage may exceed the values given above by 10%
** The distances refer to distances between bare live parts
   mechanically fixed before encapsulating:
   • of the same circuit
   • of a circuit and earthed metallic parts
   • of two separate circuits

when, in normal operation, they are used at no more than two-thirds of either their rated voltage or power as specified by the component manufacturer.

2 The following components are considered as not subject to short-circuit fault or lower resistance or higher capacitance than the rated value (when encapsulated):
   • plastic foil capacitors
   • paper capacitors
   • ceramic capacitors
   when, in normal operation, they are used at no more than two-thirds of their rated voltage as specified by the component manufacturer.

3 Optocouplers (IEC 60079-18 also refers to optocouplers and relays) used for segregation of different circuits are considered as not subject to breakdown between the segregated circuits (when encapsulated) and when:
   • the sum U of the rms values of the voltages of the circuits is not more than 1000 V
     and
   • the electric strength of the optocoupler is at least $1.5 \times U$ (when tested according to a type test procedure specified in EN 50028 and IEC 60079-18).

4 Transformers, coils and motor windings (when encapsulated) are considered as not subject to interturn short-circuits, and the transformers are considered as not subject to breakdown between windings when:
   • they comply with 'increased safety – e –' according to EN 50019 or IEC 60079-7 including those with wire diameters less than 0.25 mm and
   • they are also protected against inadmissible internal temperatures.

5 Transformers are considered as not subject to interturn short-circuits or to breakdown between windings when they comply with 'intrinsic safety – i –', clause 7.1 of EN 50020 or clause 8.1 of IEC 60079-11, except those of type 2(a) of clause 7.1.2 of EN 50020 or clause 8.1 of IEC 60079-11.

The thickness of the compound between the free surface of the compound and the components/conductors in the encapsulation shall be at least 3 mm (with an exception for 'very small components' having no free surface exceeding $2 \, cm^2$).

External connections to m-apparatus or m-components shall act as a 'seal' against the entry of the environmental atmosphere into the interior of the encapsulated apparatus. The m-standards ask for a 5 mm length (at least) of a bare electrical conductor within the compound. Bare live parts, which pass through the surface of the compound, shall be protected by one of the types of protection described in Chapter 6.

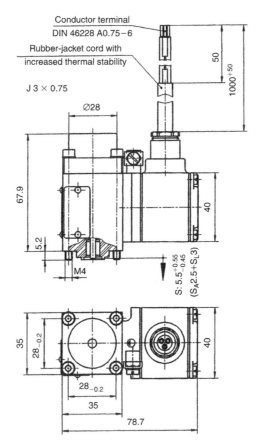

Figure 6.36   *Solenoid as a valve drive.*
Type of protection: EEx m II T4; Certificate: PTB Ex-93.C.4074 X; Dimensions in mm; Rated voltages: 6 V DC up to 250 V DC, 36 V AC up to 250 V AC; Rated currents: 4.17 A DC down to 0.057 A DC, 0.61 A AC down to 0.064 A AC; Max. dissipated power in steady state (AC/DC): 13.9 W … 22 W; Rated frequency (AC): 50 … 60 cps.

Primary and secondary cells (accumulators) are in the scope of the m-standards. Only cells, which in normal use (under the specified conditions given by their manufacturer) are not expected to release gas, do not release electrolyte or produce excessive temperature rise, may be encapsulated. The m-apparatus shall be so constructed as to allow a venting to the outside atmosphere of any gas which may be generated, or precautions shall be taken to avoid gas release or cell deformation adversely affecting the type of protection 'm'. These precautions should be based upon an agreement between the testing house or the notified body and the manufacturer.

When the charging device is not an integral part of the m-apparatus, the charging conditions shall be indicated and the apparatus shall be marked 'X'.

In addition to constructional requirements, EN 50028 and IEC 60079-18 specify requirements for type verification and testing.

Figure 6.36 shows a solenoid valve as a typical application of 'encapsulation – m –'. A rectifier bridge as a protection against incorrect input polarity and a Zener diode as an overvoltage protection form an integral part of the solenoid.

A second example of 'encapsulation' is a contactless operating plug-and-socket system with integrated electronics suitable for zone 1 application (Figs 6.37 and 6.38). The system allows power and data transmission with PROFIBUS DP/V1. The socket with an integrated 'pigtail cable' is attached to a terminal box in 'increased safety – e –'. Within the socket, the 24 V DC power

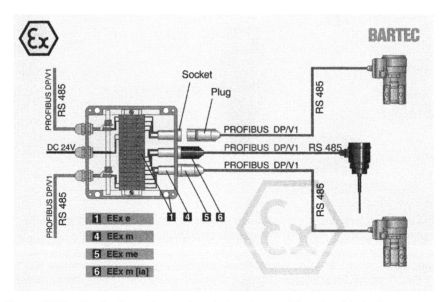

Figure 6.37 *Contactless plug-and-socket system, unidirectional inductive power transmission, bidirectional data transmission via optical fibres.*
Type of protection: socket – EEx m II T6; plug – EEx me II T6 or EEx m [ia] IIC T6;
Power input (socket): 24 V DC 80 mA; Power output (plug): 10 V DC 200 mA or 100 mA for EEx i; Data transmission: RS 485/RS 485 or RS 485/RS 485 EEx i; Baud rate max.: 1.5 Mbaud; Ambient temperature: max. +60°C; Type of ingress protection: IP 65.

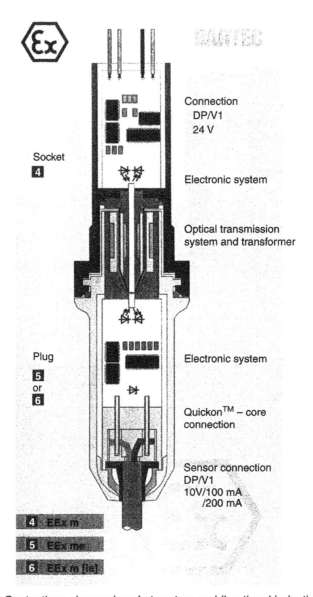

Figure 6.38   *Contactless plug-and-socket system, unidirectional inductive power transmission, bidirectional data transmission via optical fibres.*
Type of protection: socket – EEx m II T6, plug – EEx me II T6 or EEx m [ia] IIC T6; Power input (socket): 24 V DC 80 mA; Power output (plug): 10 V DC 200 mA or 100 mA for EEx i; Data transmission: RS 485/RS 485 or RS 485/RS 485 EEx i; Baud rate max.: 1.5 Mbaud; Ambient temperature: max. +60°C; Type of ingress protection: IP 65.

line is chopped to AC and fed to the primary windings of a transformer as an integral part of the socket. The secondary windings are inserted into the plug, followed by a rectifier. As an option, the DC power output can be designed as an intrinsically safe electrical circuit (with restricted power). The data transmission (RS 485) is combined electrical-optical: the bus interface is amplified as TTL level and converted into light signals which are transferred via optical fibres in two directions: socket-to-plug and plug-to-socket. The data transferred via optical fibres are once again made available as RS 485 signals. Thus, a unidirectional power transmission socket-to-plug on an 'inductive way' is combined with a bidirectional data transmission on an 'optical way'. As an internal power supply for the electronics, the socket contains a 24 V to 5 V DC–DC convertor, and the plug with its 10 V DC power line is equipped with an internal 10 V to 5 V DC–DC convertor. The electronic components are encapsulated according to EN 50028 – m –, the plug is fitted with 'e'-terminals or with terminals as a part of an intrinsically safe circuit.

6.6　Special type of protection

Standards:　– none –

In principle technical development and standardization go side by side, the latter, however, shows a certain time delay of months or even some years. So it is quite common that regulations (directives drawn up by national or international authorities, standardization) open a 'bypass' for technical solutions which do not comply (in full or in part) with established standards.

To give an example: the German Standard VDE 0170/0171/1969-01 for firedamp-proof and explosion protected electrical apparatus contains a type of protection 's' (special protection, in German: 'Sonderschutzart'), see Table 3.1. The standard does not give any technical requirements for 's'. It has been the decision and responsibility of the German testing houses (PTB, Physikalisch-Technische Bundesanstalt, and BVS, Bergbau-Versuchsstrecke) to state an equivalent safety of 's' apparatus compared with well established standards, e.g. 'flameproof enclosure – d –' or 'increased safety – e –'. In view of the great number of apparatus in service over long periods, complying with established standards, the assessment of 's' apparatus, available as prototypes or in pilot series only (and with corresponding short times in service), sets a high standard of competence for the testing houses.

The marking of such 's'-apparatus has been (Sch)s for mining equipment and (Ex)s G … for explosion protected apparatus (suitable for zone 1 applications). In many cases, protection techniques, which are covered by their own standards today, e.g. 'encapsulation – m –' or static pressurization (as a part of 'pressurization – p –'), have entered the field of explosion protection as 's'-apparatus [17].

As a next step, the directives covering the field of explosion protection within the European Community have overtaken this 'bypass' regardless of Harmonized European Norms (HN), see Table 3.7. Articles 4 and 9 of Directive 82/130/EEC (for mines endangered by firedamp and/or combustible dusts) and of Directive 76/117/EEC (for areas hazardous due to combustible gases, vapours and mists) have opened the way for a 'control certificate' as an approval for electrical apparatus not complying with the HNs (see Section 3.2.2). Directive 82/130/EEC defines a special marking for apparatus with a 'control certificate', an S in a circle in addition to the Ex marking (see Fig. 3.1(b) (the 'I' in a circle as a mark for Group I apparatus has been deleted by Directive 88/35/EEC)).

The ATEX 100a Directive, Directive 94/9/EC, specifies 'Essential Safety Requirements (ESRs)' in its Annex II (see Section 3.3).

As a general rule, explosion protected apparatus (in this book restricted to electrical apparatus) shall comply with these ESRs.

If an apparatus has been designed, constructed and successfully type tested according to established standards (e.g. according to European Standards EN … drawn up by CENELEC), the compliance with the ESRs is assumed generally (the so-called 'assumption of conformity' – to be completed – 'with the ESRs').

Where an apparatus has been designed and constructed not in accordance with established standards, the manufacturer is forced to prove compliance with the ESRs in detail. Thus, the 'bypass' is kept open by the ATEX 100a Directive. But it is a question of expenditure of time and money (and the decision of the manufacturer) to follow this route.

6.7   Increased safety

Standards: EN 50019
              IEC 60079-7
Key letter: e

6.7.1   The basic principles of 'increased safety'

This type of protection covers electrical apparatus which generate:

- no sparks/arcs
- no temperatures of components or parts exceeding the temperature limit of the corresponding temperature class (e.g. $+200°C$ for T3)

either in normal operation or under fault conditions (e.g. short-circuit conditions). One feature of e-apparatus is a dust protected and water protected enclosure, degree IP 54 according to IEC 60529, or, if the enclosure contains insulated conductive parts only, the degree of protection may be reduced to IP 44.

It should be emphasized that this fact does not exclude the penetration of combustible mixtures into the enclosure's interior. Generally, all internal parts or components shall be considered to be exposed to these mixtures. The gas exchange between external and internal atmospheres is caused by operational load cycles of the apparatus and the thermal cycles derived thereby. The following calculation will demonstrate this fact: a cage induction motor with internal free volume V_{mot} is running in an external atmosphere with a temperature

$T_1 = +20°C$ (=293 K). In steady state (full load condition) the internal air volume is heated up to $T_2 = +150°C$ (=423 K) due to the losses in the stator iron and windings, the rotor and, with a smaller contribution, the bearings due to friction. Assuming air to be an ideal gas, following the equation with:

$$pV = nRT$$

p = gas pressure
n = number of moles
$R = 8.317$ joule $\times$ mole$^{-1} \times$ K^{-1}
   ('gas constant'),

the 'low temperature' state is given by

$$pV_1 = nRT_1$$

and the 'high temperature' state is given by

$$pV_2 = nRT_2$$

When starting the motor at ambient temperature at full load and continuous operation until steady state, the air will follow the equation

$$V_2 = T_2/T_1 \cdot V_1 = T_2/T_1 \cdot V_{mot}$$

resulting in an air volume increasing of 44%, which leaves the motor's interior to the atmosphere outside. Stopping the motor and following its cooling down period to ambient temperature, the air contracts to the volume

$$V_1 = T_1/T_2 \cdot V_2 = T_1/T_2 \cdot V_{mot}$$

which decreases to 69% of the volume in the steady state. The motor starts to suck the environmental atmosphere with a certain risk of sucking a combustible mixture (not for long periods, increased safety refers to zone 1).

Generally speaking, this type of protection has several fields of application:

- cable terminals, cable connection boxes, plugs and sockets, connectors
- cage induction motors (or the stator-rotor part of a slipring asynchronous motor whose slipring has an adequate type of protection built in)
- stator-rotor part of synchronous motors (with a slipring or diode wheel with adequate type of protection built in)
- transformers, inductances, ballasts, current and voltage transformers
- heating elements for pipes, valves
- light fittings with tubular fluorescent lamps (low pressure mercury vapour lamps), incandescent lamps and blended lamps (a high pressure mercury vapour discharge is connected in series with a (tungsten) filament acting as a ballast resistor). 'Increased safety' covers the application of other lamps for which there is no danger that parts of the light source may attain, for a period longer than 10 s, a higher temperature than the limiting temperature following breakage of the bulb. Lamps containing free metallic sodium are not within the scope of the standards for explosion protection
- secondary batteries (accumulators) of the lead–acid, nickel–iron or nickel–cadmium type, e.g. for electric battery locomotives in coal mines

- busbar installations
- low voltage switchgear with internal components individually protected by an adequate type of protection, e.g. circuit breakers or contactors in a flameproof enclosure, encapsulated voltage and current transformers, electronic components protected by powder filling or controlling and monitoring equipment with intrinsically safe circuits.

For all types of electrical apparatus according to increased safety, their rated voltage is limited to 11 kV. This limitation is based upon concerns that partial discharges (PDs) in the high voltage range may adversely affect the reliability and operating lifetime of organic insulation materials [20, 30] in electrical apparatus.

What is the philosophy of 'increased safety', or, in other words, what is the difference to 'normal' industrial equipment?

There are three essential features which determine reliability and operational lifetime of an electrical apparatus:

- the temperature of (mainly) insulating materials and metallic parts
- the correlation between clearances (the shortest distance between bare live parts or between bare live parts and earthed surrounding components) in gaseous dielectrics (mainly air) and the operating voltage of the apparatus, taking into account possible overvoltages in the power network and the pollution degree of the apparatus
- the correlation between creepage distances (shortest distance along the surface of an electrical insulating material between two conductive parts of different electrical potential, i.e. exposed to an electric field), the operating voltage of the apparatus and the 'quality' of the insulating material, taking into account the pollution degree. 'Quality' refers to the tracking resistance of the insulating material, defined by the comparative tracking index (CTI), determined according to a test procedure given in IEC 60112.

These three essentials shall be considered in more detail in the following section.

6.7.1.1 The influence of temperature

Organic insulating materials are generally subject to temperature dependent deterioration during their operational lifetime. The relevant chemical processes are complicated and may be described (with a certain simplification) as slow oxidation which introduces acid groups into the polymeric insulation, the loss of plasticizer, oxygen cross-linking of polymer chains and internally catalysed depolymerization of plastic insulation.

Büssing* and Dakin* found 'lifetime laws' of the type:

$$t = t_o \cdot \exp(q/RT) \quad \text{or}$$

---

* Investigations concerning this field have been published by:
Montsinger, V. M.: *Trans. Am. Inst. Electr. Engnrs.* 49, 1930, 776–792; Büssing, W.: *Arch. Elektrotech.* 36, 1942, 333–361; Dakin, T. W.: *Trans. Am. Inst. Electr. Engnrs.* 67, 1948, 113–122; and, including electrical stresses, Busch, R.: *ETEP* 8, 1998, 105–110.

$$\ln t = \ln t_o + q/R \cdot 1/T$$
$$t = \text{aging time}$$

t_o = a constant to be determined experimentally
T = absolute temperature in K
R, q = constants (in chemical models, q may by considered as a stripping
    energy, R is the gas constant)

So, the lifetime law in a logarithmic scale is a straight line $\ln t$ versus $1/T$ with a positive gradient q/R.

The results differ in a wide range:

- increasing the temperature by an amount of 8 K shortens the lifetime to 50%
- +20 K results in a lifetime reduction to 10%.

Montsinger* found a correlation of the type:

$$t = t_o' \cdot \exp(-c \cdot \delta)$$
$$\ln t = \ln t_o' - c \cdot \delta$$
$$t = \text{aging time}$$
$$t_o' = \text{a constant to be determined experimentally}$$
$$c = \text{material constant}$$
$$\delta = T - 273, T = \text{absolute temperature,}$$
i.e. δ is the temperature measured in °C

This lifetime law, represented in logarithmic scale, is a straight line $\ln t$ versus δ with a negative gradient c. At first glance, Montsinger's law seems to contradict Büssing's or Dakin's law. But a short calculation will indicate that Montsinger's law is valid for 'small' temperatures δ and can be derived from the more general law:

$$t = t_o \cdot \exp(q/RT)$$
It is: $$q \cdot R^{-1} \cdot T^{-1} = qR^{-1} (273 + \delta)^{-1}$$

with a series expansion:

$$= q \cdot R^{-1} \cdot 273^{-1} \cdot [1 - \delta/273 + 0.5 \, (\delta/273)^2 - \cdots]$$

Neglecting all terms with exponents equal/higher than 2, it is:

$$q \cdot R^{-1} \cdot T^{-1} \sim q \cdot R^{-1} \cdot 273^{-1} - qR^{-1} \cdot 273^{-2} \cdot \delta$$

and for simplification

$$qR^{-1}T^{-1} \sim \text{const.}_1 - \text{const.}_2 \cdot \delta$$
$$\exp(q/RT) = \exp c_1 \cdot \exp(-c_2\delta)$$
$$= \text{const.} \cdot \exp(-c_2\delta)$$

So, the general law $t = t_o \cdot \exp(q/RT)$
is reduced to $t = t_o' \cdot \exp(-c_2\delta)$

for 'small' temperatures $\delta = T - 273$ with $\delta < 273$. This is exactly what Montsinger found by his experiments.

Note:

In the literature, the aging laws are often quoted as 'Montsinger's law'.

The first main feature of increased safety is the reduction of operational temperatures for (organic) insulating materials in order to achieve a higher lifetime in service and to reduce the probability of an early breakdown of the insulation in the windings of motors, generators, transformers (and inductances).

In Table 6.13 the maximum permissible overtemperatures (referring to +40°C ambient temperature) according to EN 50019 or IEC 60079-7 respectively, are compared with the overtemperatures for 'normal' industrial equipment and for motors and generators on railway vehicles. All these values are valid for indirect cooling with air. ('Direct cooling' indicates, that the coolant (this may be a liquid like mineral oil or deionized water or a gas like air or sulphur hexafluoride) is forced to flow through hollow conductors, pipes or channels forming an integral part of the windings within the main insulation. 'Indirect cooling' covers all other cooling methods with the exception of direct cooling.)

6.7.1.2   Insulation coordination – the determination of clearances

This implies the selection of electric insulation characteristics with regard to:

- voltages which can appear in the power distribution system (voltages generated by the equipment, transient overvoltages caused by switching in the system, or transients caused by external effects, e.g. lightning strokes in overhead lines). To follow the different probabilities of the occurrence of overvoltages in a power distribution system overvoltage categories have been established. Table 6.14 gives the correlation between rated impulse voltages (rated impulse voltage = an impulse withstand voltage value assigned by the manufacturer to the equipment or to a part of it, characterizing the specified withstand capability of its insulation against recurring peak voltages), overvoltage categories and the 'working voltage' of the system
- pollution at the electrical installation. The pollution may adversely affect the safe operation of an electrical component, caused by 'bridges' of solid particles between the conductive parts. In addition, some substances will become conductive in the presence of humidity. Similar to the overvoltage categories, four pollution degrees have been established in IEC 60664-1:
  pollution degree 1:   No pollution or only dry, non-conductive pollution occurs

Table 6.13   Part A   Limiting overtemperatures in K for insulated windings (indirect cooling, coolant: air, referring to an ambient temperature of +40°C and to rated service)

Thermal class[1]	A		E		B		F		H	
Method of temperature measurement	R	T	R	T	R	T	R	T	R	T
Increased safety – e –										
• insulated single layer windings	55	55	70	70	80	80	90	90	115	115
• other insulated windings	50	40	65	55	70	60	90	75	115	95
Method of temperature measurement	R	ETS	R	ETS	R	ETS	R	ETS	R	ETS
IEC 60034-1 Motors/Generators										
• AC windings ≥5000 kW	60	65^2	–	–	80	85^2	100	105^2	125	130^2
• AC windings >200 kW, <5000 kW	60	65^2	75	–	80	90^2	105	110^2	125	130^2
• AC windings ≤200 kW	60	–	75	–	80	–	105	–	125	–
• other AC windings	65	–	75	–	85	–	110	–	130	–
IEC 60726 Dry-type power transformers	60	–	75	–	80	–	100	–	125^3	–
IEC 60349-1 motors/generators for railway vehicles										
• stationary windings, rotating field windings of AC generators and synchronous motors	–	–	–	–	130	–	155	–	180	–
• all other rotating windings	–	–	–	–	120	–	140	–	160	–

R = resistance, T = thermometer, ETS = measurement with electric temperature sensors
Note: The original tables in EN 50019 and IEC 60079-7 indicate the limiting *temperatures* for insulated windings (and not the limiting *overtemperatures*). To enable an easier comparison, these values are transformed to overtemperatures taking into account an ambient temperature of +40°C.
1 The thermal classes of insulating material are defined in IEC 60085
2 There may be a matching of the limiting values for certain high voltage windings
3 Increased limiting overtemperatures are admissible for certain insulating materials of thermal class H according to an agreement between manufacturer and customer

Table 6.13   Part B   Thermal classes according to IEC 60085 – examples for insulating material

Thermal class	Insulating material	Impregnants
A	• cotton • natural silk • cellulose wool • artificial silk • polyamide fibre	• drying oil-modified natural or synthetic resin varnishes • shellac, copal and other natural resins
	• cross-linked polyester resins • polychloroprene elastomers • butadiene acrylonitril elastomers	• none –
E	• wire varnishes on the basis of polyvinyl acetate, polyurethane • epoxy or polyamide resins	• synthetic resin varnishes • cross-linked polyester resins • epoxy resins
	• acetobutyrate foil • polycarbonate foil	• none –
B	• glass fibre • varnish treated glass fibre textiles • mica products	• synthetic resin varnishes • cross-linked polyester resins • epoxy resins
	• wire varnishes based on polyterephthalate • films based on crystallized polycarbonate	• none –
F	• glass fibre • varnish treated glass fibre textiles • mica products • laminated glass fibre	• alkyd resins • epoxy resins • silicone alkyd resins
	• wire varnishes based on imide polyester or ester imide	• none –
H	• glass fibre • varnish treated glass fibre textiles • mica products • laminated glass fibre • films based on polyimide • silicone rubber • wire varnishes based on pure polyimide	• silicone resins

Table 6.14 Rated impulse voltage for equipment energized directly from the low voltage mains (according to IEC 60664-1)

Voltage line to neutral V AC or DC	Rated impulse voltage kV Overvoltage category[1]			
	I	II	III	IV
50	0.33	0.5	0.8	1.5
100	0.50	0.8	1.5	2.5
150	0.80	1.5	2.5	4.0
300	1.5	2.5	4.0	6.0
600	2.5	4.0	6.0	8.0
1000	4.0	6.0	8.0	12.0

1 The overvoltage categories are defined as follows:
  I = equipment for connection to circuits in which measures are taken to limit transients to an appropriately low level (e.g. protected electronic circuits)
  II = equipment to be supplied from the fixed installation (e.g. household and similar installations)
  III = equipment in fixed installations and for cases where the reliability and the availability of the equipment is subject to special requirements (e.g. equipment for industrial use).
  IV = equipment for use at the origin of the installation (e.g. electricity meters, primary overcurrent protection apparatus)

pollution degree 2: Only non-conductive pollution occurs except that occasionally a temporary conductivity caused by condensation is to be expected

pollution degree 3: conductive pollution occurs, or dry non-conductive pollution occurs which becomes conductive due to condensation (which is to be expected)

pollution degree 4: the pollution generates persistent conductivity caused by conductive dust or by rain or snow

So, generally, higher pollution degrees are combined with increased minimum clearances. But this is valid only for small clearances and small impulse voltages respectively. Obviously, 'great' (which means >2 mm minimum) clearances are independent of the pollution degree for given impulse voltages ('bridges' caused by pollution may not be expected)

• the type of electric field. Normally, inhomogeneous field conditions are expected. Only in the case where shape and arrangement of the conductive parts are designed to achieve an electric field with an essentially constant voltage gradient, can homogeneous field conditions reduce the clearances

• the altitude (or the atmospheric pressure) for air-insulated electrical apparatus. According to Paschen's law, the breakdown voltage of a clearance (in air and for a homogeneous field) is proportional to

distance between electrodes × atmospheric pressure

Table 6.15   Altitude correction factors (according to IEC 60664-1)

Altitude m	Normal barometric pressure $\times 10^3$ Pa	Multiplication factor for clearances (in air)
0	101.3	Reference value
2000	80.0	1.00
3000	70.0	1.14
4000	62.0	1.29
5000	54.0	1.48
6000	47.0	1.70
7000	41.0	1.95
8000	35.5	2.25
9000	30.5	2.62
10 000	26.5	3.02
15 000	12.0	6.67
20 000	5.5	14.5

High altitudes (=low atmospheric pressure) ask for increased clearances, see Table 6.15. The same correction is made for inhomogeneous fields.

Partial discharges (due to an electric field) are pressure dependent: lower pressures (or higher altitudes) will increase the partial discharge strength, and vice versa [20].

Table 6.16 gives a survey of clearances according to IEC 60664-1 and IEC 60079-7 or EN 50019 respectively, for increased safety – e –.

The second main feature of increased safety – e – is the increase of clearances compared with those of overvoltage category III. To give a rough estimation the clearances of increased safety are comparable with those of overvoltage category IV (or even higher).

6.7.1.3   Insulation coordination – the determination of creepage distances

This point is focused on the selection of electrical insulation characteristics with a special regard to the decomposition of the surface of the insulating material caused by a leakage current through a 'track' on the contaminated surface. These effects can be described as a progressive formation of conductive paths on the surface and an erosion of the material. Inorganic insulation materials like aluminium oxide, ceramics or (mineral) glass are normally not subject to these effects. However, a great part of organic insulating materials is hydrocarbon based and suffers from these effects. A method of classification of insulation materials is given in IEC 60112: the determination of the comparative tracking index (CTI) of a material by tests. High CTI values indicate a robustness against erosion due to leakage currents, low CTI values stand for poor resistance against these effects.

Table 6.16 Minimum clearances (in air, up to 2000 m above sea level, inhomogeneous field conditions)

IEC 60664-1						Increased safety – e –	
Line-neutral voltage V	System voltage V	Impulse voltage kV Overvoltage category		Minimum clearances mm Pollution degree		Minimum clearance mm	Voltage V
		III	IV	3	4		
50	12.5–48	0.8		0.8	1.6	1.6	$U \leqslant 15$
			1.5	0.8	1.6	1.8	$15 < U \leqslant 30$
						2.1	$30 < U \leqslant 60$
100	60, 66/115	1.5		0.8	1.6	2.5	$60 < U \leqslant 110$
			2.5	1.5	1.6		
150	110, 220	2.5		1.5	1.6	3.2	$110 < U \leqslant 175$
	127/220		4	3	3	5	$175 < U \leqslant 275$
300	220, 440	4		3	3	6	$275 < U \leqslant 420$
	230/400		6	5.5	5.5		
600	480, 960	6		5.5	5.5	8	$420 < U \leqslant 550$
	400/690		8	8	8	10	$550 < U \leqslant 750$
1000	1000	8		8	8	14	$750 < U \leqslant 1100$
			12	14	14		
						30	$1100 < U \leqslant 2200$
						36	$2200 < U \leqslant 3300$
						44	$3300 < U \leqslant 4200$
						50	$4200 < U \leqslant 5500$
						60	$5500 < U \leqslant 6600$
						80	$6600 < U \leqslant 8300$
						100	$8300 < U \leqslant 11\,000$

Note: Insulation coordination is given in IEC 60071-1 and IEC 60071-2 for voltages exceeding 1kV

Insulating materials are separated into four groups according to their CTI values (Table 6.17).

Generally, poor CTI values shall be compensated by an increasing creepage distance.

In addition, the reliability of an insulating material is strongly influenced by the pollution which may occur on its surface. So, creepage distances have to be increased for a higher pollution degree.

The main factor, however, is given by the electric field strength parallel to the surface of the insulating material, or, in other words, by the voltage between the conductors separated by the insulating material.

Table 6.17 Grouping of insulating materials according to their CTI values (and examples of such materials)

Material group robustness	CTI according to IEC 60 112	Insulating materials (examples)
I high	600 ⩽ CTI	Ceramics, glass Aluminium oxide Epoxy resins Polyamide Polyethylene Polypropylene Silicone rubber
II medium	400 ⩽ CTI < 600	Polystyrene
IIIA poor	175 ⩽ CTI < 400	Polycarbonate, phenoplast (with inorganic fillers)
IIIB very poor	100 ⩽ CTI < 175	Phenoplast (with organic fillers, e.g. wood, textile fibres, cellulose)

Table 6.18 gives a survey of these relations according to IEC 60664-1 for 'normal' industrial equipment compared with the specifications according to IEC 60079-7 and EN 50019 for 'increased safety – e –'. Figure 6.39 shows creepage distances at a terminal block.

The third main feature of increased safety – e – is the increase of creepage distances compared with those for pollution degree 3 and the adoption of the values for pollution degree 4 in the low voltage range. In addition, material Group IIIB is excluded.

Note:

In the high voltage range, the creepage distances are identical to those for pollution degree 3. It should be remembered that increased safety asks for IP 54 as an adequate ingress protection degree for the enclosure (with bare live parts inside).

6.7.2 Terminal compartments, cable connecting techniques and connection facilities

This heading describes an extended application of 'increased safety'. Terminal compartments are connected to the main part of the apparatus supplied with or generating electric power. In many applications, the main part

Table 6.18 Minimum creepage distances for electrical equipment subject to long-term stresses (the values given for IEC 60 664-1 are only a part of the standard's content)

Voltage V	IEC 60 664-1 (selected values) Minimum creepage distances mm						Increased safety – e – Minimum creepage distance mm			Voltage V
	Pollution degree 3			Pollution degree 4						
Material group	I	II	III²	I	II	IIIA³	I	II	IIIA¹	
16	1.1	1.1	1.1	1.6	1.6	1.6	1.6	1.6	1.6⁴	≤ 15
32	1.3	1.3	1.3	1.8	1.8	1.8	1.8	1.8	1.8⁴	15 < U ≤ 30
63	1.6	1.8	2	2.1	2.6	3.4	2.1	2.6	3.4⁴	30 < U ≤ 60
125	1.9	2.1	2.4	2.5	3.2	4	2.5	3.2	4⁴	60 < U ≤ 110
160	2	2.2	2.5	3.2	4	5	3.2	4	5	110 < U ≤ 175
250	3.2	3.6	4	5	6.3	8	5	6.3	8	175 < U ≤ 275
400	5	5.6	6.3	8	10	12.5	8	10	12.5	275 < U ≤ 420
500	6.3	7.1	8	10	12.5	16	10	12.5	16	420 < U ≤ 550
630	8	9	10	12.5	16	20	12	16	20	550 < U ≤ 750
1000	12.5	14	16	20	25	32	20	25	32	750 < U ≤ 1100
2500	32	36	40	50	63	80	32	36	40	1100 < U ≤ 2200
3200	40	45	50	63	80	100	40	45	50	2200 < U ≤ 3300
4000	50	56	63	80	100	125	50	56	63	3300 < U ≤ 4200
5000	63	71	80	100	125	160	63	71	80	4200 < U ≤ 5500
6300	80	90	100	125	160	200	80	90	100	5500 < U ≤ 6600
8000	100	110	125	160	200	250	100	110	125	6600 < U ≤ 8300
10000	125	140	160	200	250	320	125	140	160	8300 < U ≤ 11 000

1 Material group IIIB is not covered by the standards for increased safety (IEC 60079-7, EN 50019)
2, 3 Material group IIIB is not recommended for application in pollution degree 3 above 630 V and in pollution degree 4
4 For external connections, creepage distances shall have a minimum value of 3 mm

Figure 6.39   *Terminal block.*
Type of protection: EEx e II; Certificate: KEMA Ex-95.D.4411U; Rated voltage: 726 V; Rated current: 350 A; Conductor cross-section (flexible): 70–185 mm^2; Insulating material: CTI 600. The lines in dark grey indicate the creepage distances.

is protected by a different technique, e.g. flameproof enclosure, oil immersion, pressurization or powder filling. In this case, the e-terminal compartment is separated from the main part by a dividing wall. Thus, three parts of a terminal compartment have to be considered, following the current path:

- the cable entry (including fibre optics cables). As far as increased safety is concerned, the cable entries are tested (and certified) as an 'autonomous' apparatus independent of the apparatus in its entirety and which can be fitted to the apparatus during installation without further certification
- the connecting facilities between external and internal circuits, e.g. terminal blocks, which are tested (and certified) as an 'Ex component', the certificate marked with the symbol 'U'
- the bushings, carrying one or more conductors or fibre optics through the dividing wall to the main part of the apparatus (e.g. the interior of a flameproof switchgear), tested and certified as an 'Ex component' with a U certificate. In the special case of an e-terminal compartment combined with another e-apparatus or with a pressurized enclosure, bulkhead fittings may be used.

Components such as terminal blocks, bushings or busbar insulators with organic insulating materials show maximum service temperatures far below the maximum surface temperatures according to the temperature classification T1, T2 and T3 (and in some cases T4). In addition, the standards for increased safety require a 'margin of safety' for solid electrical insulating materials. The mechanical characteristics of the materials shall meet the requirements for good operational reliability at a temperature up to at least 20 K above the maximum temperature in rated service (at least +80°C). The relation

$$T_{\text{max. in service}} \leqslant T_{\text{max.}} - 20\,\text{K}$$

is valid throughout the components in increased safety (organic materials only). The majority of such materials show temperatures $T_{\text{max.}}$ in the range +100°C … +160°C, restricting the temperature in service (in increased safety) to +80°C … +140°C.

In a similar way, the maximum temperature of cables in rated service is limited to values of +70°C for the entry point and of +80°C for the branching point of the conductors in cable entries (this is a general requirement for all types of protection, given in EN 50014 and IEC 60079-0).

Higher temperatures at these points ask for installation with special heat-resistant cables (e.g. with silicon-based insulating material). In this case, the apparatus shall be marked with an 'X' for this special condition.

Caution shall be given to national installation codes demanding a certain mechanical strength and/or resistance against oil, grease and hydraulic fluids for the cable sheath which often contradicts the high temperature performance of the cable.

Remembering the maximum ambient temperature of +40°C, there is only a small gap of 30 K or 40 K respectively for the overtemperature range of cable entries. So, the (organic) insulating materials of cables and components like terminal blocks or bushings are often the 'thermal bottleneck' of terminal compartments, especially for flameproof switchgear, frequency convertors or light fittings.

The following consideration may illuminate this fact. The thermal losses of, e.g., a switchgear within a flameproof enclosure caused by components such as fuses, transformers, operating coils, ohmic losses in conductors, current dependent eddy losses in metals and hysteresis losses in ferromagnetic (steel) walls cause a heat flow directed to the enclosure walls of the flameproof cabinet. To give an example, a Group I low voltage switchgear cabinet (in German known as a 'Kompaktstation') shows a total current sum of 1600 A 3 AC at 50 cps and total heat losses of 1000 W to 2000 W. The surfaces of the front panel, and rear, top and bottom accumulate to 4 m² and act as a cooling surface. The dividing walls of the terminal compartments on right and left side generate additional hysteresis and eddy losses dependent on wall materials and the technical construction of the bushings (a three-phase bushing for one 3 AC circuit is more favourable due to its weak residual magnetic field compared with three single-pole bushings showing an uncompensated magnetic field each). On both right and left sides, the heat flow from the main part enters the terminal compartments and increases their internal losses. On the other side, the typical surface area of the enclosure of a terminal compartment – 2 m² – acts as an additional cooling surface, with the exception of the cable entry plates. For these parts, the same considerations are valid as for the bushings.

It is noteworthy that the general efforts towards increased current densities – A/mm² – in cables for economic reasons contradict the high power design of such switchgear, due to the temperature limitations for cable entries. Figures 6.40(a)–(d) show the relationship overtemperatures of entry and branching

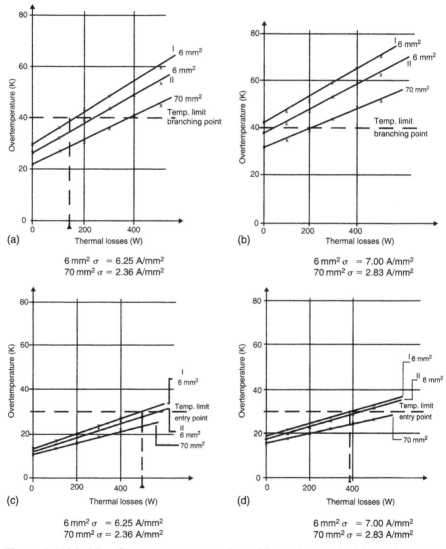

Figure 6.40(a)–(d)   *Overtemperatures at branching points and cable entries versus thermal losses. Low voltage switchgear EEx de I. The thermal losses indicate the heat dissipation of fuses, magnetic coils and transformers in the 'd' main part. Two cables (I, II) with 3 AC, 50 cps, conductor cross-section 6 mm², one cable with 3 AC, 50 cps, conductor cross-section 70 mm², at different current densities σ.*

point versus thermal losses in the flameproof part of a Group I switchgear with a terminal compartment in 'e'.

With current densities of 2.36 A/mm^2 (70 mm^2 cable) and 6.25 A/mm^2 (6 mm^2 cable), the maximum overtemperature at the branching point (40 K) sets up a limit of 160 W for the thermal losses within the main part of the switchgear. Increased current densities of 2.83 A/mm^2 and 7.00 A/mm^2 cause an overtemperature of 42 K at the branching point of cable I (6 mm^2) even at zero losses, i.e. this will be an impractical solution beyond the scope of the standards.

In a very general sense, terminal boxes, regardless of their type of protection (increased safety or flameproof), ensure an additional cooling surface and a greater distance between the thermic critical cable entries and the source of heat losses, compared with direct cable entries into the flameproof main compartment of the apparatus.

In addition to the requirements for cable entries as given in Section 6.1, they shall fulfil the ingress protection degree IP 54 according to IEC 60529 when fitted to an 'e' apparatus. Figures 6.41 and 6.42 give examples for cable entries for non-armoured and braided cables (clamping by a sealing ring and

Figure 6.41   *Cable entry with thread mount.*
Type of protection: EEx e I, EEx e II; Certificates: INIEX (former name of ISSeP) 89.B.102761, INIEX 89.B.102762; For cables with diameter 8 ... 86 mm.

Figure 6.42   *Cable entry with flange mount.*
Type of protection: EEx e I, EEx e II; Certificates: INIEX (former name of ISSeP) 89.B.102759, INIEX 89.B.102760; For cables with diameter 8 ... 86 mm.

compression element), fitted with a thread (Fig. 6.41) and a flange (Fig. 6.42) to the enclosure. High voltage cables (caution: increased safety – e – asks for a voltage limitation of 11 kV) are fitted with sealing heads (cast resin) or vulcanized (rubber) heads to ensure an appropriate potential grading.

Examples for cable entries suitable for high voltage cables are given in Figs 6.43 and 6.44 (for cast resin heads only).

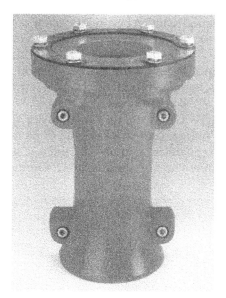

Figure 6.43   *Cable entry for high voltage cables with sealing heads (cast resin).*
Type of protection: EEx e I, EEx e II; Certificates: BVS 89.B.1098, BVS 89.C.2022;
For cables with diameter 50 ... 80 mm.

Figure 6.44   *Cable entry for high voltage cables with sealing heads (cast resin).*
Type of protection: EEx e I, EEx e II; Certificates: BVS 89.B.1107, BVS 89.C.2025;
For cables with diameter 46 ... 85 mm.

On the whole, there are two connecting techniques between external and internal or internal to internal circuits in 'e' terminal compartments. The first one is the use of terminal blocks (Figs. 6.39, 6.45 and 6.46) to avoid damaging of conductors and the impairment of contacts by temperature changes in normal operation. The contact pressure shall not be transmitted through insulating material. Figure 6.46 is inserted especially for protective earthing conductors.

In the examples given above, a screw maintains the contact pressure between conductor and clamp via a moveable metallic part. A more recent development of terminal blocks shows a clamping spring to maintain an adequate permanent contact pressure between conductor and clamp ('cage clamp technique'). Examples for multiple contact terminals in this technique are given in Figs 6.47 and 6.48.

In these terminal blocks, a metallic clamping spring (which is not intended for current transmission) is directly attached to the inner side of a permanently fixed U-shaped current bar (mainly copper), forming the current path within the terminal block. On the top, the clamping spring shows a rectangular-shaped slot embracing one 'leg' of the current bar, and, after insertion of a

Figure 6.45 *Terminal block.*
Type of protection: EEx e I/II; Certificate: LCIE 86.B.0004 U; Rated voltage: 660 V; Rated current: 138 A; Conductor cross-section max.: 35 mm^2; Insulating material: CTI 600.

Figure 6.46   *Terminal block for protective earthing conductors.*
Type of protection: EEx e II; Certificate: KEMA Ex-95.D.4412U; Conductor cross-section (flexible): 35–95 mm^2.

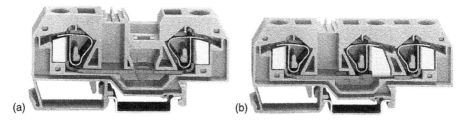

(a)   (b)

Figure 6.47(a)–(b)   *Terminal blocks for two or three conductors, 'cage clamp' technique.*
Type of protection: EEx e II; Certificate: PTB 98 ATEX 3132 U; Rated voltage: 550 V; Rated current: 68 A; Conductor cross-section: 0.5 … 16 mm^2; Insulating material: CTI 600.

cable, the conductor too, which is adjacent to the outer side of the current bar. The clamping spring generates a force perpendicularly directed to the 'leg' of this current bar, thus ensuring an adequate permanent contact pressure.

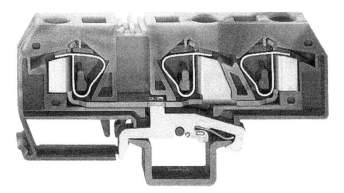

Figure 6.48 *Terminal block for three protective earthing conductors, 'cage clamp' technique.*
Type of protection: EEx e II; Certificate: PTB 98 ATEX 3132 U; Conductor cross-section: 0.5 ... 16mm².

Figure 6.49 *Connecting terminals (single units) in a high voltage cable joint box (the following data are valid for the cable joint box in its entirety).*
Type of protection: EEx e I; Certificate: BVS 87.1123, and BVS 87.1124 X, EEx e II T6, BVS 87.026, and BVS 87.027 X; Rated voltage: 11 kV; Rated current: 388 A; Conductor cross-section: 35 ... 240 mm²; Rated dynamic current: I_{dyn} 88 kA; Ingress protection code: IP 68; Insulating material: ceramics.

Though terminal blocks are used for cable connecting boxes (mainly in the low voltage range and for non-intrinsically safe remote controlling and monitoring), an extended field of application has been opened for 'e'-terminal compartments for pressurized apparatus or 'e'-main parts of apparatus separated from the terminal compartment by a dividing wall. The common way to pass the internal cables through the wall is to use bulkhead fittings. In the high voltage range, the connecting terminals are single units with ceramics for insulation (Fig. 6.49). Especially in 'e'-terminal compartments combined with a flameproof main part of the apparatus, e.g. motors or switchgear,

Table 6.19 Multiple cable bushings for transitions 'e' to 'd' or 'e' to 'p' (and 'd' to 'd' or 'd' to 'p') (by courtesy of Emil A. Peters GmbH & Co. KG, Iserlohn/Germany)

Maximum number of cables	Rated voltage V	Cross-section mm^2	Thread
40	275	0.5	M 48 × 1.5
35	275	0.75	M 48 × 1.5
35	275	1.5	M 48 × 1.5
12	750	1.5	M 30 × 1.5
12	750	2.5	M 30 × 1.5
6	750	4	M 30 × 1.5
6	750	6	M 30 × 1.5
3	750	10	M 30 × 1.5
3	750	16	M 30 × 1.5
35	750	1.5	M 48 × 1.5
25	750	2.5	M 48 × 1.5
15	750	4	M 48 × 1.5
9	750	6	M 48 × 1.5
9	750	10	M 48 × 1.5
6	750	16	M 48 × 1.5
3	750	25	M 48 × 1.5
8	1100	1.5	M 30 × 1.5
5	1100	2.5	M 30 × 1.5
3	1100	4	M 30 × 1.5
3	1100	6	M 30 × 1.5
16	1100	1.5	M 48 × 1.5
9	1100	2.5	M 48 × 1.5
8	1100	4	M 48 × 1.5
6	1100	6	M 48 × 1.5
3	1100	10	M 48 × 1.5
3	1100	16	M 48 × 1.5
6	3300	1.5	M 48 × 1.5
6	3300	2.5	M 48 × 1.5
3	3300	4	M 48 × 1.5
3	3300	6	M 48 × 1.5
3	3300	10	M 48 × 1.5
3	3300	16	M 48 × 1.5

block terminals are used for low power circuits (non-intrinsically safe). The internal cables pass through the dividing wall by cable bushings. Multiple cable bushings in the low voltage range are very common, see Table 6.19, single cable bushings are available up to a cross-section of (roughly) $70\,mm^2$. For 'low' high voltages, e.g. 3.3 kV, single cable bushings show cross-sections up to (roughly) $70\,mm^2$.

Figure 6.50 *Flameproof conductor bushings (connection between 'e' and 'd').*
Marking: Ex II 2G, EEx de II, IM2 EEx de I; Type of protection: EEx de I/EEx de II C;
Certificate: PTB 98 ATEX 1066 U; Rated voltage: up to 1100 V; Rated current: up to
400 A; Thread: up to M 42 × 1.5; Conductor diameter: up to 16 mm; Insulating
material: ceramics.

The second way – typical for 'e'-terminal compartments at a flameproof
main part of apparatus and high currents – is the direct connection of the
external cable to single conductor bushings fitted into the dividing wall of
the flameproof main part. These bushings have to comply with the 'flameproof' standards (IEC 60079-1 and EN 50018) and at one end (at least) with
the 'e'-standards. So, these bushings (Fig. 6.50) look somewhat 'asymmetrical'. Attention is given to the correct installation into the dividing walls,
i.e. the 'e'-terminal part shall not be confused with the 'non-e'-terminal part,
e.g. for the flameproof compartment of the apparatus. A typical 'e' connecting
head is shown in Fig. 6.51.

High voltage bushings are often fitted with 'pigtail cables' at the 'non-e'
end. Figure 6.52 shows high voltage bushings, the top end of which is
intended for installation in increased safety.

It is a part of the safety philosophy in power engineering that cable connections and bare conducting parts shall withstand the forces of short-
circuits without any adverse effects to the clearances.

An example is given for a busbar for low voltage power distribution: two
parallel conductors with distance r over a length l with current I in air with
relative permeability $\mu = 1$ influence one another with the force:

$$F = \mu_o \, l \, I^2 / 2\pi r$$
$$\mu_o = 1.256 \cdot 10^{-6} \ \text{Vs/Am}$$
$$(\text{'absolute permeability'})$$

Currents in opposite directions result in a repelling force, currents in same
directions result in an attractive force.

Figure 6.51   *Screw terminal. These terminals fit one cable conductor core to the screw-type bolt of a conductor bushing.*
Type of protection: EEx e II; Certificate: PTB Ex-94.C.3168 U; f = fine wire conductor; m = multiple wire conductor; Conductor cross-sections: m 25 ... 300 mm²; f 16 ... 240 mm²; Thread: M8 ... M16.

Figure 6.52   *Flameproof conductor bushings (connection between 'e' and 'd').*
Marking: Ex II 2G, EEx de II, IM2 EEx de I; Type of protection: EEx de I/EEx de II C; Certificate: PTB 98 ATEX 1067 U; Rated voltage: 6.6 kV and 11 kV; Rated current: 400 A; Conductor cross-sections: up to 300 mm²; Conductor diameter: 16 mm; Thread: M 80 × 1.5 and M 110 × 1.5; Insulating material: cast resin.

Two parallel busbars with a length $l = 1\,\text{m}$ and a distance $r = 50\,\text{mm}$ shall withstand a short-circuit current (peak value) of $100\,\text{kA}$. Then the force F will be:

$$F = 4 \cdot 10^4 \text{N}$$

(approximately 4 tons!)

In the case of a sinusoidal current:

$$I = I_o \cdot \sin \omega t$$
$$\omega = 2\pi f$$
$$f = 1/T$$

f = frequency
T = cycle duration (period)
F shows the term $\sin^2 \omega t$, containing the double frequency, 2ω or $2\pi 2f$.

To avoid resonance effects, the natural frequency f_n of the busbar system shall not be identical with $2f$.

The thermal effects caused by short-circuits can be calculated as follows: the overtemperature ΔT of a conductor of length l, resistance R, cross-sectional area A and volume V, made of material with density σ and mass m, specific heat c and specific resistance ρ, caused by the current $I(t)$, and without any heat dissipation to the environment, reads:

$$R \int I^2 (t)\, dt = c \cdot m \cdot \Delta T$$

It is:

$$R = \rho l / A$$
$$\sigma = m / V$$
$$V = A \cdot l$$

and then:

$$\frac{\rho l}{A} \int I^2 (t) dt = c\sigma Al\Delta T$$

which results in:

$$\Delta T = \frac{\rho}{c\sigma A^2} \int I^2 (t) dt$$

For sinusoidal currents with an rms value I_{rms} and a current flow time Δt (at least several half periods) this equation can be simplified:

$$\Delta T = \frac{\rho I_{rms}^2 \Delta T}{c\sigma A^2}$$

Example: in a low voltage switchgear, a busbar system (copper, $A = 20 \times 5 = 100\,\text{mm}^2$) is exposed to a $1\,\text{s}$ short-circuit current of $10\,\text{kA}$ (rms). Calculate ΔT!
For copper, it is:

$$c = 0.38\,\text{Ws/gK}$$
$$\rho = 1.7 \cdot 10^{-2} = \text{ohm} \cdot \text{mm}^2/\text{m}$$
$$= 1.7 \cdot 10^{-5}\,\text{ohm} \cdot \text{mm}$$
$$\sigma = 8.936\,\text{g/cm}^3$$
$$= 8.936 \cdot 10^{-3}\text{g/mm}^3$$

Then, ΔT follows with

$$\Delta T = 50\,\mathrm{K}$$

(at a current density of $100\,\mathrm{A/mm^2}$!). In normal service with rated current, the temperature of busbar systems is in the range $+60°\mathrm{C}$ to $+80°\mathrm{C}$. In the short-circuit condition given above, the temperature will increase up to $+110°\mathrm{C}$ to $+130°\mathrm{C}$.

The standards for increased safety – e – do not exclude plug-and-socket connections. In recent times, such connections have found a broad field of application in the low voltage and high voltage range simultaneously, overrunning the philosophy that such connections can be considered as 'safe' in a flameproof enclosure only. Figure 6.53 shows a low voltage plug-and-socket combination in 'e' as a whole with the exception of a socket-integrated 'flameproof-d-' switch according to EN 50018 as an interlock to disconnect all poles and the neutral before separation. The contacts cannot be energized when plug and socket remain separated. Very generally, IEC

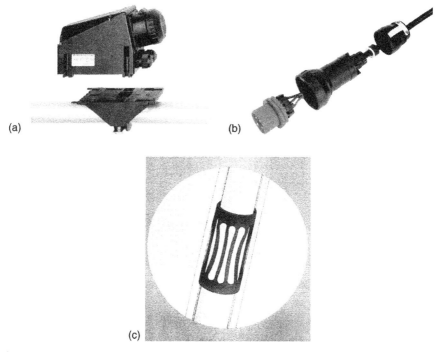

(a)

(b)

(c)

Figure 6.53 *(a) and (b) Low voltage plug-and-socket connector.*
Marking: Ex II 2G, EEx de II C T6; Type of protection: EEx de II C T6; Certificate: PTB 99 ATEX 1039 (16 A), PTB 99 ATEX 1041 (32 A); Rated voltage: up to 690 V; Rated current: 16 A, 32 A; Ingress protection code: IP 66; Types: 2P + PE, 3P + PE, 3P + N + PE; (for voltages <50 V): 2P; 3P. The 'd' in the marking is due to the inter-lock switch according to EN 50018
(c) Detail of Figs 6.53(a) and (b). The contact between the male and female part is made by a lamellar contact piece fixed in the female part.

60079-0 and EN 50014 ask for an electrical or mechanical interlock for plug-and-socket combinations in order to avoid the contacts being energized at the moment of separation, or, alternatively, plug-and-socket combinations shall be fixed together with special fasteners complying with clause 9.2 of IEC 60079-0 and EN 50014 and marked with the warning: DO NOT SEPAR-ATE WHEN ENERGIZED. In German, plug-and-sockets with an interlock are called 'Steckvorrichtungen', whereas plug-and-sockets without an interlock are called 'Steckverbinder'. In chemical plants (Group II application), plug-and-socket connections for high voltage have been established as a competitive component to the 'traditional' terminal compartments. The cables are connected directly to the apparatus to be supplied with electric power via the plug-and-socket connection. Usually, these connections are constructed to fit one single phase conductor, and consequently single-core cables are used with an earthed screening. In three-phase systems with an isolated neutral, a single insulation fault cannot create a short-circuit due to the screening covering each phase conductor over its total length. Only a line-to-earth fault can occur, resulting in low fault currents (compared with a short-circuit current) caused by the capacitances between phase conductor and earthed screening. Figure 6.54 shows a section view of a high voltage connector and Fig. 6.55 gives an example for this technique, replacing the traditional terminal compartment fitted to a cage induction motor.

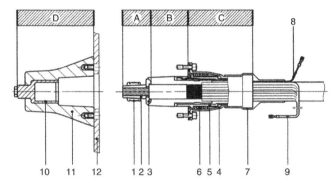

Figure 6.54 *High voltage plug-and-socket connector (single pole) (cable termination system).*
Type of protection: EEx e II T5; Certificate: PTB Ex-90.C.3183; Rated voltage: 11 kV; Rated current: up to 1250 A; Conductor cross-section: up to 630 mm^2.
System description:

A Contact system	B Insulating and potential grading part	C Housing	D Bushing
1. contact ring;		4. bell flange;	10. female
2. tension cone;		5. pressure sleeve;	contact part;
3. thrust piece;		6. pressure spring;	11. insulating
		7. heat-shrink tubing;	bushing;
		8. test lead (on request);	12. housing.
		9. cable screen;	

Figure 6.55   *High power high voltage cage induction motor with cable termination system according to Fig. 6.54 with single core cables.*

6.7.3   Motors, generators, transformers and reactors

Asynchronous motors (squirrel cage induction motors, the rotor-stator part of slipring induction motors) and the rotor-stator part of synchronous motors are within the scope of 'increased safety' due to the omission of sparks and arcs assuming an appropriate construction of rotor and windings. The standards IEC 60079-7 and EN 50019 contain mechanical and thermal requirements that follow this aim.

To ensure a certain mechanical robustness, the nominal conductor diameter of wires for windings shall be at least 0.25 mm. (Resistance thermometers and their terminals are not subject to this limitation.)

In consequence, 'e' motors, generators, transformers and reactors designed for the common rated supply voltages 500, 690 V and 1000 V are not suitable for very low power ratings. Smaller conductor diameters ask for another type of protection, e.g. 'flameproof enclosure – d –', 'encapsulation – m –' or 'powder filling – q –', the last mentioned restricted to transformers and reactors.

For rotating electrical machines, a reduction of the degree of ingress protection against solid foreign bodies (e.g. dust) and water is allowed for use in clean environments and subject to the regular supervision by skilled personnel. For Group I apparatus, IP 23, and for Group II apparatus, IP 20 are stated in the 'e' standards. IP 2X allows an internal cooling by environmental air ventilated through the enclosure. Higher degrees of ingress protection, e.g. IP 5X or IP 6X ('X' stands for an undefined degree of protection against water), ask for an internal closed-circuit ventilation (normally with air) and a heat exchanger. This may be the finned surface of the enclosure with forced outside ventilation, an air-to-air heat exchanger (in many cases an annular piping surrounding the stator windings) or an air-to-water heat exchanger.

Attention should be directed to the motor's (generator's) radial air gap. In contradiction to synchronous machines with a wide radial air gap, asynchronous machines ask for a radial air gap as small as possible to reduce the reactive power demand. On the other side, a rotor cannot be considered as a mechanical 'stiff' part but as resilient with respect to torsion showing distinct natural vibrations. Essential factors are the torque harmonics of the motor, in some cases the uneven, periodic torque of the engine powered by the motor, e.g. a piston compressor, and the interaction between the unbalanced state of the rotor and the stiffness of the bearings. If the natural vibrations of the rotor are matched with the exciting frequencies resulting from the effects mentioned above (at a 'critical' speed of the rotor), the rotor is subject to a radial deviation. So, the rotor may brush against the stator iron. In order to avoid such damage, with a certain risk of igniting a potentially dangerous gas–air mixture by frictional sparks, the radial air gaps of asynchronous machines are in the range of 0.2 … 3 mm (at standstill).

For e-rotating electrical machines, IEC 60079-7 and EN 50019 prescribe the minimum radial air gap (at standstill) according to:

$$g_{min} = \{0.15 + 780^{-1}(D-50) \cdot (0.25 + 0.75n/1000)\} \cdot r \cdot b$$

g_{min} = minimum radial air gap (in mm)

D = rotor diameter (in mm), in the formula above there are lower and upper limits with $75 \leqslant D \leqslant 750$

n = maximum rated speed (min^{-1}), with a minimum value of 1000 min^{-1}

r = ratio core length to rotor diameter according to

r = core length (in mm)/1.75·D^*

D^* = real value of rotor diameter (in mm), the condition $75 \leqslant D \leqslant 750$ is not valid here.

r is restricted to a lower limit of $r \geqslant 1.0$

b = 1.0 for machines with rolling bearings

  = 1.5 for machines with plain bearings

To illustrate the technical content of this formula, two figures (Figs 6.56 and 6.57) are given for 50 cps, motors with 2, 4 and 6 (or more) poles and for 60 cps, motors with 2, 4, 6 and 8 (or more) poles. For simplification, the synchronous speed has been given for the maximum rated speed, neglecting the motor's slippage.

An essential point considering e-motors is their thermal behaviour. For the simple case of permanent operation at rated power, the windings shall exceed neither the temperature limit of the corresponding temperature class (T1 … T6) nor the limiting temperature due to the thermal class of insulating material* given in Table 6.13, Part A.

For Group II electrical apparatus in cases where the electrical apparatus is subjected to thermal type testing, the measured temperature of any surface exposed to an explosive gas atmosphere shall not exceed the marked

---

* The values given in Table 6.13 are limiting *overtemperatures* in K. To determine the limiting *temperatures* in °C, 40 K shall be added to the overtemperatures values.

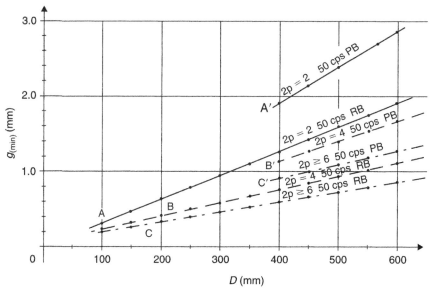

Figure 6.56   *Minimum radial air gap for asynchronous machines (at standstill)* g_{min}
versus rotor diameter D, *valid for 50 cps and* r = 1.0.
Plot A: 2 poles, rolling bearings; plot A': 2 poles, plain bearings; plot B: 4 poles,
rolling bearings; plot B': 4 poles, plain bearings; plot C: 6 (and more) poles, rolling
bearings; plot C': 6 (and more) poles, plain bearings.
RB = rolling bearings   PB = plain bearings

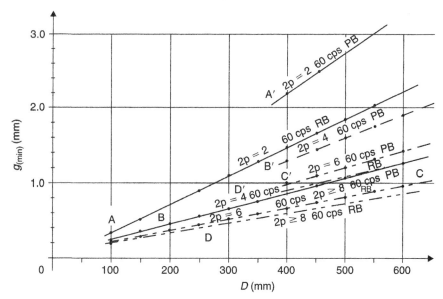

Figure 6.57   *Same as in Fig. 6.56, with frequency 60 cps.*
Plot A: 2 poles, rolling bearings; plot A': 2 poles, plain bearings; plot B: 4 poles,
rolling bearings; plot B': 4 poles, plain bearings; plot C: 6 poles, rolling bearings;
plot C': 6 poles, plain bearings; plot D: 8 (and more) poles, rolling bearings; ·
plot D': 8 (and more) poles, plain bearings.

Table 6.20 Limiting temperatures for insulated windings (for stalled motors) (referring to an ambient temperature of +40°C, and to the end of time t_E). Method of temperature measurement: resistance

Thermal class of insulating material*	A	E	B	F	H
Limiting temperature** °C	160	175	185	210	235

* According to IEC 60085
** The limiting *overtemperature* can be calculated with −40 K

temperature, or the temperature class, less 5 K for temperature classes T6, T5, T4 and T3, and less 10 K for temperature classes T2 and T1. For Group I equipment, the temperature limitations are +450°C and +150°C (see Table 4.1, Part A).

This is of general importance, with a special reference to the bare parts of a cage rotor, and establishes an additional limit for the (insulated) windings. The rotor of a squirrel cage motor shall exceed neither +300°C nor the temperature limit of the corresponding temperature class. These values shall not be exceeded even during starting.

In continuous operation, rotor and windings asymptotically attain their permanent operating temperatures at different values T_{rotor}, T_{stator}. For the steady state, a 'hotter' rotor (compared with the stator windings), $T_{rotor} > T_{stator}$, indicates a 'rotor critical' motor, whereas 'hotter' stator windings (compared with the rotor), $T_{stator} > T_{rotor}$, indicate a 'stator critical' motor. Normally, motors with a rated power exceeding 1 ... 2 kW, are 'rotor critical', smaller motors are 'stator critical'.

When the motor is stalled, the temperatures of rotor and windings start to increase and interfere with the different temperature limits, given by

- temperature class and/or
- insulation of windings.

For a short time, the limiting temperature for insulated windings given in Table 6.13, Part A, may be exceeded up to increased values (Table 6.20).

A motor stall at permanent operating temperatures causes a sudden temperature rise for rotor and stator windings and a subsequent overstep of temperature limits. In Fig. 6.58, the stator windings of a stator critical motor pass the T4 class temperature limit (minus 5 K) after a time t_{E3} at point 3. The rotor follows with a time t_{E2} at point 2. So, for the T4 rating of this motor, t_{E3} is a critical point. This is the latest moment to interrupt the motor's power supply. For a T3 rating of this motor, much more time is available: at point 1, the windings temperature crosses the short-time limiting temperature for the thermal class E after a time t_{E1}. (An additional time delay will push the windings to the T3 temperature limit.) So, t_{E1} is the relevant time for T3 rating of this motor.

Another example is given in Fig. 6.59. A rotor critical motor (windings in thermal class F) is to be rated for T2 and T3 operation. The rotor (here, for simplification, considered as a thermal 'homogeneous' component) passes the T3 class temperature limit (minus 5 K) after time t_{E3} at point 3, and somewhat later the T2 class or rotor temperature limit (+300°C minus 10 K) after time t_{E1} at point 1. Meanwhile, the stator windings have passed the

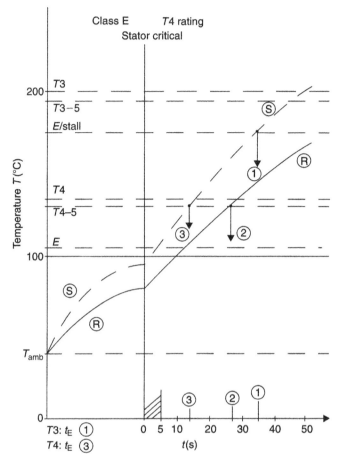

Figure 6.58   *Thermal behaviour of an E class stator critical motor, stalled at* t_o *after steady state at rated power.*
Temperature versus time for:
R: rotor; S: stator windings; *E*: limiting temperature of E class windings in continuous operation according to Table 6.13, Part A; *E*/stall: limiting temperature of E class windings according to Table 6.20; *T*3, *T*4: limiting temperatures for temperature classes T3 and T4; T_{amb}: highest ambient temperature for ex-apparatus = +40°C.

short-time limiting temperature (for thermal class F) after time t_{E2} at point 2. In consequence, the rotor critical motor (in continuous operation) remains 'rotor critical' for T3 rating (referring to time t_{E3}). But for T2 rating, the stator windings at point 2 are more time critical than the rotor at point 1. For T2 rating, this motor may be considered as 'stator critical'.

To give an impression for t_E values, the correlations in Table 6.21 may be helpful.

To ensure a time long enough to disconnect the motor from its power supply when the motor is stalled, t_E shall exceed 5 s. When current dependent protective devices are applied to initiate the disconnection, t_E shall increase with decreasing ratios of starting current to rated current.

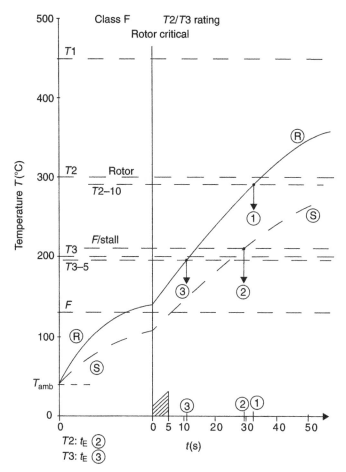

Figure 6.59 *Thermal behaviour of an F class rotor critical motor, stalled at* t_o *after steady state at rated power.*
Temperature versus time for:
R: rotor; S: stator windings; F: limiting temperature of F class windings in continuous operation according to Table 6.13, Part A; F/stall: limiting temperature of F class windings according to Table 6.20; T1, T2, T3: limiting temperatures for temperature classes T1, T2 and T3; T_{amb}: highest ambient temperature for ex-apparatus = +40°C.

Temperature sensors as an alternative to or associated with current dependent protective devices are within the scope of 'e' standards. Temperature measurements in stator windings are commonly used, but as an alternative to current dependent protective devices they are applicable in stator critical motors only.

Rotor critical motors with temperature sensors in the rotor require expensive data transmission techniques, e.g. Pt 100 resistances, data convertor with frequency modulator, and brushless data transmission to a stator-fixed

Table 6.21 t_E values of cage induction motors (thermal class F, 50 cps, for temperature classes T1/T2 and T3) (by courtesy of Siemens AG, Standard Drives Division, Erlangen/Germany)

Frame number	2p = 2				2p = 4			
	Rated power kW		t_E s		Rated power kW		t_E s	
	T1/T2	T3	T1/T2	T3	T1/T2	T3	T1/T2	T3
63	0.25	0.25	16	14	0.18	0.18	30	25
71	0.55	0.55	17	11	0.25	0.25	50	40
100 L	2.5	2.5	9	8	2.5	2.5	12	10
132 S	6.5	5.5	10	7	5.0	5.0	10	9
160 M	9.5	7.5	15	17	10	10	17	10
200 L	25	20	22	17	27	24	14	8
280 M	76	58	15	11	80	70	20	6
315 M	112	80	21	19	120	100	24	7
315	200	150	18	13	200	170	23	9
355	300	220	14	11	275	240	27	11
355	400	300	15	11	400	350	26	11

Table 6.22 Comparison of a non-explosion protected motor with (E)Ex eII design (rated synchronous speed 1500 min^{-1}, frame number 132, 50 cps, totally enclosed/fan cooled) (by courtesy of Loher GmbH, Ruhstorf/Germany)

Temperature class T …	Rated power kW	Efficiency %	cos φ	Mass kg
none (not ex-protected)	7.5	88.0	0.85	69
T3 (ex-protected)	6.8	88.0	0.86	69
T4 (ex-protected)	4.0	87	0.86	69

receiver. Such monitoring techniques are restricted to very large drives. It should be noted that – according to the 'ATEX 100a' philosophy (Directive 94/9/EC) – safety systems (e.g. the different protective devices mentioned above) are subject to approval procedures depending on the category of application.

The fact that all internal parts of a motor in 'increased safety' shall comply with the temperature limitations according to the temperature class, results in somewhat 'uneconomical' motor dimensions for classes T4, T5 and T6, or, in other words, the rated power is considerably reduced (Table 6.22) compared with 'normal industrial design'.

As an example for a large gas compressor drive unit, Fig. 6.60 shows a 3.2 MW synchronous motor, whose stator–rotor part and exciter set are

Figure 6.60 *3200 kW synchronous motor as a compressor drive.*
Type of protection (stator–rotor part of the motor and exciter set in 'e', diode wheel in a flameproof enclosure): EEx ed IIA T3; Rated voltage: 6 kV; Rated current: 356 A; Rated speed (20 poles, 50 cps): 300 min^{-1}.

protected according to EEx eII T3. The diode wheel of the exciter set is within a flameproof enclosure of type EEx d IIA T3.

6.7.4 Mains-operated luminaires

Table 6.23 gives a survey of lamps for general lighting applications in an industrial environment. The e-standards IEC 60079-7 and EN 50019 restrict the choice of lamps to the following types:

- incandescent lamps (with tungsten filament), in Table 6.23 listed as 1.1
- blended lamps (MBTF = a tungsten filament acts as a 'ballast' in series with a mercury-vapour high pressure discharge), in Table 6.23 listed as 2
- fluorescent lamps of the cold starting type with single-pin caps (Fa 6) according to IEC 60061-1 (listed under 3.1 in Table 6.23)
- other lamps for which there is no danger that parts of the light source may attain for a period longer than 10 seconds a higher temperature than the limiting temperature following breakage of the bulb. Such lamps are listed as 3.3 in Table 6.23.

Lamps containing free metallic sodium (e.g. low-pressure sodium vapour lamps) are not permissible. This is valid in general, i.e. not only a requirement for 'increased safety – e –', but for all types of protection.

Table 6.23 indicates a certain disadvantage of luminaires in 'e': lamps with high luminous densities, with high rated power and luminous flux values or with high luminous efficiency do not comply with the e-standards IEC 60079-7 and EN 50019. Incandescent lamps and blended lamps have become

Table 6.23 Lamps for general lighting in industry[1]

No.	Type of lamp	Rated power W	Luminous flux lm	Luminous efficiency lm/W	Luminous density cd/cm^2	Colour rendering property	Additional sets
1.1	Incandescent	100	1380	13.8	5 · 10^6 (filament)	medium	none
		200	3150	15.75			
		500	8400	16.8			
1.2	Halogen	500	9500	19	up to 1 · 10^7 (filament)	good	none
		1000	22 000	22			
		2000	44 000	22			
2	Blended (mercury vapour, high pressure, + tungsten filament)	160	3100	19.4	3–6	medium	none
		250	5600	22			
		500	14 000	28			
3	Fluorescent (mercury vapour, low pressure, + luminophor)						
3.1	tubular	18	1350	75	1.0–1.5	excellent	ballast, compensating capacitor, starter[2] special type
		36	3350	93			
		58	5200	90			
		20	1000	50	0.4–0.75	good	
		40	2500	62.5			
		65	4800	74			

		Power (W)	Flux (lm)	Efficacy (lm/W)		Colour rendering	Equipment
3.2	compact (folded tube)	18	1200	67	1.5–2.0	excellent	ballast, compensating capacitor, starter
		36	2900	80.6			
		55	4800	87			
3.3	**Induction lamps (HF-powered)**	**100[3]**	**8000**	**80**		**excellent**	**special type**
		150[3]	**12 000**	**80**			
		85[4]	6000	70			
		165[4]	12 000	72			
4.1	Mercury vapour (high pressure, + luminophor)	125	6300	50	7–25	medium	ballast, compensating capacitor
		400	22 000	55			
		1000	58 000	58			
		2000	125 000	63			
4.2	Metal-halide (mercury vapour, high pressure + metal halides)	250	20 000	80	500–900	good to excellent	ballast, ignition set, compensating capacitor
		1000	80 000	80			
		2000	240 000	120			
		3500	320 000	91			
5.1	Sodium vapour low pressure[5]	27	3500	130	4–10	very poor[5]	special type compensating capacitor
		35	5750	164			
		65	10 700	164			
		90	17 000	189			
		127	25 000	197			
		135	22 500	167		leakage field transformer, compensating capacitor
		185	32 000	173			

(continued)

Table 6.23 (continued)

No.	Type of lamp	Rated power W	Luminous flux lm	Luminous efficiency lm/W	Luminous density cd/cm²	Colour rendering property	Additional sets
5.2	Sodium vapour high pressure	150	14 500	97	300–600	poor	ballast, ignition set, compensating capacitor
		250	27 000	108			
		400	48 000	120			
		1000	130 000	130			
5.3	Sodium–xenon lamp (sodium vapour high pressure, + xenon)	50	3800	76	–	medium	special type
		80	6000	75			

1 Lamp types suitable for application in 'increased safety – e –' are pointed out in bold type. This table contains a selection only of lamps with a wide range of rated power. So far, this table does not claim for completeness

2 For application in 'increased safety – e –', tubular fluorescent lamps shall be of the cold starting type fitted with single-pin caps (type Fa 6 according to IEC 60061-1). These lamps do not need a starter

3 Frequency of lamp current: 250 kcps

Lamp bulb: tubular, closed loop

   (by courtesy of OSRAM GmbH, München/Germany)

4 Frequency of lamp current: 2.65 Mcps

Lamp bulb: pear-shaped, with axial cylindrical cavity

   (by courtesy of PHILIPS NV, Eindhoven/The Netherlands)

5 The very poor colour rendering property of low pressure sodium vapour lamps is due to the narrow-band emission of the Na lines (at a wavelength of 589.0 and 589.6 nm). Low pressure sodium vapour lamps are generally not within the scope of the standards for explosion protected luminaires

less important for general lighting applications due to their poor luminous efficiency. Therefore, this section focuses on fluorescent lamps.

It should be emphasized that the general requirements in the field of explosion protection, i.e. IEC 60079-0 and EN 50014, ask for constructional details for luminaires independent of their type of protection:

- the lamp shall be protected by a light-transmitting cover (which may be provided with an additional guard)
- covers giving access to the lampholder and other internal parts of luminaires shall
  (a) either be interlocked with a device which automatically disconnects all poles of the lampholder when starting the opening procedure
  (b) or be marked with 'DO NOT OPEN WHEN ENERGIZED'
  (Note: luminaires in 'intrinsic safety – i –' may not comply with this clause.)

Where it is intended that electrical components (other than the lampholder) shall remain energized after opening of the luminaire, special precautions shall be taken, e.g. the component shall comply with a type of protection, or an internal supplementary enclosure which prevents access to the components remaining energized with an IP degree of at least IP 30 according to IEC 60529 shall be installed.

Tubular fluorescent lamps in 'e' shall be fitted with single-pin caps (type Fa 6 according to IEC 60061-1).

The reason is that lamp pin and lampholder complying with IEC 60061-2 (in this case considered as a 'bore') form a 'flameproof enclosure' embracing the contact point when fitted together.

Tubular fluorescent lamps are subject to an aging effect, i.e. the lamp shows a behaviour like a rectifier.

Ballasts shall be constructed so that their limiting temperature shall not be exceeded in this case. When type tests are carried out, this effect can be simulated easily by inserting a rectifier diode in the lamp circuit.

When operated directly on the 230 V mains via an inductive ballast, only the 20 W lamp type is supplied with an adequate starting voltage. The 40 W and 65 W types ask for starting voltages in the 300–350 V range. So, leak transformers are used as a ballast for current limitation and as a 'voltage booster'. Other techniques are circuits with L–C combinations in order to supply an increased starting voltage. The starting voltage decreases with increasing ambient temperature. Due to the single-pin caps, preheating of the lamp electrodes is excluded.

A quite different technique is the application of 'electronic ballasts', commonly a combination of a rectifier on the mains side followed by a DC to AC inverter. Its frequency on the output side is adjusted to values exceeding 20 kcps. In this frequency range, the fluorescent lamps show an increased luminous efficiency compared with a 50 or 60 cps supply. In parallel, the upper limit of the human hearing is passed avoiding undesirable whistle caused by the magnetostriction of inductive components.

Inductive ballasts (or inductive components of electronic ballasts) may be designed according to the specifications for 'increased safety – e –'.

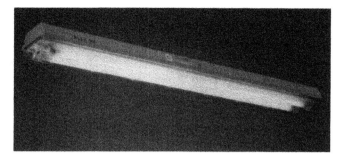

Figure 6.61　*Luminaire in 'increased safety – e –'. One or two tubular fluorescent lamps with Fa 6 sockets (single-pin cap).*
Rated power: 1 × or 2 × 58 W; Electronic ballast – Type of protection: EEx ed IIC T4; Certificate: PTB Ex-92.C.1029; Rated voltage: AC 110–254 V, DC 110–230 V; Ingress protection code: IP 66; Rated luminous flux: 1 × 58 W – 5200 lm; 2 × 58 W – 10 400 lm; Efficiency: 1 × 58 W – 83%; 2 × 58 W – 72%; Enclosure material: glass-mat reinforced polyester; Cover: polycarbonate; Weight: 8.0 kg; Dimensions (L × W × H): 1660 × 188 × 130 mm.

Semiconductors or capacitors are out of the scope of – e – and shall be protected (or the ballast in its total) in a different type of protection: e.g.

flameproof enclosure – d –
powder filling – q – or
encapsulation – m –.

The switch for disconnecting all poles of the lampholder or all components inside the luminaire as an interlock with the fastener is commonly protected according to 'flameproof enclosure – d –'. So, the marking of – e – luminaires covers the key letters of the explosion protected components inside, e.g. (E)Ex ed … or (E)Ex emd … Figures 6.61, 6.62 and 6.63 give examples of – e – luminaires for indoor and outdoor lighting.

Caution shall be given to avoid any mechanical damage of the tube of fluorescent lamps, not only when fitted inside the luminaire (and in operation), but also during transportation in a potentially explosive atmosphere.

In case of mechanical damage, the environmental atmosphere enters the tube and starts to compress the filling gas of the lamp. The filling gas is composed of (normally) argon and mercury vapour at a total pressure of some 10^2 Pa. The compression proceeds without any energy transfer to the environment as an adiabatic compression, described by Poisson's law:

$$p \cdot V^\chi = \text{constant}$$

p = gas pressure
V = gas volume
χ = adiabatic exponent
χ = c_p / c_v, i.e. the ratio between the mean moles heat of the gas, referring to constant pressure ($=c_p$) and to constant volume ($=c_v$)

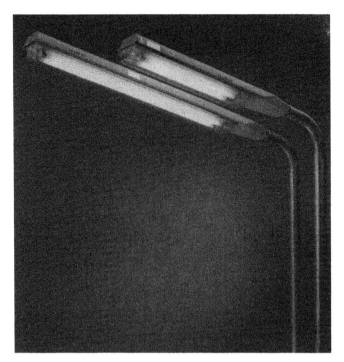

Figure 6.62 *Pole mounted luminaires in 'increased safety – e –'. Two tubular fluorescent lamps with Fa 6 sockets (single-pin cap).*
Rated power: 2 × 18 W or 2 × 36 W; Electronic ballast – Type of protection: EEx ed IIC T4; Certificate: PTB Ex-94.C.2101 (2 × 18 W), PTB Ex-95.D.2176 (2 × 36 W); Rated voltage: AC 110–254 V, DC 110–230 V; Ingress protection code: IP 66; Rated luminous flux: 2 × 18 W – 2700 lm, 2 × 36 W – 6700 lm; Efficiency: 2 × 18 W – 78%; 2 × 36 W – 78%; Enclosure material: glass-mat reinforced polyester; Cover: polycarbonate; Weight: 2 × 18 W – 6.7 kg; 2 × 36 W – 9.1 kg; Dimensions (L × W × H): 2 × 18 W – 1060 × 188 × 130 mm, 2 × 36 W – 1660 × 188 × 130 mm.

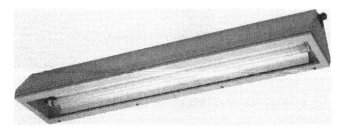

Figure 6.63 *Luminaire in 'increased safety.– e –' for paint shops and spray paint cabins. Three tubular fluorescent lamps with Fa 6 sockets (single-pin cap).*
Rated power: 3 × 65 W; Electronic ballast – Type of protection: EEx edq IIC T4; Certificate: PTB Ex-90.C.2041; Rated voltage: AC/DC 220–240 V; Ingress protection code: IP 65; Weight: 42.5 kg; Dimensions (L × W × H): 1630 × 400 × 215 mm.

The equation for 'ideal' gases:

$$p \cdot V = n \cdot R \cdot T$$

p = gas pressure
V = gas volume
n = number of moles
R = 8.317 joule $\times$ mole^{-1} $\times$ K^{-1}
T = gas temperature

can be used to eliminate one variable of state in order to get the correlation between the two residual variables.

In steady state, p_1, V_1 and T_1 describe the parameters of the filling gas. After compression, these parameters are p_2, V_2 and T_2.

It is
$$T_2 = T_1(V_1/V_2)^{x-1}$$

showing an increase of temperature after adiabatic compression, or

$$T_2 = T_1(p_2/p_1)^{\frac{x-1}{x}}$$

For the main component of the filling gas, argon, it is

$$\chi = 1.66 \text{ (for air: } \chi = 1.40)$$

Obviously, pressure ratios p_2/p_1 in the range 10–25 are sufficient to cause a temperature rise from $T_1 = 300\,\text{K}$ to $T_2 \sim 750\,\text{K}$ up to $T_2 \sim 1080\,\text{K}$, referring to argon, and may start the ignition of gas–air mixtures with low ignition energies, e.g. hydrogen–air mixtures.

A very recent development has been the introduction of 'induction lamps' into service. These fluorescent lamps are low pressure mercury vapour lamps with an additional luminophor. The difference between induction lamps and 'conventional' fluorescent lamps is a gas discharge in a 'closed loop' without the existence of electrodes. The power is supplied via the electromagnetic field of one (or two) coils into the gas discharge at frequencies of 250 kcps (manufacturer: OSRAM) or 2.65 Mcps (manufacturer: PHILIPS), the former with two coils complying with the 'e'-standards (Fig. 6.64). The advantage of such lighting systems is a very increased operating life: compared with tubular fluorescent lamps, showing 8000 to 20 000 hours, induction lamps succeed with 60 000 hours of operating life.

6.7.5   Accumulators (secondary batteries)

Accumulators in 'increased safety – e –' cover traction batteries for locomotives in coal mines (see Fig. 6.4) as well as batteries for forklifts in chemical plants or other transportation vehicles running in hazardous areas powered by their own energy storage system. Much smaller accumulators are used as a mains-independent power source for handlamps and caplamps (Fig. 6.65) used in potentially explosive atmospheres.

Accumulators shall be of the lead–acid, nickel–iron or nickel–cadmium type. For accumulators with a capacity exceeding 25 Ah (at the 5 hours rate) IEC 60079-7 and EN 50019 contain specific requirements for construction and

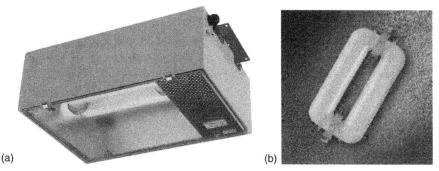

(a) (b)

Figure 6.64 *(a) Luminaire in 'increased safety – e –' for induction lamp.*
Rated power: 150 W; Ballast: special type; Frequency of lamp current: 250 kcps;
Marking: Ex II 2G EEx ed IIC T4; Type of protection: EEx ed IIC T4; Certificate: PTB
99 ATEX 2065; Rated voltage: AC/DC 220–240 V; Ingress protection code: IP 54;
Rated luminous flux: 12 000 lm; Luminous efficiency: 80 lm/W; Dimensions
(L × W × H): 652 × 400 × 205 mm
(b) Induction lamp.
Rated power: 150 W; Luminous flux: 12 000 lm; Dimensions (L × W × H):
414 × 139 × 72 mm; Tube diameter: 54 mm.

Figure 6.65 *Explosion protected caplamp.*
Type of protection: EEx e ib IIC T4; Certificate: PTB Ex-89.C.2156; Battery: recharge-
able NiCd accumulator 3.6 V, 7 Ah (component in 'increased safety'); Lamp: tungsten,
two filaments 3.75 V, 0.8 A and 0.3 A (for pilot light); (intrinsically safe electric circuit,
category ib); Light aperture: diameter 65 mm mineral glass; Operating period: approx. 8
hours; Recharging period: max. 18 hours; Weight: approx. 2.0 kg (with accumulator).

type testing. There are three 'essentials' compared with non-explosion protected accumulators:

• special precautions shall be made in order to prevent leakage currents due
  to the moistened and electrical conductive surface of cell containers caused

by the electrolyte. Battery containers shall be provided with insulating barriers so positioned to prevent nominal voltages exceeding 40 V in any compartment. Creepage distances between poles of adjacent cells and between these poles and the battery container are considerably increased compared with those given in Table 6.18 for general application in 'increased safety – e –'. This creepage distance shall be at least 35 mm. Where nominal voltages between adjacent cells exceed 24 V, these creepage distances shall be increased by 1 mm (at least) for each 2 V in excess of 24 V, e.g. 44 mm are required for 42 V and 53 mm for 60 V

- Group I batteries shall have an insulating covering for each live part to avoid leakage currents and any accidental contact
- battery containers shall be provided with ventilation openings to achieve an 'adequate' ventilation. 'Adequate' is described as a hydrogen concentration in the battery container not exceeding 2% by volume during a type test representing a charging period. For battery containers, the degree of ingress protection against solid foreign bodies and water is lowered to IP 23 according to IEC 60529.

The hydrogen generation and the decomposition of water (as an essential part of the electrolyte) during charging periods shall be considered in more detail under lead–acid accumulators.

Dominant chemical reactions during charging are oxidation of $PbSO_4$ to PbO_2, and reduction of $PbSO_4$ to Pb:

- at the positive plate:

$$PbSO_4 + 2H_2O \rightarrow PbO_2 + H_2SO_4 + 2e^- + 2H^+$$

- at the negative plate:

$$PbSO_4 + 2H^+ + 2e^- \rightarrow Pb + H_2SO_4$$

With the progress of charging, the decomposition of water starts according to:

- at the positive plate:

$$2H_2O \rightarrow 4H^+ + 4e^- + O_2$$

- at the negative plate:

$$4H^+ + 4e^- \rightarrow 2H_2$$

- summarizing as:

$$2H_2O \rightarrow 2H_2 + O_2$$

The decomposition of water is unavoidable due to the low decomposition potential of water, -0.4 V for hydrogen and $+0.8$ V for oxygen, as a total of 1.2 V. Mass and volume of hydrogen (and oxygen) generated during the charging period can be calculated according to Faraday's law:

To produce one gram equivalent (atomic weight of a substance in grams, divided by valence), an electrical charge of 96 487 As is required.

For hydrogen, 1 gram is produced by 96 487 As, corresponding to a volume of half the molecular volume (=22.4 litres, referring to a temperature of 273 K and a pressure of $1.013 \cdot 10^5$ Pa). 1 Ah $= 3600$ As generate a hydrogen mass of 0.0373 g with a volume of $0.4179 \cdot 10^{-3}$ m^3. (For oxygen with an atomic weight of 16 and valence $= 2$, 8 g are produced by $96\,487$ As. 1 Ah generates an oxygen mass of 0.2985 g with a volume of $0.20\,895 \cdot 10^{-3}$ m^3.)

The gas volume V depends on pressure p and temperature T according to $pV = n \cdot R \cdot T$ (for an ideal gas). Assuming a constant pressure, a temperature T deviating from $T_0 = 273$ K asks for a volume correction factor T/T_0, i.e. the volume V at temperature T can be calculated according to:

$$V = V_0 \cdot T/T_0 \quad (V_0 = \text{volume at } T_0)$$

Example:
With $T = +60°C$ ($=333$ K), the generation rate of hydrogen (per 1 Ah) is $0.50975 \cdot 10^{-3}$ m^3.

In the same way, the volume shall be corrected to comply with a change in pressure:

$$V = V_0 \cdot p_0/p \quad (V_0 = \text{volume at } p_0)$$

Example:
In deep coal mines, with $p = 1.1 \cdot 10^5$ Pa, the generation rate of hydrogen (per 1 Ah) reduces to $0.3848 \cdot 10^{-3}$ m^3.

It should be emphasized that these considerations refer to one single cell of an accumulator only. When cells are connected in series, the generation rates for hydrogen (and oxygen) shall be multiplied with the number of cells.

Speaking in a more general manner, the volume V of hydrogen released from N cells caused by a current I is:

$$V\,[\text{m}^3/\text{h}] = 0.4179 \cdot 10^{-3} \cdot N \cdot I$$

The loss of water can be calculated as the sum of hydrogen mass and oxygen mass generated.

1 Ah produces 0.0373 g hydrogen and 0.2985 g oxygen, resulting in a loss of 0.3358 g water.

So, the loss of water per hour, W, is given by:

$$W\,[\text{kg/h}] = 0.3358 \cdot 10^{-3} \cdot N \cdot I$$

Example:
A battery cell with a capacity of 960 Ah is charged with a current of $0.25 \cdot I_5$ ($I_5 = 5$ hours discharge current) during 2 h:

$$I_5 = 192\,\text{A}$$
$$0.25 \cdot I_5 = 48\,\text{A}$$

$W = 16.12$ g/h, resulting in a water loss of 32.24 g (and a hydrogen volume of 40.1 litres).

Hydrogen plays an important role in considerations related to the safety of accumulators in areas endangered by gas–air mixtures due to its broad range to form an explosive mixture: LEL $= 4\%$ (v/v), UEL $= 75.6\%$ (v/v), see Table 1.1.

This is the reason why standards IEC 60079-7 and EN 50019 ask for type tests, ensuring a highest allowable hydrogen concentration of 2% (v/v) in the battery container during the charging period.

Two alternative methods are stated in the standards. The first one simulates hydrogen generation during the charging period. The part of the battery container which is normally occupied by the cells shall be fitted with closed 'dummy' boxes. The lids of the boxes shall be provided with filler and vent plugs identical in form, number and location with those on the 'real' cells. The location of the boxes shall be such that the natural ventilation in the battery container remains unchanged. Into the space above the boxes, hydrogen shall be fed to the filler and vent plugs with a constant flow according to:

$$\text{hydrogen flow} \quad V_{\text{test}} \, [\text{m}^3/\text{h}] = N \cdot C \cdot 5 \cdot 10^{-6}$$

N = number of cells
C = capacity in Ah

Referring to the relation:

$$\text{hydrogen production} \quad V \, [\text{m}^3/\text{h}] = 0.4179 \cdot 10^{-3} \cdot N \cdot I$$

and corrected to an ambient temperature of +20°C (=293 K), it is:

$$V \, [\text{m}^3/\text{h}] = 0.4485 \cdot 10^{-3} \cdot N \cdot I$$

The test shall simulate the hydrogen production caused by a current I flowing after the main charging period. So, it is:

$$V = V_{\text{test}}$$
$$0.4485 \cdot 10^{-3} \cdot N \cdot I = N \cdot C \cdot 5 \cdot 10^{-6}$$
$$I \, [\text{A}] = 11.15 \cdot 10^{-3} \cdot C \, [\text{Ah}]$$

or

$$I \, [\text{A}] = 0.05575 \cdot C \, [\text{Ah}]/5$$

Obviously, the current I equals – as a very rough estimation – 5% of I_5, the 5 hours' discharge current.

For the example given above, a 960 Ah cell, $I_5 = 192$ A, and $I = 9.6$ A, V_{test} shall equal $4.8 \cdot 10^{-3} \text{m}^3/\text{h}$. With $V = 0.4485 \cdot 10^{-3} \cdot I$, $I = 9.6$ A, V will result in:

$$V = 4.306 \cdot 10^{-3} \text{m}^3/\text{h}$$

The second method uses a 'real' accumulator: the battery container shall be equipped with a battery made up of cells of the number, type and capacity it is intended to contain in service. An overcharging current shall be passed through the battery to produce hydrogen at a constant flow corresponding to the number, size, type of construction and capacity of the cells in the battery. The overcharging current is determined by

$$I_{\text{overcharging}} \, [\text{A}] = V \, [\text{m}^3/\text{h}] \, (N \cdot 0.44 \cdot 10^{-3})^{-1}$$

which is nearly identical with the formula for V, referring to +20°C (=293 K).

During both tests, the ambient temperature, the temperature of the battery container and the temperature of the cells (or of the boxes simulating the cells) shall be between +15°C and +25°C. The test is satisfactory if the hydrogen concentration does not exceed 2% (v/v).

The 'escape' of hydrogen is caused by diffusion due to the high molecular velocity. The theory of molecular cinetics gives a relation between velocity v, molecular mass μ and temperature T of a gas:

$$\tfrac{1}{2}\mu v^2 = 3/2kT$$
$$k = 1.3805 \cdot 10^{-23}\,\text{J/K}$$
$$\text{(Boltzmann's constant)}$$

With the molecular weight of a gas, m, and $\mu = m/L$:

$$L = 6.02252 \cdot 10^{23}$$
(Loschmidt's constant, sometimes quoted as Avogadro's constant)

it is

$$v^2 = 3kTL/m$$

For hydrogen, $m = 2 \cdot 10^{-3}\text{kg}$, and $T = 273\,\text{K}$, the molecular velocity $\sqrt{v^2} = 1848\,\text{m/s}$.

The mean value of the molecular velocity, $\bar{v}$, differs from $\sqrt{v^2}$ due to the distribution of molecular velocities (Table 6.24). The relation between the velocities v_I and v_II of gases with molecular weights m_I and m_II (at the same temperature T) can easily be calculated with:

$$\tfrac{1}{2} \cdot mv^2/L = \text{constant}_1 \text{ or}$$
$$mv^2 = \text{constant}_2,$$
$$m_\text{I} \cdot v_\text{I}^2 = m_\text{II}v_\text{II}^2$$
$$v_\text{I} = v_\text{II} \cdot (m_\text{II}/m_\text{I})^{0.5}$$

Table 6.24   Molecular velocities $\sqrt{v^2}$ and mean molecular velocities $\bar{v}$ of selected gases (at $T = 273\,\text{K}$)

Gas	$\sqrt{v^2}$ m/s	$\bar{v}$ m/s
Hydrogen	1848	1694
Nitrogen	492	453
Air*	485	447
Oxygen	461	425
Carbon dioxide	392	361

*Air, for simplification considered as a mixture of 79% (v/v) nitrogen and 21% (v/v) oxygen, shows an average molecular weight of 28.8

So, hydrogen (with $m = 2 \cdot 10^{-3}\,\text{kg}$) diffuses in an easy way through ventilation openings, compared with 'heavy' gases like oxygen ($m = 32 \cdot 10^{-3}\,\text{kg}$) or carbon dioxide ($m = 44 \cdot 10^{-3}\,\text{kg}$).

The density of the electrolyte in lead–acid batteries varies from $1.12\,\text{g}/\text{cm}^2$ at the end of discharge to $1.30\,\text{g}/\text{cm}^3$ at the end of charging. Unfortunately, the electrolyte shows an inhomogeneous distribution of density (higher values at the bottom of a cell, lower values at the top, showing differences in density up to $0.1\,\text{g}/\text{cm}^3$).

In order to achieve a homogeneous distribution of chemical processes during discharge/charging cycles across the surface of positive and negative plates (in other words to ensure a homogeneous current density at the electrodes), 'layers' of different densities in the electrolyte shall be removed.

One way to do so is an extended overcharging period with the production of hydrogen and the undesirable loss of water. The hydrogen bubbles 'mix' the electrolyte layers with different densities. The modern way replaces the function of hydrogen bubbles by purified compressed air (free of dust, oil, water), blown into the battery cells during the charging cycles (electrolyte circulation).

The question of overcharging of batteries is strongly related to their efficiency. Two values should be differentiated:

- the 'Ah' efficiency η_{Ah}:

$$\eta_{Ah} = \frac{\int I\,dt/\text{discharge cycle}}{\int I\,dt/\text{charging cycle}}$$

- the 'Wh' efficiency, η_{Wh}:

$$\eta_{Wh} = \frac{\int U \cdot I\,dt/\text{discharge cycle}}{\int U \cdot I\,dt/\text{charging cycle}}$$

During the overcharging period, a certain additional amount of the energy stored during the charging cycle is required and obviously strongly influences η_{Ah} and η_{Wh}.

Without electrolyte circulation, for lead–acid batteries:

$$\eta_{Ah} = 0.83 \dots 0.90$$
$$\eta_{Wh} = 0.67 \dots 0.75$$

(and for nickel–iron cells, as a guideline:

$$\eta_{Ah} = 0.72$$
$$\eta_{Wh} = 0.5 \dots 0.6)$$

With electrolyte circulation, the values for efficiency are increased for lead–acid batteries:

$$\eta_{Ah} = 0.90 \dots 0.95$$
$$\eta_{Wh} = 0.80$$

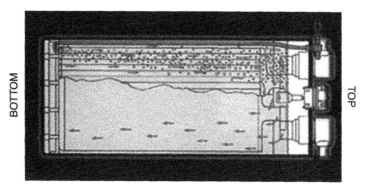

Figure 6.66 *Lead–acid battery cell with electrolyte circulation.*

due to the reduced overcharging current (only some per cent of I_5 instead of, e.g., 25% I_5).

Figure 6.66 gives an example for a lead–acid secondary cell with electrolyte circulation. Compressed air enters the cell at the top, flows downstream to the bottom via a duct and, after leaving this duct, drags the electrolyte upstream from the bottom. On the opposite side of the cell, the electrolyte flows downstream.

Constructional details of a modern lead–acid cell for locomotives in coal mines are given in Fig. 6.67. A copper expanded metal lattice parallel with the negative plate considerably lowers the internal resistance of the cell. The total view of a 540 V–480 Ah lead–acid accumulator for locomotives in coal mines is given in Fig. 6.68.

6.7.6 Current and voltage sensors

The 'classic' solution is the application of current and voltage transformers, usually as a cast resin embedded type. As a certain disadvantage of inductive transducers, their small frequency range may be considered. Other solutions for current sensors are Rogowski coils (with active or passive integrators) or the fibre-optic current sensor, based upon the Faraday effect [20].

The Rogowski coil (as a coil measuring the time-dependent azimuthal magnetic field of a current-carrying conductor) is not suitable for direct current applications, but can be designed for higher frequencies, e.g. for the output of frequency convertors. The Faraday effect is based upon the rotation of the plane of polarization (this is the plane of the electric vector) of linear polarized light in a magnetic field. The direction of light shall be identical with the axis of the magnetic field, and, in addition, the light shall spread out in matter, e.g. in glass, quartz or in an optical fibre made of plastics. The rotation angle of the plane of the electric vector, α, is strongly related to the magnetic field strength, B:

$$\alpha = c_v \cdot B \cdot l$$

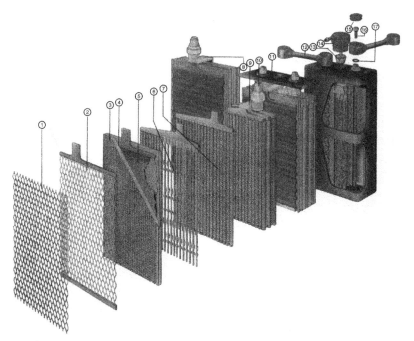

Figure 6.67   *Lead–acid battery cell for locomotives in coal mines, Group I.*
1: copper expanded metal lattice; 2: negative copper expanded metal lattice, lead
coated; 3: sheath separator, perforated, undulated; 4: microporous separator;
5: negative plate; 6: positive lead lattice; 7: positive armour plate; 8: negative plate
pile; 9: positive plate pile; 10: cell terminal; 11: welded cover; 12: cable connection,
flexible; 13, 14: floater and stopper plug of the water refilling system; 15: cable
connection cover; 16: connection screw; 17: O-ring.

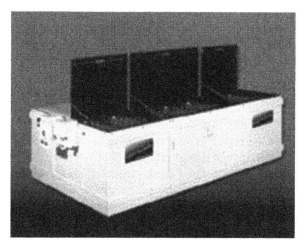

Figure 6.68   *Lead–acid accumulator for locomotives in coal mines.*
Types of protection – Battery container: EEx e I; Plug and socket: EEx de [ia] I;
Temperature sensor: EEx m I; Rated voltage: 540 V; Rated capacity: 480 Ah; IP code
of battery container: IP 23; Length: 3465 mm; Width: 1886 mm; Height: 964 mm.

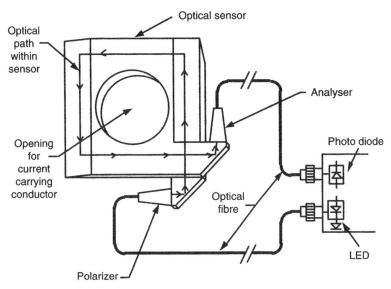

Figure 6.69 *Fibre-optic current sensor.*
Rated primary current I_r: 4 ... 4000 A; Accuracy class: 0.2; Bandwidth: >5 kcps;
Sensor output signal (measurement): analogue – 2.0 V AC at I_r, 8.0 V AC at 4 × I_r;
Sensor output signal (protection): analogue – 2.0 V AC at 10 × I_r, 20.0 V AC at
100 × I_r. An additional 1 A output is available on request.

c_v = a constant depending on the material conducting the light (to give an
impression of magnitude, c_v = 276.6°/m × tesla for light wavelength
λ = 589.3 nm (sodium line) in quartz)
l = length of the light path within the material

Figure 6.69 gives an example for an optical current sensor. The light path is
'wound' around a current-carrying conductor equidirectionally with the
azimuthal magnetic field of the current. The rotation of the plane of the electric vector is not detectable on its own and is converted to light intensity variations by a polarizer/analyser combination. A photo diode is used as a light
intensity detector. The optical sensor itself is installed in the – e – compartment, the electronics shall be protected in an adequate type of protection, e.g.
in a small flameproof – d – enclosure or in encapsulation – m –. In the special
case of an energy distribution system with combined – e – and – d – compartments, the optical fibres may enter the d-compartment to the electronics
inside via bushings complying with 'd'-standards EN 50018 or IEC 60079-1
respectively (Fig. 6.70). The 'evacuation' of the sensors into the e-compart-
ment results in additional available space in the more expensive d-compartment, compared with 'increased safety – e –'.

Other sensors, e.g. based upon the Hall effect, or capacitive or resistive
voltage dividers, are not within the scope of '– e –' and shall be manufactured
in a different type of protection.

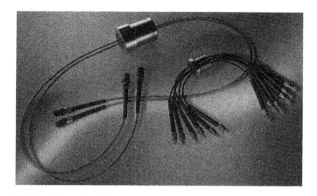

Figure 6.70 *Optical fibre bushings.*
Type of protection: EEx d IIC; Certificate: PTB Ex-94.C.1011 U; Fibre cores: max. 12 cores/bundle, 40 channels; Power limit (of optical radiation): 10 mW/mm²;
Max. temperature at place of installation: +110°C; Type and size of thread: M 16 × 1.5 to M 48 × 1.5; Diameter of sleeve: 22 mm to 46 mm; Joint length of sleeve: 15 mm, 25 mm, 40 mm; Optical fibre material: glass; Variants: solid core; solid core with plug-in connector; solid, hollow or bundled cores; solid or hollow core with plug-in connector.

6.7.7 Low voltage energy distribution systems

For some time, this has been the domain of 'flameproof enclosure – d –' and, somewhat later, of 'pressurization – p –' (see Sections 6.4 and 6.8). A more recent development results in a modular system of enclosures (mainly made of plastics), containing equipment itself complying with – e –, e.g. busbars, voltage and current transformers, measuring instruments, terminal blocks or incandescent lamps for signalling and remote controlling (E 14 size). All other components, e.g. fuses, contactors, overload protection devices, circuit breakers, capacitors, which are not within the scope of 'increased safety – e –', are individually explosion protected according to, e.g., 'flameproof enclosure – d –', 'encapsulation – m –' or 'powder filling – q –'. These components, with bare terminals or 'pigtail cables' are installed in the e-enclosures. A very simple example of this technique is given in Fig. 6.71. A (hand-operated) circuit breaker (in 'd' according to EN 50018, enclosure material: plastics) is inserted in an e-enclosure. Such an apparatus may be considered as the starting point of a development towards more complex switchboards (see Fig. 6.72, covers removed, and Fig. 6.73, covers closed, front view). This technique now competes with flameproof and pressurized switchboards, mainly in chemical plants (Group II application), notwithstanding a restricted selection of individually protected components.

A certain problem of electrical installations with a high degree of ingress protection against solid foreign bodies and water is the exchange of internal and external atmospheres (in Section 6.7.1 an example has been given). Usually, electrical apparatus is not gas-tight, and, in consequence, two parts of the environmental atmosphere should be taken into account: combustible gases (preferably in zone 1) and water. The presence of combustible gases in

Figure 6.71 *Enclosure in 'increased safety – e –' with installed circuit breaker in 'flameproof enclosure – d –'.*
Types of protection:
- enclosure   EEx ed IIC T ...
- circuit breaker   EEx de IIC
Technical data of circuit breaker:
Rated voltage: 690 V 3 AC; Rated frequency: 50/60 cps; Rated current: 0.1 A ...
22.5 A, depending on motor current range; Enclosure material: polyester resin, glass fibre reinforced.

an enclosure can be handled in a safe manner, see types of protection e, d and i. Water is a general problem: humid air enters an enclosure, and, in the cases of low thermal losses inside or inoperability, the humidity may condense. Due to the high IP protection degree, there is no way out for water. To avoid water accumulation, a 'hydraulic rectifier' at suitable (bottom) places should be inserted. A special diaphragm, comparable with those textiles for all-weather clothing, enables the egress of water (Fig. 6.74).

6.7.8   Resistance heating units

Induction heating systems, skin effect heating, dielectric heating or systems with a current passing through a liquid, an enclosure or pipework are not within the scope of 'increased safety'. However, resistance heating devices

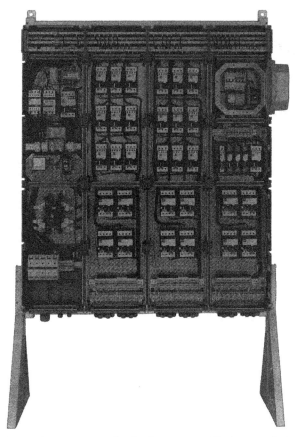

Figure 6.72   *Low voltage energy distribution system. Enclosures in 'increased safety – e –', covers removed.*
Enclosure material: polyester resin, glass fibre reinforced, or steel, galvanized, or stainless steel; Type of protection: EEx ed IIC T ...; Certificate: PTB Ex-95.D.3155 (for polyester enclosures); PTB Ex-96.D.3148 (for steel enclosures).
Enclosure dimensions, approx. (height × width × depth):
• up to 681 × 340 × 190 mm^3 (for polyester enclosures)
• up to 727 × 360 × 190 mm^3 (for steel enclosures)
IP code: IP 66
Rated voltage:
• up to 1000 V 3 AC, depending on internal components (for polyester enclosures)
• up to 690 V 3 AC depending on internal components (for steel enclosures)
Type of protection of internal components: EEx de IIC
Note: At the right side/on top, a 'flameproof enclosure – d –' is installed.

made of either a single conductor as a heating element and fed from its two ends, or a two conductor system (with low conductor resistance) fed from one end only and with the resistance heating element as a 'distributed resistor' inserted in the gap between the two parallel conductors and electrically connected with them (Fig. 6.75) may comply with the 'c'-standards if manufactured accordingly. Plug-and-socket connectors in 'e' in their entirety are

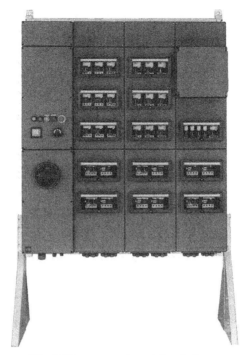

Figure 6.73 *Same as Fig. 6.72, covers closed, front view.*

Figure 6.74 *Drainage stopper and pressure compensation element.*
Type of protection: EEx e I/II; IP code (when installed): IP 68; Air flow at $\Delta p = 7000$ Pa: 14 litres/hour up to 25 litres/hour, depending on size.

used as well as terminal compartments to connect the resistance heating devices to the power system.

IEC 60079-7 and EN 50019 do not consider a heating resistor to be windings, and following this philosophy, the limitation of the wire diameter to a

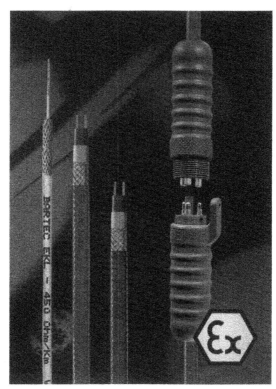

Figure 6.75 *Resistance heating cables in different versions:*
- *single conductor heater*
- *parallel conductor heaters (self-limiting heaters)*

Type of protection: EEx e II T ...; Cable connector system for heating units: Type of protection: EEx e II T5/T6; Certificate: KEMA Ex-99.E.0843 X.
alternatively:
Type of protection: EEx e II T3; Certificate: KEMA Ex-99.E.0841 X; Rated voltage: 254 V AC/DC; Rated current: 8 A, 10 A; IP code: IP 66.

lowest value of 0.25 mm and the requirement for at least two layers of insulation do not apply. The electrical insulation of the heating elements shall ensure that they cannot be in contact with the environmental atmosphere.

A very general requirement of 'e'-standards is the prevention from exceeding the limiting temperature (according to the temperature class). This shall be ensured by one of the following means:

- the limitation of rated power so that under specified conditions of installation and application in continuous operation the temperature of the heating device asymptotically climbs up to a steady value lower than the limiting temperature
- the self-limiting property of the resistance heating element, i.e. when connected to the constant voltage of the power grid, the total resistance of the heating device increases with increased temperature (i.e. there is a

positive temperature coefficient of the resistor), resulting in a continuously decreased dissipated power of the heating device and an adequate final temperature value
- an electrical protective system which, at predetermined surface temperature, isolates all live parts of the heating system. This protective system shall be independent of any other temperature control system provided for regulating the temperature under normal conditions. The 'e'-standards contain special requirements for this protective system.

6.8   Flameproof enclosure

Standards: EN 50018
             IEC 60079-1
Key letter: d

6.8.1   The basic principles of 'flameproof enclosure'

This type of protection works according to the principle that a combustible gas–air mixture is assumed to be present inside (and, of course, outside – 'flameproof enclosure – d –' is a zone 1 type of protection) due to the thermal cycles of an apparatus which generate a gas exchange between external and internal atmospheres (see the calculation in Section 6.7.1). Within a flameproof enclosure, sparks or arcs caused by, e.g., commutators, relays, sliprings, contactors, circuit breakers are permissible. In electric circuits working as a mains power supply, the energies of sparks or arcs do exceed the minimum ignition energies for gas–air mixtures (see Table 1.7 in Section 1.2.3) by orders of magnitude. Example: in a 1000 V three-phase circuit driving a 315 kW-rated power cage induction motor, switch-off arcing in a contactor (operating in air, not a vacuum contactor) causes an energy release of 1.5 ... 6.4 kJ. This should be compared with minimum ignition energies in the range of 9 ... 280 µJ!

In the same manner, temperatures exceeding the ignition temperatures (see Table 1.3 in Section 1.2.2) or the maximum surface temperatures according to the temperature class of the apparatus (see Table 4.1 in Chapter 4) are permissible inside of 'd'. This covers windings and rotors in motors, especially for 'low' temperature classes T4, T5 and T6 as well as the discharge tubes of high pressure sodium or mercury vapour lamps in case of a broken lamp bulb, or the ovens of gas chromatographs.

So, generally, the ignition of the internal gas–air mixture shall be taken into account. The enclosure itself shall fulfil two main features:

- the enclosure shall withstand the explosion pressure of the internal combustion process (as a guideline: 10 bar or 1 MPa) without any (permanent) deformation which may adversely affect the following point
- the joints between constructional parts (covers, doors) or structural openings in the enclosure shall prevent any flame transmission from the

internal to the external (combustible) atmosphere. To give a rough idea, the gap (=the clearance between corresponding surfaces, or the difference between two diameters, forming a flameproof joint), w, shall be in the range $0.1\ldots0.5\,mm$, and the width of the joint (=the shortest path through a joint from the inside to the outside, length of the flame path), L, shall be in the range $6\ldots40\,mm$ in order to prevent flame transmission. It should be emphasized that in cases of enclosures with 'abnormal' shapes, e.g. a large length to cross-section ratio, higher explosion pressures may be expected and reduced values for w may comply with flameproofness.

Domains of application for flameproof enclosures are:

- motors (especially those with commutators or sliprings, or those for heavy duty cycles) and generators (e.g. the diode wheel to excite a synchronous generator)
- semiconductors in power electronics
- switchgear and relays in non-intrinsically safe electric circuits
- incandescent and halogen lamps, high pressure discharge lamps
- analyser heads, ovens of gas chromatographs
- capacitors
- transformers
- plug-and-socket connectors
- heating systems for combustible gases, fluids and gas–air mixtures
- display screens for video transmission and remote controlling
- X-ray tubes, thyratrons, (gas) laser tubes
- non-explosion protected industrial equipment to be installed in zone 1 (or zone 2) in such a small number of items that it seems to be unprofitable to start the development of a special ex-protected type, e.g. the analyser head of a mass spectrometer.

What is the working principle of 'flameproof enclosure – d –'? How will the flame be 'cooled down' to prevent ignition of the external atmosphere? In 1815, Sir Humphry Davy discovered that a metal wire screen prevents the ignition of firedamp by the flame of an (oil) lamp. Obviously, the thermal conductivity of the screen was sufficient to cool the combustion residuals in an adequate manner. Ninety years later, Beyling [3] discovered in his experiments that ignited firedamp (methane–air mixture) causes an internal overpressure of – roughly – 6.5 bar (650 kPa) in a 'tight' enclosure, and that joints of suitable dimensions (values for $w \leqslant 0.7\,mm$ and $L = 25\ldots50\,mm$ have been reported) will prevent flame transmission to the external atmosphere. But this was not a static effect as in Davy's lamp. What is the explanation for the 'dynamical cooling' in the joints of flameproof enclosures?

After many years of research in this field, two essential facts have been discovered:

- the MESG values of gas–air mixtures are independent of the specific heat of joint materials (Table 6.25(a)) [17]. Despite a variation of 1:5 in specific heat, the MESG values remain unchanged

Table 6.25(a) Specific heat values of selected materials

Material	Specific heat J/g · K
Steel (low carbon)	0.42
Brass	0.377
Glass	0.7
Epoxy resin	1.3
Polyethylene	2.1 … 2.2

Table 6.25(b) Thermal conductivity of selected materials

Material	Thermal conductivity W/cm K
Steel (low carbon)	0.5
Brass	1.2 … 1.5
Glass	0.006
Epoxy resin	0.003
Polyethylene	0.004

- the MESG values are independent of the thermal conductivity of joint materials (Table 6.25(b)) [17]. A variation over more than two orders of magnitude ($>1:10^2$) does not affect the MESG values.

So, the conclusion will be:

the 'cooling effect' of gases along their path within joints is by no means caused by any kind of heat transfer to or heat storage in a material.

The MESG values of gas–air mixtures depend on the width L of the joint. In IEC 60079-1 A, the MESG values refer to $L = 25$ mm (and this is the reference value for defining MESG values in general). In trials, L has been varied between 0 (zero!) and 50 mm [2]. The main result obtained has been (Fig. 6.76):

Even with a zero width of a flameproof joint, there will be no flame transmission if the gap is suitably reduced.

So, cutting edge-shaped joints do prevent flame transmission to the external atmosphere. For the purpose of manufacturing electrical equipment, this fact may be unimportant, but it is very helpful towards an explanation of the working principle of a flameproof joint.

In a next step, caloric values of combustible gases and mixtures shall be compared (Table 6.26).

For gas–air mixtures, the MESG exponentially decreases with increasing power density σ (Fig. 6.77). It seems very likely that the gas mixture itself – without the 'assistance' of any joint material – is subdued to a cooling effect.

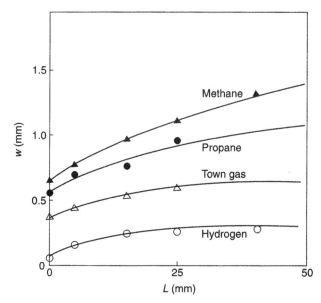

Figure 6.76   *Experimental safe gap values* w *of gas–air mixtures (referring to an initial pressure* p_o = 1.0 · 10^5 Pa) *versus joint width L.*
Note: The components of town gas are: 45 ... 55% (v/v) H_2; 6 ... 10% (v/v) CO; 25 ... 33% (v/v) CH_4; inert residues.

To give an example, $1\,m^3$ of a stoichiometric CH_4–air mixture (with initial parameters pressure p_o = 1.0 · 10^5 Pa, temperature T_o = 293 K = +20°C) shall be ignited in a constant volume. The energy obtained is 3.19 · 10^6 J at a peak explosion pressure $p_1 = \pi \cdot p_o$, p_1 = 8.2 · 10^5 Pa. With this peak pressure, the maximum temperature T_1 of the combustion can be roughly calculated. For constant volume, it is:

$$T_1 = T_o p_1/p_o = \pi \cdot T_o, \quad \text{resulting in } T_1 = 2400\,K$$

The potential energy $p \cdot V$ = 8.2 · 10^5 J corresponds to 25.7% of the combustion heat only, the remainder is due to radiation losses (the walls of the combustion vessel), thermal conductivity losses and dissociation energy.

During the combustion process the gas flows off passing the joint without significant heat transfer to the surfaces.

The gas changes its state in an adiabatic way. Following Poisson's law:

$$p \cdot V^\chi = \text{constant}$$

p = gas pressure
V = gas volume
χ = adiabatic exponent
$\chi = c_p/c_v$, i.e. the ratio between the mean moles heat of the gas, referring to constant pressure ($=c_p$) and to constant volume ($=c_v$)

and the equation for an 'ideal' gas:

$$p \cdot V = n \cdot R \cdot T$$

Table 6.26 Caloric values of gases and gas mixtures (according to [17], [55], [56])

Gas	Stoichiometric point % (v/v) gas	Caloric value[1] of gas J/m³	Caloric value[1] of gas mixture J/m³	Reaction time[2] Δt ms	Power density[3] σ kW/cm³	μ[4]	Pressure ratio π[5]	MESG value[6] mm
Part A	Gas–oxygen mixtures			7	7		7	7
H₂	66.7	$1.01 \cdot 10^7$	$6.74 \cdot 10^6$			1.50		
CH₄	33.3	$3.35 \cdot 10^7$	$1.12 \cdot 10^7$			1.00		
C₂H₂	28.6	$5.31 \cdot 10^7$	$1.52 \cdot 10^7$			1.165		
C₂H₄	25.0	$5.59 \cdot 10^7$	$1.40 \cdot 10^7$			1.00		
C₂H₆	22.2	$6.01 \cdot 10^7$	$1.33 \cdot 10^7$			0.90		
C₃H₈	16.7	$8.71 \cdot 10^7$	$1.45 \cdot 10^7$			0.856		
Part B	Gas–air mixtures[8]		9					
H₂	29.6	$1.01 \cdot 10^7$	$2.98 \cdot 10^6$	30	0.099	1.175	7.9	0.29
CH₄	9.5	$3.35 \cdot 10^7$	$3.19 \cdot 10^6$	220	0.0144	1.00	8.2	1.14 (1.17)
C₂H₂	7.75	$5.31 \cdot 10^7$	$4.10 \cdot 10^6$	40	0.102	1.04	10.3	0.37
C₂H₄	6.54	$5.59 \cdot 10^7$	$3.65 \cdot 10^6$	85	0.043	1.00	9.3	0.60 (0.65)
C₂H₆	5.66	$6.01 \cdot 10^7$	$3.41 \cdot 10^6$	150	0.022	0.974	9.1	0.83 (0.91)
C₃H₈	4.03	$8.71 \cdot 10^7$	$3.52 \cdot 10^6$	160	0.022	0.960	9.4	0.92

1 Two different caloric values shall be differentiated:
- the 'lower' caloric value, *not* incorporating the energy of condensed water as a combustion residual
- the 'upper' caloric value, *incorporating* the energy of condensed water

In this table, all values listed refer to the 'lower' caloric value

All values refer to $T = 293\,K$ and $p = 1 \cdot 10^5 Pa$ (before ingition!)

(continued)

Table 6.26 (*continued*)

2 In this table, the reaction time Δt is the time difference between ignition and the time of the peak pressure (as an integral value), measured in a cylindric vessel with:

diameter   480 mm
length   850 mm
volume   0.155 m^3

3 The power density (as an integral value) is the ratio $\sigma = H_L/\Delta t$
   H_L = 'lower' caloric value as defined in 1

4 μ is the volume ratio of the gas mixture before (b) and after (a) combustion
   μ = volume (b)/volume (a) (see Section 1.2.1)

5 The pressure ratio π is the ratio $\pi = p_1/p_o$
   p_1 = peak pressure of the reacting gas mixture in a gas-tight volume
   p_o = initial pressure of the gas mixture

6 MESG values for gas–air mixture, referring to $p_o = 1.0 \cdot 10^5$ Pa
   $T = 293$ K
   $L = 25$ mm (see Table 1.6 in Section 1.2.2)

7 No values available

8 It should be emphasized that these values refer to the gas only (and *not* to a gas–oxygen or gas–air mixture). So, the values in Part A and Part B of this table are identical

9 As a guideline – with the exception of hydrogen, methane and acetylene – the 'lower' caloric value of 1 m^3 gas–air mixture is comparable with 1 kWh ($=3.6 \cdot 10^6$ J)

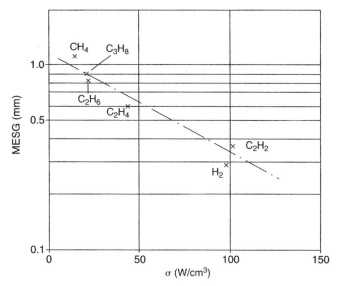

Figure 6.77 *MESG (maximum experimental safe gap) values of gas–air mixtures versus power density σ (according to Table 6.26).*

p = gas pressure
V = gas volume
n = number of moles
$R = 8.137 \text{ joule} \times \text{mole}^{-1} \times \text{K}^{-1}$
T = gas temperature

the state of the gas at the end of the expansion process (p_2, V_2 and T_2) can be calculated:

$$T_1/T_2 = (p_1/p_2)^{\frac{\chi-1}{\chi}}$$

(p_1, V_1 and T_1 = the state at the time of peak explosion pressure, $p_1 = 8.2 \cdot 10^5 \text{Pa}$ $V_1 = 1\,\text{m}^3, T_1 = 2400\,\text{K}$)

The combustion residuals are mainly CO_2, H_2O (as a vapour) and N_2.
   With the adiabatic exponents for

$$N_2\ \chi = 1.40$$
$$CO_2\ \chi = 1.34$$
$$H_2O \text{ (vapour)}\ \chi = 1.28$$

the combustion residuals show a mean value of $\chi = 1.35$. With $p_2 = 1.0 \cdot 10^5 \text{Pa}$ (the pressure of environmental atmosphere), it follows $T_2 = 1390\,\text{K}$.
   By adiabatic expansion when passing the joint the combustion residuals have decreased their temperature by $1000\,\text{K}$ and enter the environmental atmosphere as a gas jet incapable of causing an ignition.
   This simple model indicates that a flameproof joint works in the same way as a rocket. A chemical reaction (combustion) within the combustion chamber

generates heat which increases the gas pressure in a defined volume. With neglectable heat transfer, the gas expands through the nozzle into the environmental atmosphere, converting the 'potential energy' inside the combustion chamber to kinetic energy, accompanied by a considerable temperature decrease. A flameproof enclosure with a volume of $1\,m^3$ at a temperature of $T = 298\,K$ $(= +25°C)$, filled with a stoichiometric CH_4–air mixture, contains $0.095\,m^3$ CH_4, corresponding to $0.095 \cdot T_0/T = 0.087\,m^3$ (referring to $T_0 = 273\,K) = 3.89$ moles or $62\,g$ CH_4 respectively. For air, $0.905\,m^3$ corresponds to $0.8291\,m^3$ (referring to $T_0 = 273\,K) = 37.1$ moles or $1070\,g$, resulting in a total mass of $1.132\,kg$. The potential energy in this volume (at peak pressure) is $8.2 \cdot 10^5\,J$. When this energy is converted to kinetic energy without any losses, a jet velocity of $1200\,m/s$ will result, i.e. a supersonic jet.

More detailed studies concerning the 'working principle' of flameproof enclosures are reported in [33], [40] and [42].

Before considering individual types of electrical apparatus, flameproof joints as an essential element of 'd' shall be dealt with. The contents of EN 50018 and IEC 60079-1 are different, depending on the edition of the standard. The requirements are listed in Tables 6.27, 6.28 and 6.29. The values given in these tables are constructional requirements and ensure the flameproofness of an enclosure except for 'extreme' shapes, e.g. long U-shaped pipes for electric heaters. In such cases, the gap shall be reduced to an appropriate value.

The 'd'-standards IEC 60079-1 and EN 50018 give detailed requirements for holes in joints, threaded and cemented joints, for the application of gaskets and O-rings, for surface roughness of joints and for light transmitting parts.

To give an idea of the 'behaviour' of an explosion inside a flameproof enclosure, two diagrams, pressure versus time, $p(t)$, may be helpful (Figs 6.78 and 6.79). In Fig. 6.78, curve 1 represents the time dependence of a 9.75% (v/v) methane–air mixture in comparison with curve 2 for a 12.5% (v/v) 'mixed gas'–air mixture, the 'mixed gas' composed of 58% (v/v) methane and 42% (v/v) hydrogen, MESG = 0.80 mm. These two diagrams were attained in a flameproof terminal compartment (Group I). According to Table 6.26, the pressure ratios of CH_4–air or H_2–air mixtures are nearly at the same value. The considerably smaller values of curves 1 and 2 (and the somewhat smaller peak pressure of curve 1 compared to that of curve 2) are due to the gas losses escaping through the flameproof joints of the enclosure.

Note:

The pressure ratios given in Table 6.26 are valid for a gas-tight enclosure.

In Fig. 6.79, three curves $p(t)$ are compared (attained in a flameproof terminal compartment, but not identical with that in Fig. 6.78):

- curve 3:　20% (v/v) 'mixed gas'–air mixture, the 'mixed gas' composed of 40% (v/v) natural gas (mainly methane) and 60% (v/v) hydrogen

Table 6.27 Minimum width of joint and maximum gap for enclosures Groups I, IIA and IIB (according to EN 50018 and IEC 60079-1/4th edition)

Type of joint	Minimum width of joint L mm	Maximum gap (in mm) for volumes V (in cm³) of enclosure											
		V ≤ 100			100 < V ≤ 500			500 < V ≤ 2000			V > 2000		
		I	IIA	IIB	I	IIA	IIB	I	IIA	IIB	I	IIA	IIB
Flanged, cylindrical or spigot[2] joints	6	0.30	0.30	0.20	–	–	–[1]	–	–	–[1]	–	–	–[1]
	9.5	0.35	0.30	0.20	0.35	0.30	0.20	–	–	–[1]	–	–	–[1]
	12.5	0.40	0.30	0.20	0.40	0.30	0.20	0.40	0.30	0.20	0.40	0.20	0.15
	25	0.50	0.40	0.20	0.50	0.40	0.20	0.50	0.40	0.20	0.50	0.40	0.20
Cylindric joints of shafts of rotating electric machines — Plain bearings	6	0.30	0.30	0.20	–	–	–[1]	–	–	–[1]	–	–	–[1]
	9.5	0.35	0.30	0.20	0.35	0.30	0.20	–	–	–[1]	–	–	–[1]
	12.5	0.40	0.35	0.25	0.40	0.30	0.20	0.40	0.30	0.20	0.40	0.20	0.20
	25	0.50	0.40	0.30	0.50	0.40	0.25	0.50	0.40	0.25	0.50	0.40	0.25
	40	0.60	0.50	0.40	0.60	0.50	0.30	0.60	0.50	0.30	0.60	0.50	0.30
Rolling bearings	6	0.45	0.45	0.30	–	–	–[1]	–	–	–[1]	–	–	–[1]
	9.5	0.50	0.45	0.35	0.50	0.40	0.25	–	–	–[1]	–	–	–[1]
	12.5	0.60	0.50	0.40	0.60	0.45	0.30	0.60	0.45	0.30	0.60	0.30	0.20
	25	0.75	0.60	0.45	0.75	0.60	0.40	0.75	0.60	0.40	0.75	0.60	0.30
	40	0.80	0.75	0.60	0.80	0.75	0.45	0.80	0.75	0.45	0.80	0.75	0.40

Note:

The *width* of a joint is the shortest path through a joint from the inside to the outside of a flameproof enclosure (in a very general sense, it is the length of the flame path)

The *gap* of a joint is the distance between the corresponding surfaces of a flameproof joint. For cylindrical surfaces, the gap is the diametral clearance (difference between the two diameters)

(continued)

Table 6.27 (continued)

1 Not within the scope of EN 50018 and IEC 60079-1/4th edition
2 Spigot joints are composed of a cylindrical joint of width d and a flanged joint of width c. The flame has to pass first of all through the cylindrical part, followed by the flanged part. The two parts may be considered as a 'series connection of joints'. The total width is $L = c + d$ with: $c \geqslant 3.0$ mm. A bevel at the transition cylindrical to the flanged joint is permissible with a length $f \leqslant 1.0$ mm in the axial as well as in the radial direction. In the case of 'spigot joints' whose cylindrical part fulfils the requirements of Table 6.27, the flanged part may not comply with the content of Table 6.27

Table 6.28 Minimum width of joint and maximum gap for enclosures Group IIC (according to EN 50018 and IEC 60079-1/4th edition)

Type of joint	Minimum width of joint L mm	Maximum gap (in mm) for volumes V (in cm³) of enclosure			
		$V \leqslant 100$	$100 < V \leqslant 500$	$500 < V \leqslant 2000$	$V > 2000$
Flanged joints[2,7]	6	0.10	—[1]	—[1]	—[1]
	9.5	0.10	0.10	—[1]	—[1]
Spigot joints[3]	12.5	0.15	0.15	0.15	—[1]
	25	0.18[4]	0.18[4]	0.18[4]	0.18[4]
	40	0.20[5]	0.20[5]	0.20[5]	0.20[5]
Cylindrical and spigot joints[6]	6	0.10	—[1]	—[1]	—[1]
	9.5	0.10	0.10	—[1]	—[1]
	12.5	0.15	0.15	0.15	—[1]
	25	0.15	0.15	0.15	0.15
	40	0.20	0.20	0.20	0.20

	L				
Cylindrical joints of shafts of rotating electrical machines with rolling bearings	6	0.15	—[1]	—[1]	—[1]
	9.5	0.15	0.15	—[1]	—[1]
	12.5	0.25	0.25	0.25	—[1]
	25	0.25	0.25	0.25	0.25
	40	0.30	0.30	0.30	0.30
Flanged joints[2,7]	$6 \leq L < 9.5$	0.10	—[1]	—[1]	—[1]
	$9.5 \leq L < 15.8$	0.10	—[1]	—[1]	—[1]
	$15.8 \leq L < 25$	0.10	0.10	0.04	—[1]
	$25 \leq L$	0.10	0.10	0.04	0.04

Note:

The *width* of a joint is the shortest path through a joint from the inside to the outside of a flameproof enclosure (in a very general sense, it is the length of the flame path)

The *gap* of a joint is the distance between the corresponding surfaces of a flameproof joint. For cylindrical surfaces, the gap is the diametral clearance (difference between the two diameters)

1   Not within the scope of EN 50018 and IEC 60079-1/4th edition

2   Flanged joints are not permissible for combustible acetylene–air mixtures

3   Spigot joints are composed of a cylindrical joint of width d and a flanged joint of width c. The flame has to pass first of all through the cylindrical part, followed by the flanged part. The two parts may be considered as a 'series connection of joints'. The total width is $L = c + d$ with:

   $c \geqslant 3.0\,\text{mm}$ for Groups I, IIA, IIB
   $c \geqslant 6.0\,\text{mm}$ for Group IIC
   $d \geqslant 0.50 \times L$ for Group IIC

   A bevel at the transition cylindrical to the flanged joint is permissible with a length $f \leqslant 1.0\,\text{mm}$ in the axial as well as in the radial direction

4   The maximum gap may be increased to 0.20 mm when $f \leqslant 0.50\,\text{mm}$

5   The maximum gap may be increased to 0.25 mm when $f \leqslant 0.50\,\text{mm}$

6   In the case of 'spigot joints' whose cylindrical part fulfils the requirements of Table 6.28, the flanged part may not comply with the content of Table 6.28

7   In IEC 60079-1/4th edition, this part of Table 6.28 has been extended

Table 6.29 Part I Minimum width of joint and maximum gap for Group I enclosures (according to IEC 60079-1/3rd edition + Amendment 1 + 2)

Type and width L of joint (in mm)	Maximum gap (in mm) for volumes V (in cm³) of enclosure	
	$V \leqslant 100$	$V > 100$
Flanged and spigot[2] joints		
$6 \leqslant L < 12.5$	0.30	$-^1$
$12.5 \leqslant L < 25$	0.40	0.40
$25 \leqslant L$	0.50	0.50
Cylindrical joints of operating rods and spindles		
$6 \leqslant L < 12.5$	0.30	$-^1$
$12.5 \leqslant L < 25$	0.40	0.40
$25 \leqslant L$	0.50	0.50
Cylindrical joints of shafts of rotating electrical machines		
• Plain bearings		
$6 \leqslant L < 12.5$	0.30	$-^1$
$12.5 \leqslant L < 25$	0.40	0.40
$25 \leqslant L < 40$	0.50	0.50
$40 \leqslant L$	0.60	0.60
• Rolling bearings		
$6 \leqslant L < 12.5$	0.45	$-^1$
$12.5 \leqslant L < 18.75$	0.60	0.60
$18.75 \leqslant L < 25$	0.60	0.60
$25 \leqslant L$	0.75	0.75

Table 6.29 Part IIA Minimum width of joint and maximum gap for Group IIA enclosures (according to IEC 60079-1/3rd edition + Amendment 1 + 2)

Type and width L of joint (in mm)	Maximum gap (in mm) for volumes V (in cm³) of enclosure		
	$V \leqslant 100$	$100 < V \leqslant 2000$	$V > 2000$
Flanged and spigot[2] joints			
$6 \leqslant L < 9.5$	0.30	$-^1$	$-^1$
$9.5 \leqslant L < 12.5$	0.30	$-^1$	$-^1$
$12.5 \leqslant L < 25$	0.30	0.30	0.20
$25 \leqslant L$	0.40	0.40	0.40

(continued)

Table 6.29 Part IIA *(continued)*

Type and width L of joint (in mm)	Maximum gap (in mm) for volumes V (in cm³) of enclosure		
	$V \leqslant 100$	$100 < V \leqslant 2000$	$V > 2000$
Cylindrical joints of operating rods and spindles			
$6 \leqslant L < 12.5$	0.30	$-^1$	$-^1$
$12.5 \leqslant L < 25$	0.30	0.30	0.20
$25 \leqslant L$	0.40	0.40	0.40
Cylindrical joints of shafts of rotating electrical machines			
• Plain bearings			
$6 \leqslant L < 12.5$	0.30	$-^1$	$-^1$
$12.5 \leqslant L < 25$	0.35	0.30	0.20
$25 \leqslant L < 40$	0.40	0.40	0.40
$40 \leqslant L$	0.50	0.50	0.50
• Rolling bearings			
$6 \leqslant L < 12.5$	0.45	$-^1$	$-^1$
$12.5 \leqslant L < 25$	0.50	0.45	0.30
$25 \leqslant L < 40$	0.60	0.60	0.60
$40 \leqslant L$	0.75	0.75	0.75

Table 6.29 Part IIB Minimum width of joint and maximum gap for Group IIB enclosures (according to IEC 60079-1/3rd edition + Amendment 1 + 2)

Type and width L of joint (in mm)	Maximum gap (in mm) for volumes V (in cm³) of enclosure		
	$V \leqslant 100$	$100 < V \leqslant 2000$	$V > 2000$
Flanged and spigot² joints			
$6 \leqslant L < 9.5$	0.20	$-^1$	$-^1$
$9.5 \leqslant L < 12.5$	0.20	$-^1$	$-^1$
$12.5 \leqslant L < 25$	0.20	0.20	0.15
$25 \leqslant L$	0.20	0.20	0.20
Cylindrical joints of operating rods and spindles			
$6 \leqslant L < 12.5$	0.20	$-^1$	$-^1$
$12.5 \leqslant L < 25$	0.20	0.20	0.15
$25 \leqslant L$	0.20	0.20	0.20

(continued)

Table 6.29 Part IIB (*continued*)

Type and width L of joint (in mm)	Maximum gap (in mm) for volumes V (in cm³) of enclosure		
	$V \leqslant 100$	$100 < V \leqslant 2000$	$V > 2000$
Cylindrical joints of shafts of rotating electrical machines			
• Plain bearings			
$6 \leqslant L < 12.5$	0.20	—[1]	—[1]
$12.5 \leqslant L < 25$	0.25	0.20	0.15
$25 \leqslant L < 40$	0.30	0.25	0.20
$40 \leqslant L$	0.40	0.30	0.25
• Rolling bearings			
$6 \leqslant L < 12.5$	0.30	—[1]	—[1]
$12.5 \leqslant L < 25$	0.40	0.30	0.20
$25 \leqslant L < 40$	0.45	0.40	0.30
$40 \leqslant L$	0.60	0.45	0.40

Table 6.29 Part IIC   Minimum width of joint and maximum gap for Group IIC enclosures (according to IEC 60079-1/3rd edition + Amendment 1 + 2)

Type and width L of joint (in mm)	Maximum gap (in mm) for volumes V (in cm³) of enclosure				
	$V \leqslant 100$	$100 < V \leqslant 500$	$500 < V \leqslant 1500$	$1500 < V \leqslant 2000$	$2000 < V \leqslant 6000^3$
Flanged joints[4]					
$6 \leqslant L < 9.5$	0.10	—[1]	—[1]	—[1]	—[1]
$9.5 \leqslant L < 15.8$	0.10	0.10	—[1]	—[1]	—[1]
$15.8 \leqslant L < 25$	0.10	0.10	0.040	—[1]	—[1]
$25 \leqslant L$	0.10	0.10	0.040	0.040	0.040
Spigot joints[2,5]					
$12.5 \leqslant L < 25$	0.15	0.15	0.15	0.15	—[1]
$25 \leqslant L < 40$[6]	0.18	0.18	0.18	0.18	0.18
$40 \leqslant L$[7]	0.20	0.20	0.20	0.20	0.20
Spigot joints[8]					
$6 \leqslant L < 12.5$	0.10	0.10	—[1]	—[1]	—[1]
$12.5 \leqslant L < 25$	0.15	0.15	0.15	0.15	—[1]
$25 \leqslant L < 40$	0.15	0.15	0.15	0.15	0.15
$40 \leqslant L$	0.20	0.20	0.20	0.20	0.20
Cylindrical joints of operating rods and spindles					
$6 \leqslant L < 9.5$	0.10	—[1]	—[1]	—[1]	—[1]
$9.5 \leqslant L < 12.5$	0.10	0.10	—[1]	—[1]	—[1]
$12.5 \leqslant L < 25$	0.15	0.15	0.15	0.15	—[1]

(*continued*)

Table 6.29 Part IIC (*continued*)

Type and width L of joint (in mm)	Maximum gap (in mm) for volumes V (in cm³) of enclosure				
	$V \leqslant 100$	$100 < V \leqslant 500$	$500 < V \leqslant 1500$	$1500 < V \leqslant 2000$	$2000 < V \leqslant 6000^3$
$25 \leqslant L < 40$	0.15	0.15	0.15	0.15	0.15
$40 \leqslant L$	0.20	0.20	0.20	0.20	0.20
Cylindrical joints of shafts of rotating electrical machines with rolling bearings					
$6 \leqslant L < 9.5$	0.15	$-^1$	$-^1$	$-^1$	$-^1$
$9.5 \leqslant L < 12.5$	0.15	0.15	$-^1$	$-^1$	$-^1$
$12.5 \leqslant L < 25$	0.25	0.25	0.25	0.25	$-^1$
$25 \leqslant L < 40$	0.25	0.25	0.25	0.25	0.25
$40 \leqslant L$	0.30	0.30	0.30	0.30	0.30

Note:
The *width* of a joint is the shortest path through a joint from the inside to the outside of a flameproof enclosure (in a very general sense, it is the length of the flame path).

The *gap* of a joint is the distance between the corresponding surfaces of a flameproof joint. For cylindrical surfaces, the gap is the diametral clearance (difference between the two diameters)

1 Not within the scope of IEC 60079-1/3rd edition + Amendment 1 + 2
2 Spigot joints are composed of a cylindrical joint of width d and a flanged joint of width c. The flame has to pass first of all through the cylindrical part, followed by the flanged part. The two parts may be considered as a 'series connection of joints'. The total width is $L = c + d$
   For Group IIC only, it is:
   $c \geqslant 6\,mm$
   $d \geqslant 0.5 \times L$
   A bevel at the transition cylindrical to the flanged joint is permissible with a length $f \leqslant 1.0\,mm$ in the axial as well as in the radial direction
3 Enclosures of volume larger than 6000 cm³ and with any one dimension larger than 1 m are subject to special requirements upon which agreement should be reached between the manufacturer and the testing laboratory
4 Flanged joints are not permitted for explosive acetylene–air mixtures except if the gap is $\leqslant 0.040\,mm$ for $L \geqslant 9.5\,mm$ up to an enclosure volume of 500 cm³
5 For Group IIC, it is:
   $c \geqslant 6\,mm$
   $d \geqslant 0.5 \times L$
6 The maximum gap of the cylindrical part may be increased to 0.20 mm if $f \leqslant 0.5\,mm$
7 The maximum gap of the cylindrical part may be increased to 0.25 mm if $f \leqslant 0.5\,mm$
8 In the case of spigot joints whose cylindrical part fulfils the requirements of Table 6.29, Part IIC, the flanged part may not comply with the content of Table 6.29, Part IIC. For a gap $\leqslant 0.040\,mm$ up to 6000 cm³ is permitted and for diametral clearance of cylindrical parts this value may be 0.060 mm

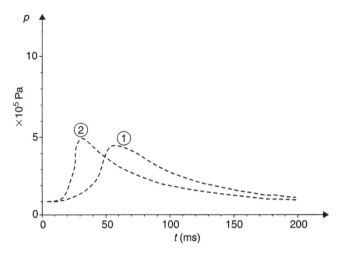

Figure 6.78   *Explosion pressure* p *versus time* t *(ignition at* t = 0*).*

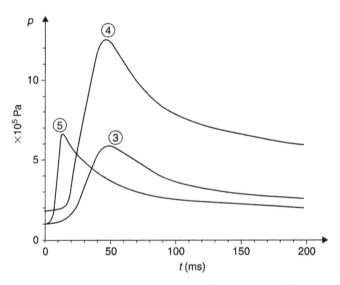

Figure 6.79   *Explosion pressure* p *versus time* t *(ignition at* t = 0*).*

- curve 4:   same gas–air mixture as for curve 3, but with increased precompression ($1.9 \cdot 10^5$ Pa) of the gas–air mixture before ignition.
  Obviously, the peak pressure occurs at the same time after ignition, independent of precompression. The peak pressures $p_{max.}$
  (as absolute pressure values, referring to $p = 0$) of curves 3 and 4
  show exactly the same ratio as the pressures $p_{initial}$ of the gas–air
  mixtures before ignition, i.e. the pressure ratio $\pi = p_{max.}/p_{initial}$ is
  constant, or, in other words, $p_{max.} = \pi p_{initial}$

Table 6.30   Constants K_G of selected gas–air mixtures (referring to $p_o = 1.013 \cdot 10^5$ Pa, $T = 273$ K) according to [2]

Gas, in stoichiometric ratio to air	K_G bar $\cdot$ m/s
Acetylene	1415
Ammonia	10
Butane	92
Carbon disulphide	105
Ethane	106
Hydrogen	550
Methane	55
Methanol	75
Pentane	104
Propane	100
Toluene	94

Note: 1 bar $= 1.0 \times 10^5$ Pa

- curve 5:   44.5% (v/v) hydrogen–air mixture. The peak pressure is somewhat higher than $p_{max.}$ of curve 3, due to the much faster reaction of H_2–air mixtures and the hereby lowered gas losses through the flameproof joints.

When gas–air mixtures are ignited in different volumes of the same shape and at identical conditions (gas temperature, initial pressure), the explosion is 'slower' with increasing enclosure volume V [2]:

$$dp/dt/\text{max.} \times V^{1/3} = K_G$$

$p(t)$ = explosion pressure, time dependent
dp/dt = pressure rise velocity
V = volume of enclosure
K_G = a constant, typical for the gas

This behaviour is known as the 'cubical law'. In a double-logarithmic scale, this law is represented by a straight line with a negative gradient:

$$\log dp/dt/\text{max.} = -1/3 \log V + \text{const.}$$

In Table 6.30, K_G values for selected gas–air mixtures are summarized.

The explosion pressure depends on the density ρ of a gas–air mixture. Starting with $p \cdot V = n \cdot R \cdot T$ for an 'ideal' gas,

with: p = gas pressure
  V = gas volume
  n = number of moles
  $R = 8.137$ joule $\times$ mole$^{-1} \times$ K^{-1}
  T = gas temperature

and $\rho = n/V$, it follows

$$p = \frac{\rho}{RT}$$

Keeping T = constant, ρ is proportional to p. The above-mentioned relation between explosion pressure p_1 and initial pressure p_o, $p_1 = \pi \cdot p_o$ is based on this fact.

Keeping p = constant and changing the temperature from T_1 to T_2, it follows:

$$\rho_2 = \rho_1 T_1 / T_2$$

At low ambient temperatures, e.g. a natural gas compressor station in arctic regions, the explosion pressure will increase in winter (say $-60°C$ or $T_2 = 213\,K$) by 38% compared with summer conditions ($+20°C$ or $T_1 = 293\,K$), an important fact when designing flameproof apparatus for installation in cold regions. In addition, some mechanical properties of the enclosure material may deteriorate drastically at such low temperatures.

The lowest ambient temperature for explosion protected electrical apparatus is normally rated at $-20°C$ or $T_2 = 253\,K$. Compared with the type test conditions (see Section 8.1) for reference pressure determination in a laboratory at $+20°C$ or $T_1 = 293\,K$, the explosion pressure is increased by 16% only.

In many cases there will be a certain demand for protective coatings (paints) against corrosion for the joint surfaces. This question should be handled very carefully. Many coatings contain particles (e.g. aluminium) which may adversely affect the flameproof properties of the joint. Tests should be made to ensure the coating's appropriateness for this application.

Caution shall be given to combustible liquids and organic insulating materials in flameproof enclosures. Combustible liquids, e.g. oil (as a hydrocarbon-based insulating material), are often used as impregnating agents in capacitors. The molecular structure of hydrocarbons can be cracked mainly by two effects: arcing due to faults within the insulating material, or partial discharges due to a high electric field strength in a dielectric. In the case of arcing, the main gaseous products of decomposition of hydrocarbons are hydrogen, H_2, propylene, C_3H_6, ethylene, C_2H_4, methane, CH_4, acetylene, C_2H_2, and carbon monoxide, CO [30].

Partial discharges generate hydrogen (as a major part) and methane. Most of these combustible gases show – in a mixture with air – lower MESG values compared with the gases for Group I or IIA (or even IIB). So, faults may initiate the generation and accumulation of an internal explosive atmosphere. This fact, in general, is within the scope of 'd'. But the problem is that there are two different combustible atmospheres: outside the atmosphere upon which the classification of the electrical apparatus has been based, e.g. Group I or IIA, and inside a completely different atmosphere with low MESG values. The flameproofness of the apparatus is adversely affected. EN 50018 excludes the use of liquids, if their decomposition generates gaseous components with MESG values not complying with the group of the flameproof enclosure. A way out of such difficulties may be a special type test of the enclosure for non-transmission, with the decomposition components (and air) inside and an environmental atmosphere as given in the standard for type testing.

A similar problem arises from the use of organic insulating materials as a constructional part of the arc chute, e.g. for circuit breakers and contactors operating in air. In the case of hydrocarbon-based insulating materials, their carbon content is in the range 30% (by weight) to 75%, the hydrogen content is in the range 2.5% (by weight) to 10%, to give a rough idea. The radiation of the arc decomposes hydrocarbons to hydrogen, acetylene (and some other gaseous components of minor importance) and soot. To give an example: an insulating material with 8% (by weight) hydrogen content and a density of $1 \, g/cm^3$ may be decomposed. A volume of 0.1 litre of this (solid) insulating material, corresponding with 100 g, has 8 g hydrogen as an integral part. But these 8 g correspond with 4 moles of hydrogen, resulting after decomposition in a volume of 4×22.4 litres = $0.09 \, m^3$ at 'normal conditions'. Starting with 0.1 litre, the volume of the gaseous decomposition components has been increased by nearly three orders of magnitude. Assuming a gas-tight enclosure, suitable for an internal overpressure of $10 \times 10^5 \, Pa$ (=10 bar), the content of hydrocarbon-based insulating material exposed to arcs shall be limited to 1% of the enclosure volume. In EN 50018, this restriction to 1% of the enclosure volume is given for insulating materials with a CTI (comparative tracking index) value of lower than CTI 400 M (the 'M' indicates the addition of a wetting agent to the test fluid in the CTI testing procedure according to IEC 60112), referring to arcs in air and in electric circuits with a rated current exceeding 16 A, valid for Group I equipment.

EN 50018, as an alternative way, and IEC 60079-1 generally ask for CTI 400 M quality or better in this case.

6.8.2 Techniques for cable entries

In general, cable entries according to IEC 60079-1 and EN 50018 shall withstand the explosion pressure inside the enclosure and shall ensure the flameproofness by an adequate design and construction of all mechanical parts. In addition, the general requirements stated in IEC 60079-0 and EN 50014 shall be fulfilled. On the whole, to pass a flameproof enclosure with electrical power or electrical and optical signals four methods have been established (Figs 6.80 and 6.81):

1 A terminal compartment in 'increased safety – e –' according to IEC 60079-7 or EN 50019 is installed 'electrically in front' of the flameproof enclosure. The cable entry shall comply with the 'e'-standards. Via terminal blocks and cable (or fibre) bushings the electrical power or the signals are fed to the interior of the flameproof enclosure, or, as an alternative, by the use of single conductor bushings (for high power transmission only). These methods are described in more detail in Section 6.7.2. Looking back at the

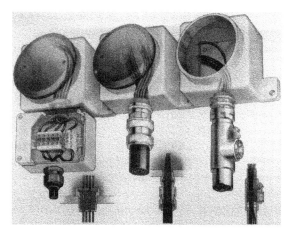

Figure 6.80   *Techniques for cable entries into a flameproof enclosure (left to right):*
- *with 'e'-terminal compartment and terminal blocks, cable bushings fitted into the partition wall between 'e'- and 'd'-enclosure*
- *direct cable entry into the flameproof enclosure*
- *conduit technique, sealing the 'd'-enclosure by a 'stopping box'.*

Figure 6.81   *Flameproof plug-and-socket connectors ('cable couplers'), fitted into a flameproof enclosure (EEx dI equipment for coal mines).*

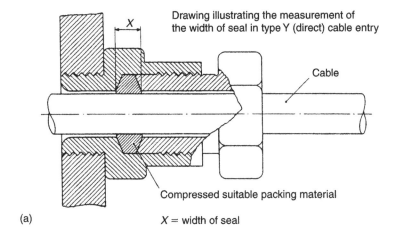

Drawing illustrating the measurement of
the width of seal in type Y (direct) cable entry

Cable

Compressed suitable packing material

(a) X = width of seal

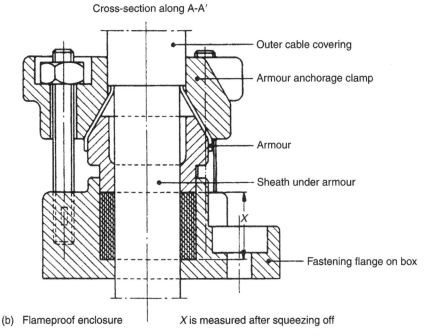

Cross-section along A-A'

Outer cable covering

Armour anchorage clamp

Armour

Sheath under armour

Fastening flange on box

(b) Flameproof enclosure X is measured after squeezing off

Figure 6.82(a) to (c)   *Examples of direct cable entries into a flameproof enclosure, according to IEC 60079-1.*

history of explosion protection, the 'e'-terminal compartment has been favoured in Germany for a long time.

2 The cable for power or signal transmission passes the flameproof enclosure 'on the direct way' via a cable entry complying with the 'd'-standards. This technique has its origins in the United Kingdom and France (including their historical spheres of economical influence), and in the countries of the former 'Eastern bloc'. Examples for direct cable entries are given in Fig. 6.82.

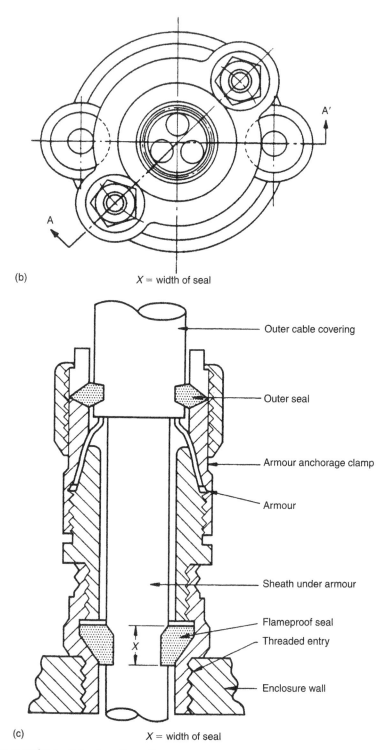

(b) X = width of seal

Outer cable covering

Outer seal

Armour anchorage clamp

Armour

Sheath under armour

Flameproof seal

Threaded entry

Enclosure wall

(c) X = width of seal

Figure 6.82(a) to (c)   *(continued)*

It should be remembered that IEC 60079-0 and EN 50014 'General require-ments' limit the maximum temperatures at the entry and branching points:

entry point: +70°C
branching point: +80°C

As an exception, cables with an increased temperature resistance shall be inserted.

Note:

In some countries, e.g. the United Kingdom, terminal compartments in 'd' have been installed in front of the 'd'-enclosure. Notwithstanding this separate terminal compartment, the requirements for this cable entry technique are identical with those for 'direct cable entries'. For these 'd'-terminal compartments, the same thermal considerations are valid as for 'e'-terminals, given in Section 6.7.2. The cable entry point is shifted away from the main enclosure with all its thermal losses inside, the terminal compartment inserted hereby acts as an additional 'cooling surface'.

Especially for Groups IIB and IIC, a cable with a direct entry into a flameproof enclosure shall have a minimum length. Regarding the cross-sections of cables, the cables are not gas-tight due to the space between the single wires of a conductor, found mainly in multiple-wire conductors, very seldom in fine-wire conductors. At very short cable lengths (as an order of magnitude, 0.1 m) this space allows flame transmission from the flameproof enclosure to another 'd'-enclosure. In practice, a cable length of 1 m (as a rough idea) erases this problem.

---

3 The 'conduit technique'. Plastic or rubber insulated single conductor wires inserted in steel conduits are used for power or signal transmission. The flameproof enclosure shall be fitted with an appropriate thread to insert the conduit (with at least five threads). The diameter range of the conduits is 12.7 mm to 76.2 mm for most applications. To avoid flame transmission through the conduit system from one flameproof enclosure to the other, 'stopping boxes' shall be inserted in the conduit system as near as possible to the flameproof enclosures. After installation, a setting compound (=resin) shall be cast into the stopping box. After hardening, the com-pound acts as a flameproof sealing element. During the casting process, the compound (resin) is supported by mineral wool in order to prevent flow-off. Obviously, this technique is applicable for fixed apparatus only, i.e. Group II installations (Fig. 6.83). This technique has its origin in North America, especially in the United States.

4 Plug-and-socket connectors, 'cable couplers', see Fig. 6.81. These parts are inserted into an opening of the flameproof enclosure and act as a flame-proof 'seal'. This also applies to the unplugged state of plug and socket. Such means are favoured where apparatus is frequently removed, e.g. in coal mines (Group I application).

Figure 6.83   *Low voltage energy distribution system. Flameproof enclosures are interconnected by a conduit system.*

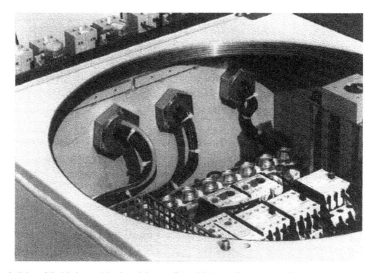

Figure 6.84   *Multiple cable bushings, fitted into a flameproof enclosure (cover removed).*

Electrical and optical connections between two flameproof enclosures separated by a partition wall are made by the insertion of (multiple) cable bushings or fibre bushings in an opening or thread of the partition wall. This technique is identical with the means of connection between 'e'- and 'd'-compartments (see Section 6.7.2, Table 6.19, and Section 6.7.6, Fig. 6.70, and, in addition, Fig. 6.84).

6.8.3   Motors and generators

Compared with rotating machines in 'increased safety – e –', there are no additional requirements for the windings or for the gap between stator core and rotor. The 'interior' of flameproof motors and generators is identical

Table 6.31 Comparison between 'e', 'd' and non-explosion protected industrial purpose cage induction motors (4 poles, 50 cps/totally enclosed/fan cooled) (by courtesy of Loher GmbH, Ruhstorf/Germany)

Type of protection	Temperature class	Rated power kW	Efficiency η %	cos ϕ	Weight kg
(a) Frame number 132					
none (industrial)	–	7.5	88.0	0.85	69
eII	T3	6.8	88.0	0.86	69
eII	T4	4	87	0.86	69
dIIC	T4	7.5	88	0.85	100
(b) Frame number 200					
none (industrial)	–	30	92.5	0.87	254
eII	T2	27	92.7	0.85	254
eII	T3	24	92.7	0.83	254
dIIC	T4	30	92.5	0.87	310

with that of normal industrial equipment. Rated power, efficiency and cos ϕ are identical with the values of non-explosion protected motors, even for 'low' temperature classes T4 or T5 (T6 is rather exceptional, e.g. carbon disulphide asks for such a 'low' temperature class). Table 6.31 gives a comparison between 'e', 'd' and non-explosion protected cage induction motors with identical frame numbers. Normally, a flameproof cage induction motor shows cylindrical and threaded joints only:

- for shafts and bearings (cylindrical)
- for end shields (cylindrical)
- for bushings and connection plate (single conductor or multiple cable bushings) to the terminal compartment (cylindrical, threaded)

(see Fig. 6.85). So, a dIIC design (for hydrogen, acetylene and carbon disulphide) can be fitted without great mechanical complications (with the exception of 'high' frame numbers, due to the radial bearing clearance).

In general, there are two ways to assemble shaft joints:

- the shaft joints are *outside of* the bearings (see Fig. 6.85(a))
- the shaft joints are *inside of* the motor's enclosure (see Fig. 6.85(b)).

The first method incorporates the bearings into the flameproof enclosure, i.e. the bearings may not comply with the temperature range according to the temperature class for the motor. Especially for classes T5 and T6, rolling bearings may exceed the adequate temperature limit in continuous operation. A certain disadvantage of such an arrangement is the fact that the joints are exposed to water and dust in case of failure of the sealing rings with subsequent heavily increasing erosion and an undefined gap which may exceed the safe values according to the equivalent group or subgroup (I, IIA, IIB or IIC).

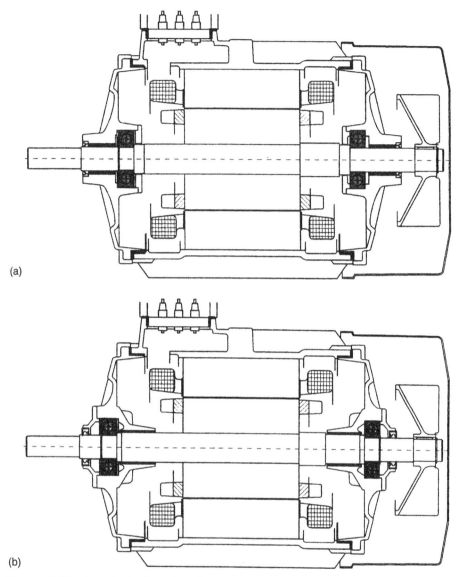

(a)

(b)

Figure 6.85   *Longitudinal section drawings of flameproof cage induction motors, indicating the flameproof joints:*
(a) Bearings inside the flameproof motor housing
(b) Bearings outside the motor housing.

In the second case, the bearings are exposed to the environmental atmosphere, i.e. they are no longer a part of the flameproof enclosure. They shall comply with the temperature limitations according to the temperature class. So, this solution is adequate for Group I application and for Group II, classes T1–T3, sometimes even for class T4. Table 6.31 indicates a somewhat higher weight of flameproof motors in comparison with 'e'- or non-explosion

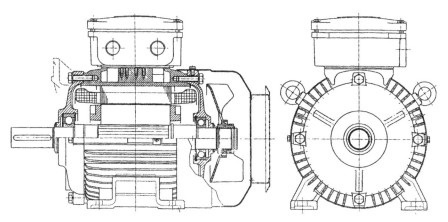

Figure 6.86 *Longitudinal section drawing of a cage induction motor with identical frame number (132) to that shown in Fig. 6.87: motor not explosion protected.*

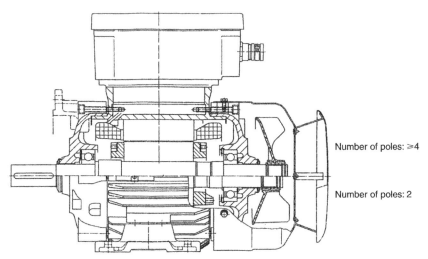

Number of poles: ⩾4

Number of poles: 2

Figure 6.87 *Longitudinal section drawing of a cage induction motor with identical frame number (132) to that shown in Fig. 6.86: flameproof motor. Type of protection: EEx de IIC T4 (terminal box in 'e').*

protected motors. This increased weight is caused mainly by the reinforced design of the end shields with additional material expenditure for the flame-proof joints at the shaft. Figures 6.86 and 6.87 give a direct comparison of a non-exprotected motor with a 'd'-motor of the same frame number (132) in longitudinal section drawings. Figure 6.88 completes this survey with a flameproof motor with 'external' bearings and 'internal' flameproof joints for the shaft.

A special problem for flameproof motors is adequate cooling. Obviously, internal cooling by ventilation is not possible. So, flameproof motors are

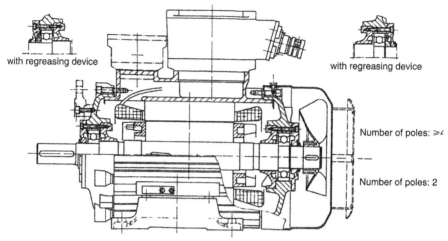

Figure 6.88   *Longitudinal section drawing of a flameproof cage induction motor, bearings outside the flameproof enclosure. Type of protection: EEx de IIC T4.*

totally enclosed. Most of them are surface cooled: an internal closed-circuit ventilation transports the heat losses to the finned surface of the motor. An external airstream (by forced ventilation) removes the dissipated heat. When increasing the motor size, by a linear factor a, its volume is increased with a^3, its surface area with a^2 only. So, the ratio volume to surface increases with a, opposite to an effective cooling of the motor. Motors for continuous operation in the high power range or for heavy duty cycles (frequent acceleration of great inertia masses) ask for more effective cooling methods. Two very different techniques have been established. The first one replaces the 'surface' of the motor by an annular assembly of tubes as an internal air-to-external air heat exchanger (Figs 6.89 and 6.90). In the case of a frequency converter-fed motor, the rapidly decreasing internal airflow with decreasing speed restricts the speed range at the lower end to values not too far away from the rated speed. (This will be no problem for the external fan: separated from the motor shaft and powered by a motor with a constant speed it ensures a constant external airflow for cooling.)

The second technique uses water as a coolant. Following the historical development, the stator was at first water cooled, followed some years later by a combined water cooling of rotor and stator (Fig. 6.91). The water enters into the rotor shaft, a hollow shaft with an axially inserted tube via rotating sealing elements, cools down the bearing at the driving end, flows back through the hollow rotor shaft, overtaking the heat losses in the rotor and cooling the bearing at the rear end, leaves the rotor shaft via sealing elements and enters the coolant chambers of the stator, now overtaking the losses in windings and in the stator core. The advantages of such motors are a very compact design, the total independence of the cooling effect from motor speed, a completely smoothed surface of the motor without any tubes or fins, a considerable advantage in plants loaded with dust deposits, and – due to the damping effect

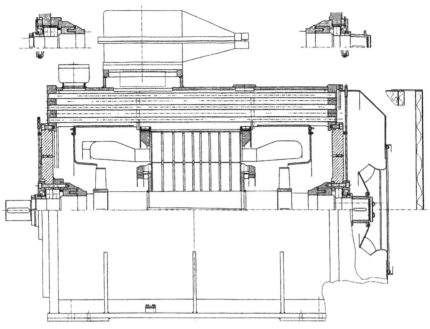

Figure 6.89   *Longitudinal section drawing of a flameproof cage induction motor, with an annular-shaped air-to-air heat exchanger.*

Figure 6.90   *Total view of a motor according to Fig. 6.89 (with plain bearings).*
Rated power: 1500 kW; Rated voltage: 6 kV; Type of protection: EEx de IIB T4.

of the water chambers surrounding the stator – a reduced engine noise. Figure 6.92 shows a water-cooled (stator and rotor) cage induction motor.

The reduced noise of water-cooled motors has opened a field of application far away from the domain of explosion protection: as an electric drive in suburban trains.

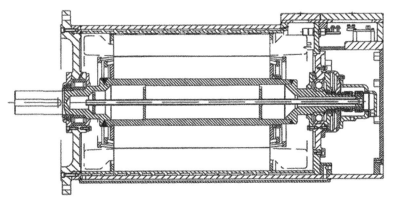

Figure 6.91 *Longitudinal section drawing of a flameproof cage induction motor, shaft and stator housing, water cooled, with pole changing.*
Number of poles: 12, 4; Rated power: 135 kW, 400 kW; Rated voltage: 1 kV, 1 kV; Rated current: 155 A, 267 A; Rated speed: 492 min^{-1}, 1485 min^{-1}; Rated frequency: 50 cps, 50 cps; Thermal class: H, H; Cooling water flow: ⩾15 litres/min; Inlet temperature of cooling water: ⩽+40°C; Type of protection: EEx dl; as an option: terminal compartment according to 'increased safety'; EEx de I; Certificate: BVS 91.B.1056 X/2. N.

Figure 6.92 *Total view of a water-cooled flameproof cage induction motor for coal mines.*
Rated power: 630 kW; Rated voltage: 3.3 kV; Type of protection: EEx de I.

To ensure correct cooling conditions, the water flow and/or the temperature difference between water outlet and water inlet shall be monitored continuously, acting on an electrical interlock with the motor's power supply, e.g. by intrinsically safe electric circuits.

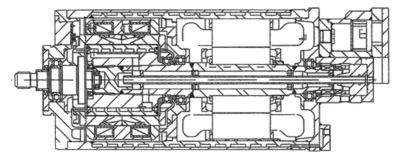

Figure 6.93 *Longitudinal section drawing of a flameproof cage induction motor with integrated eddy current clutch, stator housing water cooled.*
Motor data:
Number of poles: 2; Rated power: 55 kW; Rated voltage: 1 kV; Rated current: 37 A;
Rated speed: 2950 min^{-1}; Rated frequency: 50 cps; Thermal class: H;
Clutch data:
Rated voltage: 300 V DC; Current: ≤1.75 A; Thermal class: F;
Others:
Cooling water flow: ≥23 litres/min; Inlet temperature of cooling water: ≤+40°C; Type of protection: EEx dI; as an option: terminal compartment according to 'increased safety' EEx de I; Certificate: BVS 96.D.1060 X.

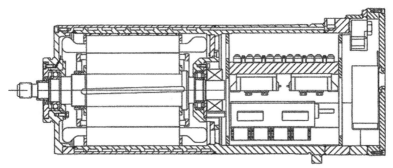

Figure 6.94 *Longitudinal section drawing of a flameproof cage induction motor with integrated frequency convertor, stator housing water cooled.*
Rated voltage: 1 kV 3 AC; Current: ≤54 A; Rated frequency (power line): 50 cps;
Rated power, motor: 75 kW; Motor speed: 0–3600 min^{-1}; Cooling water flow:
≥12 litres/min; Inlet temperature of cooling water: ≤+40°C; Type of protection: EEx dI; Certificate: BVS 98.D.1054 X.

Following the philosophy of explosion protection, the piping for cooling water shall be considered as 'empty' and, as a part of the environment, filled with the gas–air mixture according to the grouping of the motor. (This is the practice when type testing such a motor for flameproofness.)

The introduction of water cooling with its great capacity for power dissipation facilitates the integration of additional equipment for speed or starting power control into the motor. Figure 6.93 shows a water-cooled cage induction motor with an integrated eddy current clutch and Figs 6.94 and 6.95

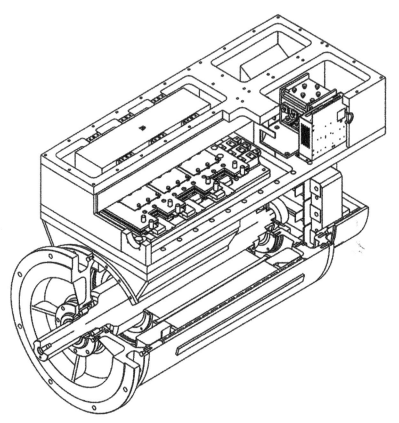

Figure 6.95   *Flameproof cage induction motor, with integrated 'piggyback' frequency convertor.*
Rated voltage: 1 kV 3 AC; Current: ⩽260 A; Rated frequency (power line): 50 cps; Rated power, motor: 400 kW; Motor speed: 0–1800 min^{-1}; Cooling water flow: ⩾20 litres/min; Inlet temperature of cooling water: ⩽+40°C; Type of protection: EEx dI, EEx de I, EEx d [ia] I, EEx de [ia] I, EEx d [ib/ia] I, EEx de [ib/ia] I, depending on terminal compartment and remote control and monitoring circuits; Certificate: BVS 99.D.1051 X.

demonstrate the integration of frequency convertors into motors, the latter in a 'piggyback' design. Figure 6.96 presents a double-rotor cage induction motor as a main drive for mining equipment.

Other motor-related equipment, e.g. electromagnetic brakes, is designed using the same basic principles as for motors. An example is given in Fig. 6.97, showing a flameproof – d – brake motor. A flameproof incremental encoder with a hollow shaft (which enables 'riding' on the driving axle by use of a torque support) is given in Fig. 6.98. It monitors speed or position of a driving axle.

Submersible centrifugal pumps are powered by a cage induction motor due to its maintenance-free design. An example of a flameproof drive unit is

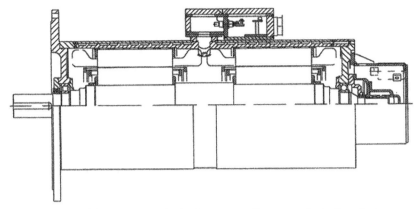

Figure 6.96 *Double-rotor cage induction motor, flameproof, shaft and stator housing water cooled.*
Rated power: 800 kW; Rated voltage: 1 kV; Rated current: 2 × 272 A; Rated speed: 1480 min⁻¹; Rated frequency: 50 cps; Thermal class: H; Cooling water flow: ⩾20 litres/min; Inlet temperature of cooling water: ⩽+40°C; Type of protection: EEx dI; as an option: terminal compartment according to 'increased safety', EEx de I; Certificate: BVS 96.D.1130 X/1. N.

Figure 6.97 *Flameproof brake motor, totally enclosed, non-ventilated. Motor and brake housing flameproof, terminal compartments according to 'increased safety'.*
Motor:
Rated power (for frame number 112): 1.85 kW; Type of protection: EEx de IIB T4; Certificate: BAS Ex 83 1409;
Brake:
Type of protection: EEx de IIC T5; Certificate: BVS 81.001;
Technical data of brake depend on motor speed.
Thermal capacity: 270 … 400 kJ/h; Input power: 56 … 82 W DC; 62 … 88 VA AC; Response time: ON: 80 … 170 ms; OFF: 80 … 110 ms.

Figure 6.98   *Incremental encoder with 70 mm diameter hollow shaft.*
Counts per turn: Z = 1024, 2500; Speed, max. (min^{-1}) (for Z < 1200):
7.2 · 10^6/Z ⩽ 7000; (for Z > 1200): 15·10^6/Z ⩽ 7000; Switching frequency: 120 kcps,
250 kcps; Logic level: HTL or TTL; Supply voltage: 9–26 V DC (HTL); 5 V DC (TTL);
Rise time: ⩾10 V/μs; Driving torque: 60 N cm; IP protection code: IP 56; Weight:
6.2 kg; Type of protection: EEx de IIC T6; Certificate: PTB Ex-99.E.1129.

shown in Fig. 6.99. The stator housing and terminal compartment are separated and sealed off from the environment. The drive unit is water cooled (or, better, liquid cooled): a portion of the pumped liquid is circulated from the pump housing up between the cooling jacket and the stator housing and carries away the thermal losses generated by the motor. Where external cooling is required, the cooling jacket can be sealed off from the pump housing and connected to a separate cooling system. As an option, the 'hydraulic end' of the pump is prepared for zinc anodes for corrosion protection and flushing of outer seal and wear rings.

In many applications, temperatures of stator windings and bearings of motors are continuously monitored during operation by Pt-100 resistance thermometers or PTC (positive temperature coefficient) temperature sensors.

6.8.4   Energy distribution systems

In the low voltage range, three directions of development have been established:

- the flameproof enclosure of components, e.g. motor protection switches, contactors, relays or circuit breakers. Usually, the enclosure is made of plastic materials. The terminals comply with 'increased safety – e –' and the components are installed in an 'e'-enclosure, forming a single part in a modular energy distribution system. This technique has found its major domain of application in Group II installations (see Section 6.7.7)

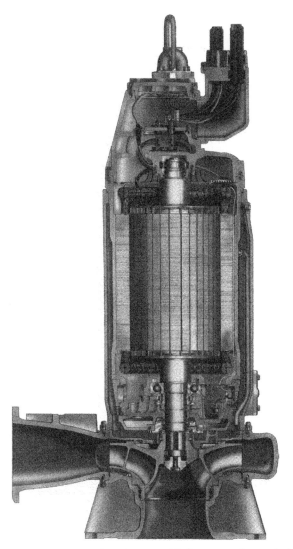

Figure 6.99 *Submersible centrifugal pump with flameproof cage induction motor.*
Motor:
Rated power: 560 kW; Rated voltage: 400 V; Rated current: 990 A; Rated frequency: 50 cps; Rated speed: 990 min^{-1};
Pump:
Impeller diameter: 715 mm; Pump capacity: 0.667 m^3/s; Height of lift: 59 m; Efficiency: 82.9% (Referring to water, max. +40°C, rated operational point); Type of protection: EEx d IIB T3; Certificate: INERIS 92.C 5015 X.

- a flameproof enclosure containing 'normal industrial', i.e. non-explosion protected, components as a single unit or as a part of a modular system acts as an energy distribution system, mainly in Group II installations. Formerly, such modular systems with flameproof enclosures (containing circuit breakers, contactors, fuses, lighting transformers), connected to

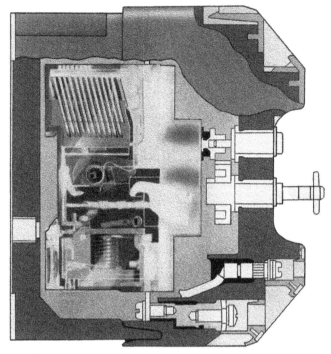

Figure 6.100   *Circuit breaker as an Ex component.*
Type of protection: EEx de IIC; Rated voltage: 690 V 3 AC; Rated frequency:
50/60 cps; Rated current: 0.1 A ... 22.5 A, depending on motor current range;
Enclosure material: plastics; Terminals according to 'increased safety – e –'.

a busbar system in 'increased safety – e –' and with individual terminal
compartments in 'e', are widely used in German coal mines as a low voltage distribution system [17]*.
• modern equipment for low (and high) voltage energy distribution in coal
  mines (Group I installations) is characteristic of large-scale integration of
  components in one single flameproof enclosure (in German known as
  'Kompaktstation'), or, in some cases, in two separated, but mechanically
  and electrically connected enclosures.

Flameproof components (Fig. 6.100) show a 'body' of insulating material,
acting both as a mechanical support for terminals, magnetic coils, switching
contacts, operating rods or spindles and as a flameproof enclosure for the

---

*In the middle of the 1980s, attempts were made for a generally revised design of mod-
ular low voltage energy distribution systems for coal mines. Flameproof enclosures are
rapidly interchangeable by the use of plug-and-socket connections to the busbar system
(in 'e'): Baltzer, Schröder: Schlagwettergeschützte Niederspannungs-Schaltanlage mit
abgesetzten eigensicheren Steuerungs- und Überwachungseinrichtungen, Siemens AG,
1985 (Original: German) [Firedamp-proof low voltage switchgear with intrinsically safe
remote control and monitoring units]

Figure 6.101 *Flameproof enclosure, cover with threaded joint.*
Type of protection: EEx d IIC T4/T5/T6 (temperature classification depending on power losses of internal components); Certificate: PTB Ex-94.C.1065; Enclosure material: steel.
In this example, the flameproof enclosure is fitted with a terminal compartment in 'increased safety – e –'. As an alternative, flameproof cable entries or conduits may be fitted.
Enclosure dimensions in mm

Width ×	Height ×	Depth	Weight (kg)	Rated current (A)
180	240	190	20	135
235	235	268	37	200
360	360	268	63	400
360	480	268	76	630
360	480	325	81	630
480	480	268	101	630
480	480	325	112	630
480	480	410	127	630
730	730	465	300	800

Rated voltage: 1000 V AC/DC; IP-code: IP 55.

contact elements. To avoid short-circuit arcing (flash-over) between current paths at different electric potentials, each contact element as a part of a single current path is enclosed separately by the insulating material (in German known as 'Einzelkontaktkapselung').

A typical flameproof enclosure for Group II installation is shown in Fig. 6.101. A cover with a threaded joint enables quick access for replacement

Figure 6.102 *Low voltage energy distribution system with EEx dII C and EEx e II modules in a chemical plant.*

of parts and adjustment of, e.g., overcurrent protection relays. The example of Fig. 6.101 shows a terminal compartment in 'increased safety – e –', equipped with additional push-button switches (in 'd') and indicating lamps.

Such a flameproof enclosure may be considered as a 'building stone' of a modular energy distribution system containing extensive low voltage installations (Fig. 6.102) in a chemical plant.

The very different philosophy of low voltage switchgear design for coal mines may be obvious in Fig. 6.103. The flameproof main part with two doors (supported by swing levers) contains contactors for 10 motor drives, divided in four groups protected by fuses, powered by three feedings each with an isolating switch (Fig. 6.103(b)). The isolating switches are mechanically interlocked with the doors (which can be opened only in the combined OFF position of the switches) and comply with the supplementary requirements for switchgear given in EN 50014 and IEC 60079-0 'General requirements'. Both terminal compartments (right side for incoming feeders, left side for outgoing lines) comply with 'increased safety – e –'. For remote controlling and monitoring, an enclosure containing intrinsically safe electric circuits with related coupling elements to the non-intrinsically safe circuits is fitted in front of the door of the right side terminal compartment.

Frequency convertors for motor drives with variable speed have shown reliability and cost reduction for some time. An example for application in coal mines is shown in Fig. 6.104. A frequency convertor with impressed current in the DC link is installed in two separated flameproof enclosures:

- at the right side, an isolating switch (mechanically interlocked with the door), fuses, vacuum contactor, mains reactor, mains controlled convertor, self-controlled convertor (at the load side), commutating capacitors and the control unit are installed. The door is supported by swing levers to enable quick access to internal components

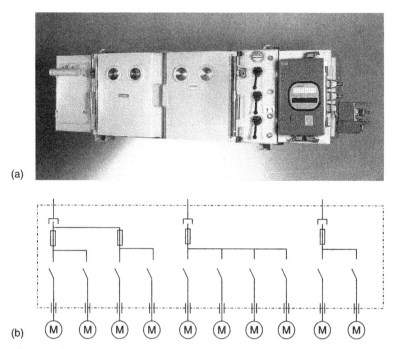

(a)

(b)

Figure 6.103 *(a) Low voltage switchboard for coal mines. A flameproof main compartment (with two doors) is fitted with a terminal compartment at right side (incoming power lines) and at left side (outgoing lines).*
Rated voltage: 1000 V; Rated current: 1200 A; Dimensions: Height × width × depth: 900 × 2800 × 900 mm³; Volume of the flameproof compartment: 1.0 m³; Weight (without internal electric parts): 1400 kg; Type of protection: EEx de I.
(b) Single-pole block diagram of the flameproof compartment given in Fig. 6.103(a).

- at the left side, the DC link reactor and capacitors are installed. The door is fixed by screws, complying with the requirements for special fasteners in EN 50014 or IEC 60079-0. The terminal compartment for the mains supply and the outgoing line complies with 'increased safety – e –'. A compartment separated from all other enclosures contains remote control and monitoring devices with intrinsically safe electric circuits.

A very different design of a low voltage energy distribution system is given in Fig. 6.105, allowing an extended scale of integration. A single flameproof enclosure with one cover at the high voltage mains side (rear in the figure) and two doors at the low voltage lines end (front in the figure) contains:

- the high voltage terminals
- a 1250 kVA transformer
- two 1050 V busbar systems, these components complying with 'increased safety – e –'
- eight places for low voltage slide-in modules.

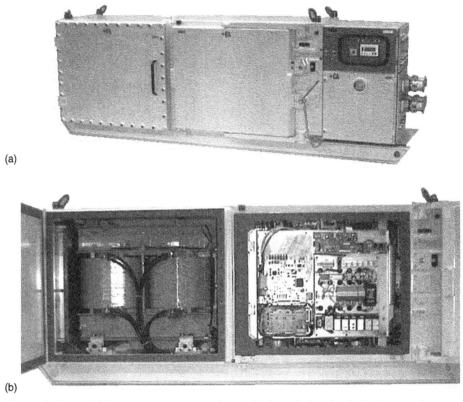

(a)

(b)

Figure 6.104 *(a) Frequency convertor for coal mines (total view). The 'e'-terminal compartment at the right side is used for both incoming and outgoing lines.*
Rated voltage: 1000 V; Max. motor rated power: 511 kW; Cooling: water cooled; Dimensions: Height × width × depth: 1000 × 2860 × 800 mm³; Added volume of flameproof compartments: 1.5 m³; Weight (total): 3700 kg; Type of protection: EEx de I.
(b) Frequency convertor according to Fig. 6.104(a), doors of flameproof compartments opened.

At the low voltage side, plug-and-socket connectors are applied for the outgoing lines. So, no low voltage terminal compartments are fitted.

The two doors at the low voltage end are mechanically interlocked with an extending device for a slide-in module each, which is fitted to the low voltage busbar system via a plug-and-socket connector. In the OFF position, the module is isolated from the busbar system. When all modules are set to their OFF position, the low voltage side doors may be opened. High voltage terminals, transformer and busbar systems (in 'e'!) remain energized.

Three types of slide-in modules are available:

- an 1100 V, 450 A vacuum contactor module (Fig. 6.106) with a motor-powered reversing switch
- an 1100 V, 2 × 80 A module with two vacuum contactors

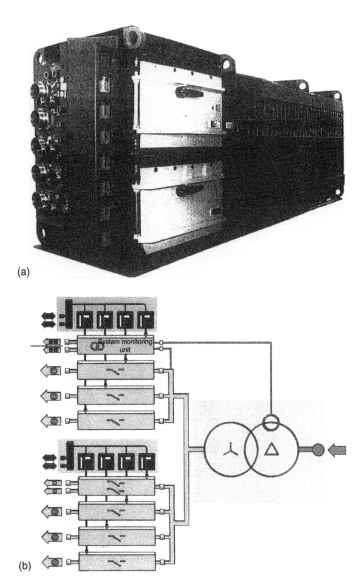

Figure 6.105 *(a) Flameproof low voltage energy distribution system with integrated transformer.*
Primary rated voltage: 5.0 or 6.0 or 10 kV; Transformer: Rated power: 1250 kVA; Rated frequency: 50 cps; Secondary rated voltage: 1050 V and 233 V (12 kVA); Low voltage switching cabinet: 8 slide-in modules (max.); Contactor module: Rated current: 450 A; Breaking capacity: 20 kA rms; Double contactor module: Rated current: 2 × 80 A; Breaking capacity: 1.6 kA; Lighting module: Rated power/voltages: 11 kVA/233 V, 1 kVA/24, 42, 127, 230 V; Flameproof plug-and-socket connectors at the low voltage side.
Dimensions: Height × width × depth: 1250 × 4450 × 980 mm^3; Type of protection: EEx d [ia/ib] ia/ib m I; Certificate: INERIS 99.D.7001 X.
(b) Block diagram of the energy distribution system shown in Fig. 6.105(a).

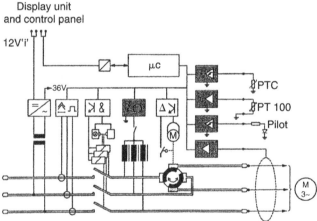

Figure 6.106   *Rear and front view and block diagram of a 450 A contactor module for the energy distribution system according to Fig. 6.105(a).*
Contactor: Vacuum contactor
Due to the high breaking capacity of the contactor (20 kA rms), fuses for short-circuit protection are unnecessary.

- a lighting module with a 230 V 1 AC, 50 A and a separated 24-42-127-230 V, 5 A output.

A similar low voltage distribution unit (without transformer) is shown in Fig. 6.107. Four slide-in modules can be inserted.

For remote controlling and monitoring, the energy distribution units according to Figs 6.105 and 6.107 are fitted with coupling elements to intrinsically safe electric circuits.

As a solution typical for the high voltage range, transformers (at their primary side) may be fitted to a combination isolating switch and contactor or

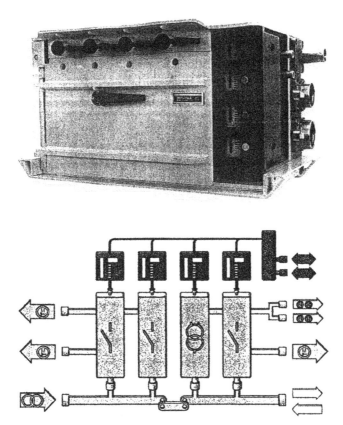

Figure 6.107 *Low voltage switching cabinet for coal mines, with four slide-in modules (total view and block diagram). Incoming and outgoing lines via flameproof plug-and-socket connectors.*
Rated voltage/frequency: 1000 V/50 cps or 1140 V/60 cps; Rated current (total): 900 A; Rated current per outgoing line: 450 A; Dimensions: Height × width × depth: 620 × 1150 × 620 mm³; Weight: 550 kg; Type of protection: EEx d [ia/ib] ia/ib ml; Certificate: INERIS 96.D.7006 X.

power circuit breaker, the latter alternatively equipped with vacuum switching tubes (Fig. 108(a)) or with SF_6-quenching chambers.

Vacuum circuit breakers show a very small stroke due to the high 'insulating capacity of vacuum'. In a gas, the mean free path length l_F of molecules, atoms or ions is reciprocal to the gas pressure p: $l_F \sim 1/p$.

So, at very low gas pressure, a high mean free path length will result accompanied by neglectable impact ionization probability. To give an example, a 2 millimetre spacing in air (at 10^5 Pa) will withstand a voltage of some kV, whereas in a vacuum (say 0.1 Pa) the same spacing will withstand (roughly) 120 kV. In other words, the vacuum circuit breaker (or contactor) operates on the 'left branch' of Paschen's law. This law correlates a certain breakdown voltage U_B to the product gas pressure, p, times contact spacing, d: $U_B = f(p \cdot d)$, for a given gas. In homogeneous electric fields, the function

(a)

Figure 6.108   *(a) High voltage switching cabinet and transformer for coal mines.*
High voltage switching cabinet:
Rated voltage: 6.6 kV; Rated frequency: 50/60 cps; Rated current: 630 A;
fitted alternatively with:
• SF$_6$ contactor
• SF$_6$ circuit breaker
• vacuum circuit breaker

		SF$_6$	
	Contactor	Circuit breaker	Vacuum circuit breaker
Rated current	400 A	400 A	630 A
Breaking capacity	10 kA	12.5 kA	25 kA
Making capacity	25 kA	30 kA	63 kA

Dimensions:
Height × width × depth: 1410 × 790 × 1050 mm^3; Weight: 1300 kg; Type of
protection: EEx de I
Transformer:
Primary rated voltage: 5.0 or 6.0 or 10 kV; Rated frequency: 50/60 cps; Rated power:
1400 kVA; Rated secondary voltage: 1050 V
Dimensions: Height × width × depth: 1470 × 980 × 2400 mm^3; Weight: 5600 kg;
Type of protection: EEx d I
High voltage incoming line via cable coupler to switching cabinet.

U_B versus $p \cdot d$ is V shaped, with a U_B minimum for most gases at $p \cdot d = 10°$ Pa·m (as an order of magnitude). The gradient dU_B/d (pd) is positive for increasing values and extremely negative for values lower than $p \cdot d$ those indicating U_B minimum. The general advantage of vacuum tubes is a very low arc voltage, resulting in very low power and heat dissipation when opening an electric circuit [15], [16].

In Fig. 6.108(a) the door is mechanically interlocked with the isolating switch (the door cannot be opened in the ON position of the isolating switch, the switch cannot be set to the ON position when the door remains open). The contactor or power circuit breaker is mounted on a slip-in module. Two earthing

(b)

Figure 6.108 *(b) High voltage switching cabinet for coal mines.*
Rated voltage: 11 kV; Rated frequency: 50/60 cps; Rated current: 630 A; Breaking capacity: 25 kA; Making capacity: 63 kA; Weight: 2000 kg; Type of protection: EEx d I.

isolators, one each at the input and output side of the power circuit breaker, mechanically interlocked with both isolating switch and door, ensure an adequate grounding of internal components when the door is opened.

The mains line is fed via a flameproof cable connector to the high voltage unit. Figure 6.108(b) shows a high voltage unit, containing:

- a flameproof busbar compartment with an isolating switch
- a flameproof main compartment with a vacuum power circuit breaker on a slip-in module
- a flameproof terminal compartment for auxiliary circuits
- an additional terminal compartment, with IP code IP 54, for auxiliary circuits
- a flameproof terminal compartment for the mains supply (at the right side)
- and a terminal compartment (alternatively as a flameproof enclosure or as an IP 54 enclosure complying with 'increased safety – e –') for the outgoing line at the rear.

The other constructional features correspond with the cabinet as shown in Fig. 6.108(a).

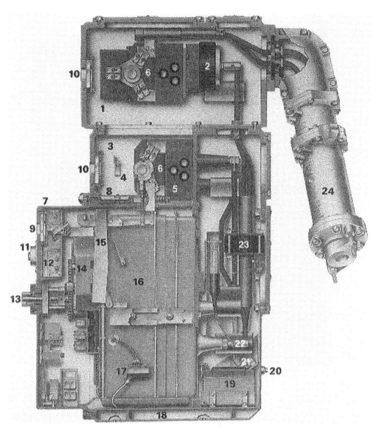

Figure 6.109   *Flameproof switchgear for Group I application, with SF$_6$ power circuit breaker.*
1: Circuit isolator chamber; 2: Overload, short-circuit and ammeter current transformers; 3: Busbar isolator chamber; 4: Isolator earth bar with fixed contacts; 5: Moulded-in busbars; 6: Isolator blades with moving contacts; 7: Hinged front door; 8: Door interlocked with isolator; 9: Flameproof window for instruments, indicator lights and operation counter; 10: Flameproof windows for visual isolation; 11: Fault reset and/or earth fault test push-buttons under covers; 12: Static protection unit; 13: Manual mechanism charging shaft; 14: Motorized mechanism charging drive; 15: Dropdown slide rails for circuit breaker withdrawal; 16: SF$_6$ circuit breaker chamber; 17: Auxiliary wiring harness multi-pin plug and socket; 18: Integral skid underbase; 19: Voltage transformer; 20: Earth terminal; 21: Fuses; 22: Split clamp connectors; 23: Earth fault current transformer; 24: Incoming adaptor and half coupler.

Rated voltage	7.2 kV
Rated current	400 A
Short-circuit rating	150 MVA at 6.6 kV
Breaking capacity	13 kA
Dimensions:	
Height × width × depth	1370 × 1020 × 785 mm^3
Weight	717 kg.

A flameproof mining switchgear with SF_6 power circuit breaker for 7.2 kV is shown in Fig. 6.109. When the SF_6 (sulphur hexafluoride) circuit breaker interrupts the current, an arc is formed and the SF_6 gas becomes ionized in the usual manner. As the alternating current decays to zero, the free electrons are rapidly captured by the SF_6 molecules to form negatively charged ions. These ions having considerable mass are almost motionless and are of no use as current carriers. Consequently, the arc path loses its conductivity, and its dielectric strength quickly builds up to prevent the arc being re-established after the current zero.

Figure 6.110 shows a sectional drawing of an SF_6 power circuit breaker.

A typical example for a 7.2 kV switchboard for coal mining, oil and petrochemical industries is given in Fig. 6.111. Each panel contains an SF_6 power circuit breaker.

6.8.5   Mains-operated luminaires

The standards for 'flameproof enclosures – d –', IEC 60079-1 and EN 50018, do not specify special requirements for flameproof luminaires. So, these shall comply with IEC 60079-0 or EN 50014 'General requirements', showing two main aspects:

1 Lamps containing free metallic sodium (e.g. sodium vapour low pressure lamps, see 5.1 in Table 6.23, Section 6.7.4) are not permitted.
2 With the exception of intrinsically safe luminaires, covers of luminaires giving access to the lampholder and other internal parts shall either be interlocked with a device which automatically disconnects all poles of the lampholder as soon as the cover opening procedure begins, or be marked with the warning 'DO NOT OPEN WHEN ENERGIZED'. In the first case, where it is intended that some parts (other than the lampholder) remain energized after opening of the luminaire, special precautions shall be taken, e.g.:

   • the parts shall comply with an independent type of protection listed in IEC 60079-0 or EN 50014
   • clearances and creepage distances between phases (poles) and to earth complying with the requirements of IEC 60079-7 or EN 50019 'increased safety – e –'
   • an internal supplementary enclosure which prevents access to the components remaining energized with an IP degree of at least IP 30 according to IEC 60529 shall be fitted.

Flameproof luminaires open a wide range of applications, based upon the fact that all types of lamps – with the exception of sodium vapour low pressure lamps – can be used (see Table 6.23, Section 6.7.4), especially those with:

• high rated power and luminous flux (high pressure lamps in general)
• high luminous densities, e.g. halogen and metal halide lamps for floodlights

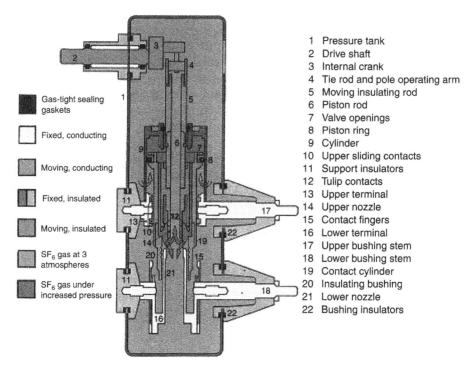

1	Pressure tank
2	Drive shaft
3	Internal crank
4	Tie rod and pole operating arm
5	Moving insulating rod
6	Piston rod
7	Valve openings
8	Piston ring
9	Cylinder
10	Upper sliding contacts
11	Support insulators
12	Tulip contacts
13	Upper terminal
14	Upper nozzle
15	Contact fingers
16	Lower terminal
17	Upper bushing stem
18	Lower bushing stem
19	Contact cylinder
20	Insulating bushing
21	Lower nozzle
22	Bushing insulators

Legend:
- Gas-tight sealing gaskets
- Fixed, conducting
- Moving, conducting
- Fixed, insulated
- Moving, insulated
- SF_6 gas at 3 atmospheres
- SF_6 gas under increased pressure

Figure 6.110   *SF_6 power circuit breaker, longitudinal section drawing.*
Rated voltage: 11 kV; Rated current: 400 A; Short-circuit rating: 250 MVA at 11 kV;
Breaking capacity; 13 kA.

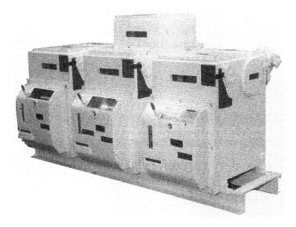

Figure 6.111   *Three panel 7.2 kV switchboard with underbase for coal mines and Group*
II applications. Two outgoing feeder units with an incoming unit situated in the middle.
Rated current: 630 A; Short-circuit rating: 150 MVA at 6.6 kV; Breaking capacity:
13.1 kA; Busbar current rating: 800 A;
Dimensions and Weight: Height × Width × Depth:
Incoming unit: 1175 × 940 × 1010 mm³, 1040 kg;
Feeder unit: 915 × 765 × 800 mm³, 818 kg; Type of protection: EEx dI/EEx dII A T …

- high luminous efficiency (metal halide and sodium vapour high pressure lamps)
- increased operating lifetime (induction lamps).

In spite of the fact that tubular fluorescent lamps are used predominantly in 'e'-luminaires complying with IEC 60079-7 or EN 50019 (in most cases with a T4 rating), flameproof luminaires with such lamps do offer an advantage with respect to temperature classification. Luminaires in 'increased safety – e –' allow the ingress of a potentially explosive atmosphere, and consequently all internal components – including the lamp – shall go along with the temperature limitations according to the temperature class. For flameproof luminaires, the enclosure and the corresponding surface temperatures determine the T rating. So, due to the considerably enlarged surface area and with constant thermal losses inside compared with an adequate 'e'-luminaire, flameproof luminaires allow a temperature classification for T5 or T6, even for increased ambient temperatures (+50°C or +55°C instead of +40°C).

A well-known construction of a flameproof luminaire uses a glass tube as luminaire body, fitted with two end covers forming a flameproof enclosure, type of protection (E)Ex dII C T5 or T6 respectively.

When tubular fluorescent lamps are used, special caution shall be taken to avoid any mechanical damage of the tube both in operation and during transportation in a potentially explosive atmosphere (the reasons are given in Section 6.7.4).

In the following, examples for zone 1 luminaires are described in brief. The floodlight in Fig. 6.112 is designed for outdoor use (illumination of large areas or floodlighting of large objects in chemical and off-shore plants). High pressure metal halide and sodium vapour lamps (both 250 W) can be used. Two pendant light fittings for high pressure discharge lamps (400 W and 250 W) are shown in Figs 6.113 and 6.114. As an alternative, the latter can be fitted with 85 W or 165 W induction lamps (see Table 6.23, 3.3, Section 6.7.4), or with compact (folded tube) fluorescent lamps. Figure 6.115 gives an example of a rectangular-shaped floodlight for halogen (250…1000 W) or high pressure discharge lamps (mercury vapour and metal halide 250 W or 400 W, sodium vapour up to 600 W).

In addition to the domain of general purpose lighting in industrial plants, flameproof luminaires have found fields of application as warning lights (in most cases a flashlight) or special purpose illumination, e.g. vessel light fittings. A flashlight is shown in Fig. 6.116. Vessel light fittings (Fig. 6.117) are fixed at inspection glasses of windows outside of vessels, tanks, bunkers, silos, stirring apparatus and chemical reactors and ensure an adequate illumination of the interior.

6.8.6   Heating systems

In chemical plants, in steam and gas turbine powered compressor and power stations and pressure control facilities for natural gas, combustible fluids and gases need to be preheated in order to comply with the process parameters,

Figure 6.112   *Flameproof floodlight.*
Lamp power rating: 250 W; Lamp type*: HIT-DE, HST-DE
Rated luminous flux of lamps:
HIT-DE: 19 000 lm; HST-DE 25 000 lm;
Light efficiency

- with diffuser   46%
- without diffuser   72%

IP protection code: IP 65; Diameter, max.: 320 mm; Height: 325 mm; Weight: 25 kg;
Type of protection: EEx de IIC T3 (terminal compartment in 'e'); Certificate: INIEX
85.103.431 (former name of ISSeP);
*Key for lamp type marking:
1: (2): 3
(1) T: Tubular fluorescent lamp; TC: Tubular fluorescent compact lamp; H: High pressure; HM: High pressure mercury; HI: High pressure iodide; HS: High pressure
sodium; L: Low pressure; LS: Low pressure sodium; Q: Quartz (halogen) lamp; .T:
Tubular lamp; .E: Elliptical form; (2) -D: Double tube compact lamp; -DE: Double
ended; -L: Long compact lamp; -SE: Self ballasted electronic; -U: U-shaped fluorescent lamp; -Us: U-shaped fluorescent lamp, short; -EL: Compact fluorescent lamp for
external electronic ballast; -with key letter T: Tube diameter (in mm); (3) Lamp rated
power (without ballast), in W.

Figure 6.113   *Pendant light fitting.*
Lamp rating: 250 W, 400 W; Lamp type˙:
HME, HIE, HIT, HSE, HST; IP protection
code: IP 65; Diameter, max.: 380 mm;
Height: 676 mm; Weight: 40 kg; Type of
protection: EEx de IIC T4/T5 (terminal
compartment in 'e'); T4 rating: 400 W
lamps; T5 rating: 250 W lamps; Certificate:
ISSeP 93.C.103.1074 X.

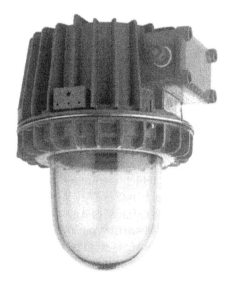

Figure 6.114   *Pendant light fitting.*
Lamp rating: up to 250 W; Lamp type*:
TC-DEL up to 42 W; HME, HIE, HSE up
to 250 W; Induction lamp: 85 W and
165 W; Rated luminous flux of induction
lamps: 85 W: 6000 lm; 165 W: 12 000 lm;
IP protection code: IP 67; Diameter, max.:
320 mm; Height: 420 mm; Weight: 14.5 kg
or 16.2 kg, depending on lamp and
ballast; Type of protection: EEx de IIC
T4/T5/T6 (terminal compartment in 'e');
T4 rating: HME 125/250 W; HIE, HSE
250 W; T5 rating: HIE, HSE 100 W, 150 W;
T6 rating: induction lamps 85 W, 165 W,
TC-DEL; Certificate: PTB 98 ATEX 1122;
Marking: II 2 G EEx de IIC T …

Figure 6.115   *Flameproof floodlight.*
Lamp rating: up to 1000 W for halogen lamps, up to 600 W for discharge lamps; Lamp type*: QT, HME, HIT, HST; IP protection code: IP 67;
Dimensions:
Width × height × depth: 489 × 400 × 390 mm³; Weight: 37.5 kg for QT lamp; 43.5 kg for H ... -lamp;
Type of protection: EEx de IIB T2/T3/T4 (terminal compartment in 'e');
T2 rating: QT 1000 W; T3 rating: QT 500 W; HME/HIT/HST 400 W, HST 600 W; T4 rating: QT 250 W, HME/HIT/HST 250 W; Certificate: PTB Ex-93.C.2107 (for QT, HME and HIT lamps); PTB Ex-93.C.2108 X (for HST lamps).

Figure 6.116   *Flameproof flashlight.*
Flash energy: 5 J; Flashing rate: 1 pulse per second; Cyclic duration factor: 100%; Supply voltage: 24/42/110/230/240 V AC; 12/24/48/60/80 V DC; Overall height: 305 mm; Protective cage diameter: 75 mm; Weight: 1.3 kg; Type of protection: EEx d IIC T6; Certificate: PTB Ex-86.2028.

e.g. fuel oil shall comply with a specified viscosity which cannot be guaranteed in winter using outdoor storage. In many cases, waste energy is applied for preheating purposes via heat exchangers, e.g. the exhaust of a gas turbine or steam from a bleeder turbine. However, during the starting period of such a plant and even in normal operation, electric preheating is required.

So, these heaters have to comply with the standards for explosion protected electrical equipment. In zone 1, flameproof heaters have found a wide range of application. The heating element shows a single-ended or a double-ended (in this case U-shaped) metallic tube as a flameproof enclosure of a winding embedded in inorganic insulating material. To achieve higher heating power,

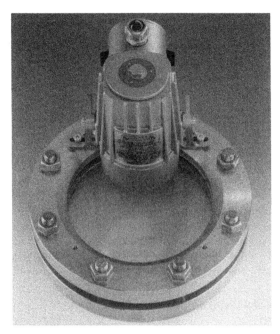

Figure 6.117 *Flameproof vessel light fitting (cable entry via conduit).*
Lamp type: Halogen, spot 35 W and 50 W; Halogen, flood 35 W, 50 W and 100 W;
Rated voltage: 120 V; Socket: E 26; IP protection code: IP 65; Diameter, max.:
130 mm; Height: 212 mm; Weight: 3.2 kg; Conduit size: NPT ½″; Type of protection:
Class I, Div. 1 and 2; Explosion group: C, D/Wet conditions; Temperature class: T3 B
for 50 W and 100 W; T4 A for 35 W; Certificate: UL 9 N 77.

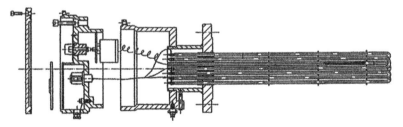

Figure 6.118 *Electrical heater, formed by a bundle of single U-shaped heating
elements. A flameproof enclosure contains the thermostatic controller and the
interconnection wiring of the single heaters. The terminal box complies with
'increased safety – e –'.*

several heating elements are combined to form a heating system (see Fig. 6.118).
The terminals of the windings are fitted in a terminal compartment as a flameproof enclosure and comply with the standards for 'increased safety – e –'. To
avoid any undesirable overtemperatures exceeding the rated value, the energy

Figure 6.119   *Natural gas preheater.*
Medium: Natural gas (as gas turbine fuel); Flow rate: 9300 m³/h; Design pressure: 2.5 MPa (=25 bar); Design temperature: +160°C; Operation pressure: 1.9 MPa (=19 bar); Inlet temperature: +40 ... +80°C; Temperature differential: 30°C; Rated power: 150 kW; Rated voltage: 460 V 3 AC; 60 cps; Type of protection: EEx d IIB T3; IP code: IP 55.

or power balance shall be maintained:

$W = c \cdot m \cdot \Delta T$
W = energy absorbed by the material to be preheated
c = specific heat of this material
m = mass
ΔT = temperature rise to be attained (temperature differential $\Delta T = T_{out} - T_{in}$)

Dividing the equation given above by time t will result in:

$P = W/t = c \cdot m/t \cdot \Delta T$ (neglecting thermal losses)
P = electric power of the heating system
m/t = mass flow of the material to be preheated

This equation demonstrates two basic possibilities for process control. Keeping P constant (P = rated power), the mass flow m/t shall be kept constant. To avoid exceeding values for ΔT conflicting either with the temperature classification of the heating system or resulting in a thermal decomposition of the fluid to be heated, the mass flow shall be monitored continuously. When it falls below a given limit, the heating system shall be switched off. In cases not allowing a constant mass flow, P shall be matched to m/t, e.g. by a thyristor controlled power supply.

For reasons of safety, an additional temperature monitoring system fitted to the heating elements shall be installed superposing the process control unit.

Figure 6.119 shows a natural gas preheater for the thermal conditioning of the fuel gas of a gas turbine power station. Three thyristor units of 50 kW each ensure a constant temperature differential $\Delta T = T_{out} - T_{in} = 30$ K.

6.8.7   Remote monitoring and controlling

In the history of explosion protection of electrical apparatus, remote monitoring and controlling has become more and more the domain of 'intrinsic safety – i –'.

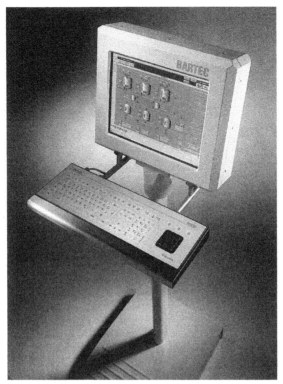

Figure 6.120   *Remote station with a 18.1" TFT display.*
Resolution: 1280 × 1024 pixels; Colours: 16.7 millions; Brightness: 180 cd/m^2;
Contrast: 150:1; View angle: left/right 60°; Transmission distance: up to 300 m;
Dimensions: Width × height × depth: 500 × 430 × 150 mm^3; Weight: approx. 28 kg;
Type of protection: EEx d [ia] IIB T4.

Nevertheless, there are some aspects which will preserve a certain field of application for flameproof apparatus:

- the very low power level for intrinsically safe electrical circuits, especially for Groups IIB and IIC (see Section 6.9) excludes some applications
- the need for barriers and other coupling devices for i-circuits when entering the non-hazardous area and the costs involved hereby generate some restrictions
- special apparatus asks for power supplies out of the scope of 'intrinsic safety' or contain internal circuits not complying with the 'i'-standards
- in many cases, transducers operate in intrinsically safe circuits. They are combined with a convertor unit, providing the power for the transducer and signal conditioning. These convertors are fitted into a flameproof enclosure, sometimes with an additional 'e'-terminal compartment.

The following examples of remote monitoring and controlling equipment may make clear what has been said above. A remote station with a 18.1" TFT colour display in a flameproof housing is shown in Fig. 6.120. This station

Figure 6.121   *Flameproof enclosure for a television camera (protective cover optional).*
Type of protection: EEx de IIC T6 ('e'-type terminal compartment);
Dimensions (without protective cover): Length × width × height:
$480 \times 200 \times 140\,mm^3$; Enclosure weight: 7.5 kg; Rated voltage (for camera power supply): up to 400 V; Max. dissipated power for T6 rating: 70 W;
IP code: IP 65

is connected to the VGA port of a personal computer in the non-hazardous area. Television cameras are brought into zone 1 by fitting them inside a flameproof enclosure (Fig. 6.121). The EEx de IIC T6 rating covers all fields of application in chemical plants, the oil and gas industry and in off-shore facilities. A Coriolis mass flowmeter with a transducer operating in an 'ib' electrical circuit combined with a flameproof convertor is shown in Fig. 6.122. The convertor has a weight of 4.5 kg, whereas the transducer – depending on mass flow and pressure – shows weights in the range 4–450 kg, the latter referring to a mass flow of $700\,t/$hour and a pressure of 1.6 MPa. The mode of operation of the transducer shall be considered in more detail in Fig. 6.123. The Coriolis force (after Gaspard Gustave de Coriolis, French engineer and physicist) $\mathbf{F_c}$ is a force of inertia acting on a body of mass m, which is moving with velocity $\mathbf{v}$ relative to a rotating system with a (constant) angular velocity $\bar{\boldsymbol{\omega}}$

$$\mathbf{F_c} = 2m[\mathbf{v} \times \bar{\boldsymbol{\omega}}]$$

(vector cross product!)

The flow Q to the flowmeter is divided into two parallel flow streams entering one of two omega-shaped measuring loops 1 and 2.

Figure 6.122 *Total view of a Coriolis mass flowmeter.*
Measuring range: from 0.8 kg/h to 700 t/h (for water);
Measurement tolerance: ±0.2% of measured value; + zero point error;
Process temperature range: −25°C ... +150°C; Process pressure limit: up to
4.0 MPa (=40 bar);
Convertor:
Type of protection: EEx d [ia] IIB/IIC T3/4/5/6 or EEx de [ia] IIB/IIC T3/4/5/6;
Certificates: BVS 98.E.2024 X; DMT 00 ATEX E 050 X; Power supply: 230 V AC 50/60
cps; 24 V DC; Outputs: 2 × 0/4–20 mA, galvanically isolated (for EEx e ... or EEx ia ...)
(HART communication optional); Weight: 4.5 kg; IP code: IP 68;
Transducer:
Type of protection: EEx ib IIC T3/4/5/6; Certificate; PTB Ex-89.C.2134 X; IP code: IP 65;
Overall dimensions (for 700 t/h mass flow):
Width × height × depth: 1750 (including flanges) × 1717 × 400 mm^3.

These measuring loops 1 and 2 fitted symmetrically are excited by oscillations against one another, referring to the axis of rotation X and Y respectively.
At mass flow zero, the legs RS 1 and RS 2 of the measuring loop 1 and the legs
RS 1 and RS 2 of the measuring loop 2 are moving on a part of a circular
trajectory, in opposite directions. With a mass flow deviating from zero, the

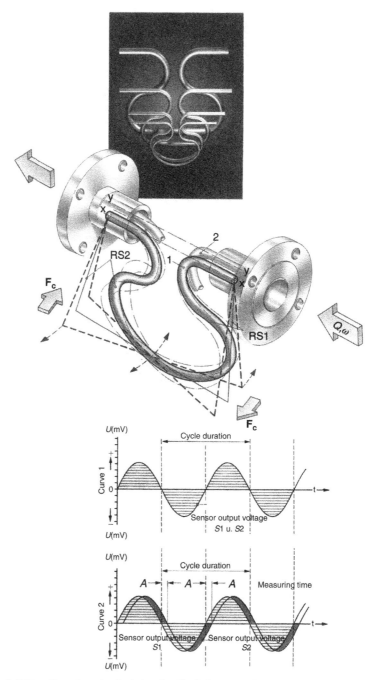

Figure 6.123   *Functional principle of a Coriolis mass flowmeter.*
A = shift to zero transitions; m = mass; ω = angular velocity; Q = flow;
F_c = Coriolis force.

mass flow in RS 1 is accelerated from a lower to a higher velocity on the circular trajectory. So, the mass m of the mass flow is subjected to the Coriolis force F_c. On the other side, the mass flow in RS 2 is slowed down from a higher to a lower velocity on the circular trajectory. The Coriolis force acts on both 'legs' of RS 1 and RS 2 with an identical amount, but in opposite directions. In other words, RS 1 shows a time delay in its oscillations, referring to zero flow conditions, and RS 2 is running in front (see the figures at the bottom of Fig. 6.123, curve 1 – mass flow zero – and curve 2 – mass flow > zero). Inductive couplers at both the measuring loops show a phase shift in their output voltages $S1$ and $S2$ which can be correlated to the Coriolis force and to the mass flow.

6.8.8 Analysers

Analysers are based upon a very wide range of modes of operation: infrared absorption for molecules with a dipole momentum, oxygen detection by electrochemical sensors or magnetic sensors owing to the paramagnetism of O_2, thermal conductivity, flame ionization detectors (due to the electrical conductivity of ionized hydrocarbons in a hydrogen-supplied flame), 'pellistors', i.e. electrically heated platinum coils with a catalytic reaction between a (combustible) gas and the oxygen of air, the ovens of gas chromatographs and the analyser heads of mass spectrometers, separating the ions of different mass in a quadrupole field or in combined electric and magnetic fields. This list – far from complete – gives an impression of the variety of types of analysers, whose explosion protection is focused on three types of protection: flameproof – d –, pressurization – p – (especially with the 2nd edition of EN 50016 or IEC 60079-2 4th edition, allowing leakage of combustibles in the analyser housing, see Chapter 7), and intrinsic safety – i –, e.g. for electrochemical sensors.

In the following, modes of operation are described in brief, combined with examples of corresponding analyser heads. In contradiction to a great number of gases showing diamagnetism (i.e. their relative permeability μ is slightly smaller than exactly 1), e.g.:

$$CO_2 \ \mu = 1\text{--}1.0 \cdot 10^{-8}$$
$$H_2 \ \mu = 1\text{--}2.3 \cdot 10^{-9}$$

oxygen shows paramagnetism with μ slightly exceeding 1:

$$O_2 \ \mu = 1 + 1.9 \cdot 10^{-6}$$

and air, due to the oxygen content, with $\mu = 1 + 0.37 \cdot 10^{-6}$.

This small deviation from $\mu = 1$ (for vacuum) forms the basis for oxygen analysers (Fig. 6.124). The gas to be analysed passes a measuring chamber. In this chamber a dumbbell-shaped quartz hollow body is pivoted by tension wires. The two cylindrical parts of the diamagnetic dumbbell dive into the inhomogeneous magnetic fields of four pole shoes of a permanent magnet. Due to $\mu > 1$, the oxygen molecules are pulled into the magnetic fields of the

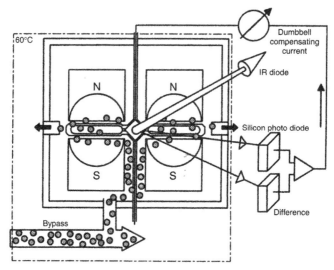

Figure 6.124 *Functional principle of an oxygen analyser.*

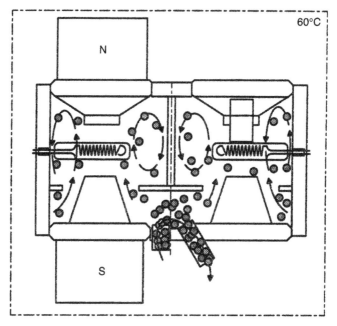

Figure 6.125 *Funtional principle of an oxygen analyser.*

permanent magnet, resulting in a partial pressure gradient and a force upon the dumbbell. It is twisted around the pivot axis. Via mirrors fitted on the dumbbell, an infrared light beam is deviated from its neutral position and generates an output signal of a photodiode.

In a modified version (Fig. 6.125), the gas passes a double-chamber system. In both chambers temperature-dependent resistors generate output signals

Table 6.32   Thermal conductivity of gases (referring to
$1 \cdot 10^5$ Pa (=1.0 bar absolute) and $T = 273$ K)

Substance	Chemical symbol	λ μW/cm $\cdot$ K
Helium	He	1430
Neon	Ne	461
Argon	Ar	164
Hydrogen	H_2	1710
Oxygen	O_2	245
Nitrogen	N_2	240
Air	–	241
Ammonia	NH_3	216
Carbon monoxide	CO	231
Carbon dioxide	CO_2	145
Methane	CH_4	303
Ethane	C_2H_6	183
Propane	C_3H_8	151
Butane	C_4H_{10}	135

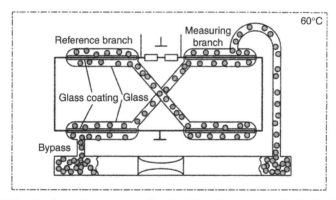

Figure 6.126   *Analyser based upon the deviating thermal conductivity of gaseous components.*

which are compensated to zero in case of a lack of oxygen. Oxygen molecules are pulled into the inhomogeneous magnetic field of a permanent magnet. The 'magnetic wind' generated hereby cools down the electrical heated resistors and forms an unbalanced output signal.

A gaseous component (e.g. H_2) with a thermal conductivity (see Table 6.32) deviating from that of the main component (e.g. air) can be detected by a resistor bridge (Fig. 6.126). Two branches are purged with the gas to be analysed, the remaining branches of the bridge are environed by a reference gas. The different thermal conductivity of the gaseous component causes a

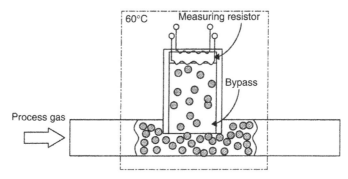

Figure 6.127   *Analyser based upon the deviating thermal conductivity of gaseous components.*

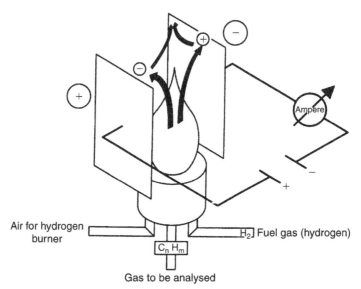

Figure 6.128   *Flame ionization detector.*

temperature gradient compared with the reference branch and an unbalanced resistor bridge, resulting in an output signal proportional to the gas concentration.

A different analyser method based upon the same effect (deviating thermal conductivities of gaseous components) is shown in Fig. 6.127. The gas to be analysed diffuses into the measuring cell. Here, a thermal conductivity sensor made of three superimposed silicon chips shows a balanced (zero) output of two thin film resistors fitted on a membrane on the chip in the middle of this stack. One of these thin film resistors is exposed to the gas to be measured. Due to its thermal conductivity, the pair of thin film resistors show an unbalanced signal output.

Hydrocarbons can be detected by passing a hydrogen–air flame in an electrical direct current field (FID, flame ionization detector, see Fig. 6.128). Due

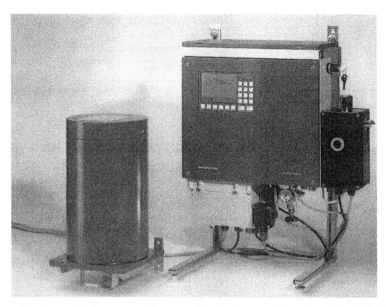

Figure 6.129   *Modular gas analyser system. Flameproof analyser head (left) and central unit (protected by pressurization), right.*
Analyser head – Type of protection: EEx d IIC T4; Certificate: BVS 97.D.2021 X; Inlet pressure: $\leqslant 10^4$ Pa (=0.1 bar); Flow rate: max. 100 l/h; Enclosure diameter: 240 mm; Enclosure height: 406 mm; Weight: approx. 26 kg; IP code: 54; Power consumption: $\leqslant 75$ W; Central unit: see Fig. 6.25.

to the very hot H_2 flame, the hydrocarbons are ionized in the flame, and the ion flow results in an electric current.

Figure 6.129 shows a modular analyser system. The flameproof analyser head (left) is suitable for zone 1 installation as well as the central unit (right), ex-protected by 'pressurization – p –'. The analyser head covers a wide range of analyser methods, e.g. such as described in Figs 6.124–6.128 and others based upon infrared absorption or electrochemical sensors for oxygen.

Many molecules show absorption of radiation in the infrared range due to oscillations (see Table 6.33), e.g. hydrocarbons such as C_2H_2, C_2H_6, CH_4 etc., NH_3, N_2O, C_2H_5OH, CO_2, CS_2, CO, SO_2, SF_6, NO, H_2O (as a vapour). The radiation from an infrared source (Fig. 6.130) passes – via a chopper wheel – alternatively a measuring cell with the gas to be analysed or a cell with a reference gas and enters an absorber cell. A membrane capacitor detects the small pressure variations in the absorber cell due to the alternating infrared light path.

An extended range of application – covering the detection of (maximal) four gaseous components – can be achieved by inserting a filter wheel into the infrared light path according to Fig. 6.131. The absorption lines of different molecules are specific for these molecules, and by selection of the 'right' absorption lines up to four different gases can be detected and identified.

Table 6.33 Main infrared absorption lines of hydrocarbons (referring to a transparency <50%)

Substance	Chemical symbol	Wavelength of absorption line peak μm					
Methane	CH_4	7.7					
Ethane	C_2H_6	12	7	3.5	2.4		
Butane	C_4H_{10}	13.5	10.4	7.7	7.2	6.8	3.4
Pentane	C_5H_{12}	13.7	10.9	7.2	6.8	3.4	
Hexane	C_6H_{14}	7.1	6.8	3.6			
Heptane	C_7H_{16}	13.9	7.1	6.8	3.4	2.3	
Octane	C_8H_{18}	13.9	11.4				
Nonane	C_9H_{20}	13.9	7.0				
Decane	$C_{10}H_{22}$	13.9	7.1				
Benzene	C_6H_6	9.8	6.8				

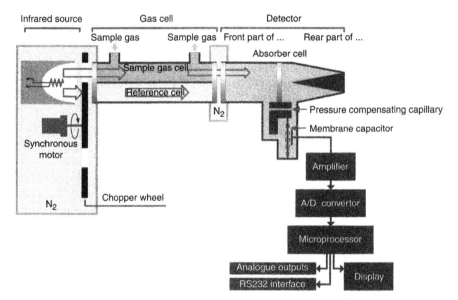

Figure 6.130 *Infrared absorption analyser (functional principle).*

Figure 6.132 demonstrates the assembly of such a modular analyser system. The different analysers are based upon infrared absorption (see Figs 6.130 and 6.131), thermal conductivity, ultraviolet absorption (e.g. for C_6H_6, Cl_2, O_3, Hg (as a vapour), SO_2, H_2S, NO_2, and toluene $C_6H_5–CH_3$), paramagnetism of oxygen and electrochemical cell for oxygen.

To avoid any propagation of explosions starting in the interior of the analyser's enclosure through the gas piping, flame arrestors shall be fitted at the gas inlet and outlet. These flame arrestors may be, e.g., concentric tubes

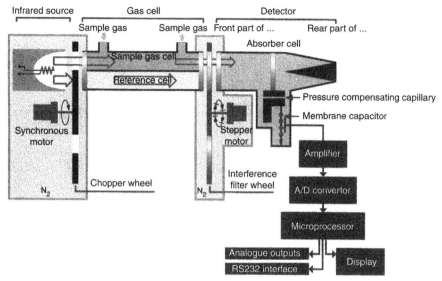

Figure 6.131 *Infrared absorption analyser (funtional principle).*

with diameter differences according to the joint dimensions given in the d-standards EN 50018 or IEC 60079-1. The more recent editions of these standards allow (as 'breathing and draining devices') the application of crimped ribbon elements, sintered metals, pressed metal wire or metal foam elements as flame arrestors.

Note:

The application of such devices as a pressure relief in case of internal explosions is out of the scope. Fitted in gas pipings, the gas shall not be capable of forming an explosive mixture with air. The pressure inside the piping shall not exceed the atmospheric pressure by more than 10 per cent.

Sintered metals show a strong correlation between open porosity and 'effective gap'. (The 'effective gap' w_{eff} of a flameproof component is the highest MESG (Maximum Experimental Safe Gap) value of combustible gas mixtures (see Section 1.2.2), for which that component ensures flame-proofness. For gas mixtures with MESG $< w_{eff}$, flame transmission occurs.) (see [19]).

The determination of open porosity (or fluid permeability) as given in ISO Standard 4022 is not expensive, and by this way the permeability of a sintered metal element can be correlated with its effective gap indicating the grade of flameproofness.

Sintered metal elements as a part of a flameproof enclosure are often used for gas warning detectors for combustible gases. As a cost effective gas sensor, an electrically heated platinum coil covered with a catalyst and as a part of a

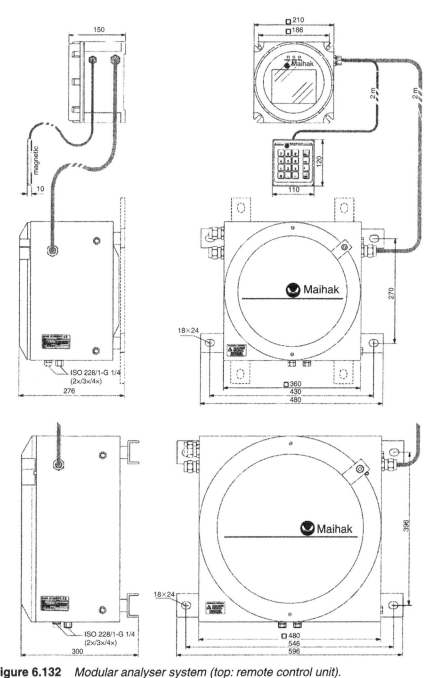

Figure 6.132   *Modular analyser system (top: remote control unit).*
Flameproof analyser heads – (middle and bottom)
(dimensions 360 × 360 × 276 mm³ and 480 × 480 × 300 mm³); Marking: II 2G EEx
d ia IIC T6 or II 2G EEx d ia [ia] IIC T6; Type of protection: EEx d ia IIC T6 or EEx d
ia [ia] IIC T6; Certificate: TÜV 97 ATEX 1207 X; IP code: IP 65.

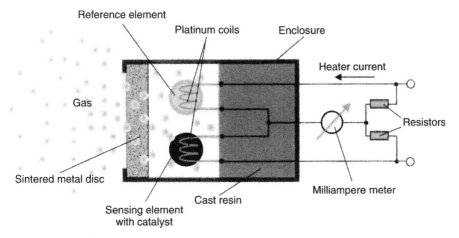

Figure 6.133   *Gas detector with 'pellistors', functional principle.*

resistor bridge (Fig. 6.133) generates an output signal due to an unbalanced bridge, caused by the temperature rise of the 'active' platinum coil in the presence of a combustible gas, which reacts with the oxygen content of air forced by the catalyst. The sintered metal element enables an adequate diffusion of the gas into the measuring cell and acts as a flame barrier. In fact, the electrical circuits of such a 'pellistor' bridge are intrinsically safe in most cases. But the 'active pellistor' may form a 'hot spot' and ask for a flame barrier against the environmental atmosphere. A flameproof gas warning detector with a 'pellistor' is shown in Fig. 6.134.

The same protection method can be applied for gas sensing elements with deviating modes of operation, e.g. a semiconducting element with an integrated heating element. Powered with a constant current, the semiconductor shows a constant temperature. Combustible gases penetrating a sintered metal element by diffusion come into contact with the semiconductor's surface and cause a shift of its electrical conductivity and output signal (Fig. 6.135).

6.8.9   Special equipment for coal mining

Most of electrical apparatus for coal mines, e.g. light fittings, motors and switchgear is closely related to Group II equipment. Deviations are caused by environmental conditions in mines asking for increased mechanical strength, increased dust tightness or resistance to special chemical agents (oils, greases and hydraulic liquids) or are based upon the need to strictly avoid frictional sparks resulting in a limited content of aluminium, magnesium and titanium for light metal enclosures.

Nevertheless, increasing power demands initiated a large-scale integration of electrical components into the machine bodies and created unique designs for mining equipment. Two examples, a shearer-loader and an automatic directional drilling system, shall be considered in more detail in

Figure 6.134   *Flameproof gas warning detector (combustible gases).*
Marking: II 2G EEx de IIC T4/5/6; Type of protection: EEx de IIC T4/5/6; Certificate:
DMT ATEX E 006 X; Dimensions: 80 × 146 × 55 mm³; Weight: 0.6 kg; IP code:
IP 65; Ambient temperature range/T ratings:

−50°C ... +40°C	T6
−50°C ... +55°C	T5
−50°C ... +85°C	T4

Detector is suitable for gas–air mixtures ⩽100% LEL, see Table 1.1.

the following. In Fig. 6.136 (as a photomontage) the machine frame of a shearer-loader is shown in front view and, additionally, with covers removed. A steel frame (as 'machine body') takes up the mechanical forces and contains four flameproof modules (from left to right):

- high voltage unit (rated voltages 3.3 kV and 1 kV) with mainly an isolating switch (at 3.3 kV level) and two vacuum contactors (forward/rearward) of the cutting motors of 300 kW each

Figure 6.135 *Flameproof gas warning detector.*
Type of protection: EEx de IIC T6; Certificate: BVS 98.E.2051; Dimensions: $80 \times 55 \times 75\,mm^3$; Weight: 0.66 kg; IP code: IP 54; Ambient temperature range: $-20°C \ldots +60°C$; Sensor: semiconducting element; Detector is suitable for gas–air mixtures ≤50% LEL, see Table 1.1.

(a)

(b)

Figure 6.136 *Flameproof compartments in the machine frame of a shearer-loader. (a) Front view (b) Covers removed.*
Type of protection: EEx dI.

- low voltage cabinet (rated voltage 1 kV) with two covers containing an isolating switch, contactors and thyristor control units for two DC haulage motors of 55 kW each, contactors for two hydraulic units of 7.5 kW each and the lumpbreaker motor (60 kW)

Figure 6.137   *Shearer-loader according to Fig. 6.136 in operation.*
Technical specifications – Machine height: 1412 mm; Length (between cutting drum shafts): 10 700 mm; Length of machine body: 6700 mm; Cutting range: +3125 mm to −435 mm; Cutting drum diameter: 2000 mm; Cutting depth: 850 mm; Cutting drum speed: 28 min^{-1}; Tractive force, max.: 370 kN; Speed, max.: 15.3 m/min; Installed rated motor power: 725 kW; Machine weight: 40 t.

- remote control and monitoring unit, with a transformer to supply the auxiliary voltages and the interfaces to external intrinsically safe electrical circuits as well as a wireless transmitter – receiver for remote radio control and a cable connection compartment (at the rear side, not shown at the photo) as a terminal for the cables to the different motors and sensors. The intrinsically safe circuits comply with EN 50020, Group I, category ib.

Figure 6.137 shows this shearer-loader in operation at Walsum Colliery, Deutsche Steinkohle AG, near Duisburg/Germany.

In the fields of coal mining, rock excavation, tunnelling and mineral exploration there is a considerable demand for improved drilling precision, especially for:

- pilot drilling for large diameter boreholes without the risk of expensive errors
- freeze boring can be done without constant check measurements and alignment drilling runs
- more accurate and reliable information will be obtained from exploratory or prospecting drilling.

To achieve such results, an automatic directional drilling system is fitted between drilling head and bore rod (Fig. 6.138). An inclinometer system detects deviations from the vertical line, and on this basis steering electronics generate correction signals for solenoid operated valves in a hydraulic system. Four steering skids covering a great part of the system's overall length and fitted in 90° azimuthal steps are actuated by hydraulic cylinders with a force depending on the pressure modulation of the solenoid valves. The

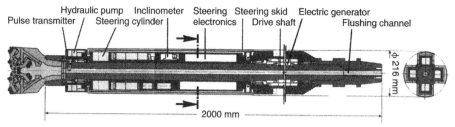

Figure 6.138   *Automatic directional drilling system. All electrical components are fitted into flameproof compartments inside the steel body.*
Length (without drilling head): 2000 mm; Diameter: 216 mm; Speed: 60 min^{-1};
Drilling depth: 600 m; Drilling pressure: 18 t; Weight (without drilling head): approx. 300 kg; IP code: IP 68; Generator: 48 V DC 2 A.

steering skids prevent the rotation of the drilling system with the bore rod by supporting the casing on the bore hole wall. In the vertical position, the drilling system is adjusted by the steering cylinders (each impinged with identical hydraulic pressure) exactly in the bore hole axis. Deviations cause adjustment signals from the inclinometers via the steering electronics to the solenoid valves and the hydraulic system, resulting in a radial displacement of the casing until the correction is completed and vertical drilling is restored. The directional drilling system itself does not depend on an external power supply. An electric generator and a hydraulic pump both powered by the drive shaft form a self-supporting power source. A 'wireless' measurement transmission system for remote monitoring of the system (especially transmitting the inclination values) during drilling has been developed. An electrohydraulic pulse transmitter narrows the flushing channel of the directional drilling system and thus produces a pressure increase in the form of a brief pressure pulse in the flushing system, monitored by a differential pressure receiver at the surface installations. At intervals of five minutes, a complete data set consisting of 11 time-coded pressure pulses is transmitted to the drilling platform. Via the pressure transducer, these signals are converted to electric signals and decoded into eight independent value channels. The automatic drilling system can be used with conventional drilling equipment. Figure 6.139 shows the breakthrough at the target point with the automatic directional drilling system.

6.8.10   Plug-and-socket connectors

Requirements are given in EN 50014 as well as in IEC 60079-0 'General requirements', covering all types of protection. Plugs and sockets shall be mechanically or electrically (or otherwise) interlocked to avoid disconnection when energized, or shall be fixed together by special fasteners and additionally be marked with a warning label 'DO NOT SEPARATE WHEN

Figure 6.139 *Breakthrough at the target point with the automatic directional drilling system.*

ENERGIZED'. (In Germany, plugs and sockets with an interlock are known as 'Steckvorrichtungen', without an interlock as 'Steckverbinder'.)

In the low power range (not exceeding 10 A and 250 V AC or 60 V DC) there are less stringent requirements.

The standards for 'flameproof enclosure – d –', EN 50018 and IEC 60079-1 give additional requirements for plugs and sockets not fixed together by means of special fasteners:

- the widths and the gaps of flameproof joints (of the flameproof enclosures of plugs and sockets and cable couplers) shall be determined by the volume which exists at the moment of separation of the contacts (other than those for earthing or bonding or those in intrinsically safe electrical circuits)
- the flameproof properties of the enclosure shall be maintained in the event of an internal explosion both when the plugs and sockets (or cable couplers) are connected together and at the moment of separation of the contacts (other than those for earthing or bonding or those in intrinsically safe electrical circuits).

An example of a flameproof plug and socket is given in Figs 6.140 and 6.141. The socket – in the non-operating mode – is covered by a rotating 'shutter'

Figure 6.140   *Plug-and-socket connector.*
Marking: II 2G/2D EEx ed IIC T6; Type of protection: EEx ed IIC T6; Certificate: LCIE 99 ATEX 6027 X; Cable termination compartment in 'increased safety – e –';
Rated voltage: up to 500 V AC; Rated current: 20 A; Versions with: 2, 3, 4 or 5 contacts; Rated frequency: 50 cps; Breaking capacity according to EN 60947-3: AC 22; IP code: IP 66/67.

made of insulating material. When inserting the plug, the 'shutter' (rotating with a given angle) opens the paths to the socket contacts, thus forming a flameproof joint with the plug contacts. Each contact forms a small flameproof 'chamber' on its own.

The plug and socket according to Fig. 6.142 shows a 'coaxial' arrangement of the contacts of a three-phase system. The central multi-contact pin is intended for six pilot and auxiliary circuits. Plug and socket are fixed together by means of a special fastener.

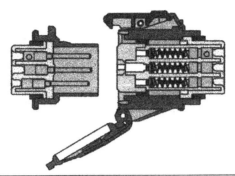

*Cut view – socket moulding and contacts. The locked
safety shutter achieves a 'd' explosion-proof chamber.*

As soon as a plug is mounted by rotation on the socket in the rest position, the 'd' explosion-proof chamber is maintained between the outer diameter of plug contacts and the contact holes in the socket safety shutter.

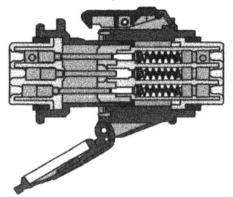

Cut view – plug in the rest position.

The plug can then be connected and disconnected under full load, and its switching capability is that of utilization category AC22.

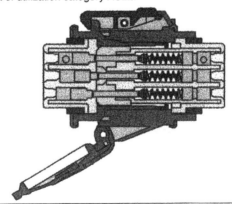

Cut view – plug fully mated. The explosion-proof chamber is maintained.

Figure 6.141   *Plug-and-socket connector, functional principle.*

Figure 6.142 *Plug-and-socket connector.*
Type of protection: EEx dl; Certificate: INERIS 95.D.7027 X; Rated voltage: 1100 V
3 AC; Rated current: 450 A; Cable conductor cross-section: up to 185 mm^2;
Auxiliary contacts: 4 mm^2; Cable conductor cross-section: 6 mm^2; (Top to bottom):
two plugs for different cable diameters; socket with mounting flange; longitudinal
section of a plug.

A Group I plug-and-socket connector is shown in Fig. 6.143. Three power
pins and one or three auxiliary pins are connected together. A plug-and-socket
connector for the high voltage range according to Fig. 6.144 contains three
power and three auxiliary pins.

Figure 6.143   *Plug-and-socket connector (plug, socket and plug coupler).*
Type of protection: EEx dl; Certificates – Plug: MECS 93 C 5502 U; Socket: MECS
93 C 5503 U; Plug coupler: MECS 93 C 5504; Rated voltage: 1100 V 3 AC; Rated
current: 300 A; Cable conductor cross-section: up to 120 mm^2.

Figure 6.144   *Plug-and-socket connector.*
Type of protection: EEx dl; Certificates – Plug: MECS 96 D 5084 U; Socket: MECS
96 D 5085 U; Plug coupler (not shown in the picture): MECS 96 D 5088; Rated voltage:
3.3 kV 3 AC; Rated current: 350 A; Cable conductor cross-section: up to 120 mm^2.

6.8.11   Limits of application

Obviously, type of protection 'flameproof enclosure – d –' opens a wide
range of applications for zone 1 equipment. But it should be mentioned that

there are some limits resulting from the rules of machine building or derived from basic chemistry and physics in the field of gaseous reactions and combustion. Three main aspects will be elucidated in this Section:

- size limitation of flameproof rotating electrical machines due to the radial clearance of bearings
- nitrogen oxide production in quenching chambers operating in air
- flame transmission initiated by high power arcing.

The radial clearance of plain and rolling bearings asks for a certain gap between shaft and casing bore. With fixed glands, as shown in Figs 6.85, 6.87, 6.88 and 6.89, the gap shall comply with the maximum values given in Tables 6.27–6.29. In many cases, great shaft diameters, e.g. for motors with eight, 10 or even more poles, are accompanied by radial clearances not complying with the corresponding gaps. Thus, the application of fixed glands is limited (only to give a rough idea) to rated motor powers in the range 3...4 MW. Flameproof motors exceeding this power range require a different gland design according to Fig. 6.145. This floating gland follows the radial movement of the shaft, remaining concentric to the shaft and by this way going along with the requirements for the maximum gap given in the standards. However, the application of floating glands for Group IIC motors (or generators) is not within the scope of EN 50018 or IEC 60079-1.

Switching contacts operating in air, e.g. contactors or circuit breakers, cause arcs which initiate a chemical reaction between the two main constituents of air, nitrogen and oxygen, resulting in the production of nitrogen oxides, generally speaking N_xO_y. In low power circuits this effect is negligible, but with increasing voltage, current and switching rate considerable amounts of nitrogen oxides are present [16, 17]. Due to the poor 'ventilation' (to give an example: a typical 1 kV switching cabinet for – say – 1000 A rated current is equipped with a door of 800 mm width and 600 mm height at a mean gap

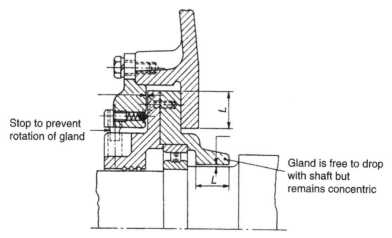

Stop to prevent rotation of gland

Gland is free to drop with shaft but remains concentric

Figure 6.145   *Floating gland (according to IEC 60079-1).*

value of 0.3 mm. A typical $0.6 \, m^3$ volume is fitted with a ventilation opening (due to the gap) of $(2 \times 800 + 2 \times 600) \times 0.3 \, mm^2 = 840 \, mm^2$ only!) of flameproof enclosures, these components cannot escape, and subsequent switching causes an accumulation in the interior of switching cabinets. Together with the water vapour content of air, nitrogen oxides react to nitric acids such as HNO_2 or HNO_3, which adversely affect the surface resistance of organic insulating materials. In Figs 1.9 and 1.10(a) the N_xO_y generation is related to current, phase angle and motor power in a 1000 V 3 AC system. The mass of N_xO_y (given in moles) is in good linearity to the energy (in joules) of the switching arc [16, 17]:

- for the low voltage range ($\leqslant$1000 V):
  $2 \cdot 10^{-8}$ moles/J ... $5 \cdot 10^{-8}$ moles/J
- for the high voltage range (6 kV):
  $6 \cdot 10^{-8}$ moles/J ... $1.1 \cdot 10^{-7}$ moles/J
  (referring to atmospheric air pressure).

A stalled 315 kW rated 1 kV motor, e.g., causes an arc energy of $6.34 \cdot 10^3$ J as a mean value during the switch-off cycle in a contactor (operating in air). The corresponding N_xO_y mass is $3.17 \cdot 10^{-4}$ moles. At atmospheric pressure and at a temperature of $+70°C$ (=343 K) in a flameproof switching cabinet with a free volume of $0.05 \, m^3$, the N_xO_y concentration is 0.0177% (v/v) for only one switch-off cycle. Thus, after 100 switch-off cycles, the nitrogen oxide concentration accumulates to 1.77% (v/v). Contaminated surfaces of organic insulating materials may cause a leakage current and thus a subsequent formation of conductive paths on the surface. By this way, a breakdown of an insulating surface may be initiated, resulting in a thermal decomposition of the insulating material (mainly hydrocarbons). The effect will be the same as described in Section 6.8.1 (thermal decomposition of organics in electric arcs).

To avoid these effects, it should be recommended that for:

- voltages $\geqslant$690 V AC and
- currents in the range of some 10^2 A (and higher) and for
- switching-off cycles at a rate of $\geqslant$10/hour,

vacuum or SF_6 contactors shall be used in flameproof switchgear. In the power lines, arcing in air is then excluded, and the small arc energy in auxiliary circuits can be neglected as well as the N_xO_y production caused hereby.

The effect of nitrogen oxides on the MESG values of combustible gas–air mixtures has been investigated [56]. With addition of NO_2 to a gas–air mixture, the MESG values increase with increasing NO_2 content (see Fig. 1.3(a)–1.3(c), Fig. 6.146), and the 'vertex' of the MESG versus gas concentration parabola shifts slightly to higher gas concentrations due to the higher oxygen content in NO_2 compared with air and by this way in the gas–air–NO_2 mixtures as a whole. As far as the flameproofness of an enclosure is concerned, the MESG values are shifted to 'the safe side' and do not adversely affect the safety of the enclosure.

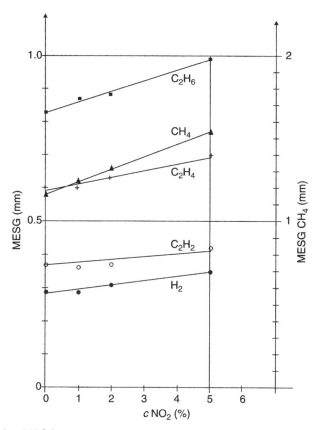

Figure 6.146 *MESG values of gas–air–NO₂ mixtures versus NO₂ concentration* c *given in % (v/v).*

In the early 1980s, in the United States of America and in Germany the electrical breakdown in electrical fields during gas–air explosions has been investigated independently [48, 54, 55].

Using a very simple model (neglecting dissociation effects), the flame temperature of an exploding gas–air mixture in a constant volume V can be calculated. For an ideal gas it is:

$$p_o \cdot V = \Sigma \, m_o R T_o \quad \text{before ignition}$$

and then, in the moment of peak pressure, $p_p V = \Sigma \, m_p R T_p$

V = volume of combustion chamber
$\Sigma \, m$ = sum of moles
subscript o = before ignition
subscript p = at peak pressure, after ignition
$R = 8.137 \, \text{joule} \times \text{mole}^{-1} \times \text{K}^{-1}$
p = gas pressure
T = gas temperature

Table 6.34   Flame temperatures* for gas–air mixtures at the stoichiometric point ($p_o = 1.0 \cdot 10^5$ Pa (=1 bar absolute), $T_o = 293$ K)

Gas	Pressure ratio π	T_p K	Simple model: T_p with $\mu = 1$ K
H_2	7.9	2720	2320
CH_4	8.2	2400	2400
C_2H_2	10.3	3140	3020
C_2H_4	9.3	2720	2720
C_2H_6	9.1	2590	2660
C_3H_8	9.4	2650	2760

* The temperature values given in this table refer to a combustion in *a constant volume* and are not identical with combustion temperatures given for constant pressure combustion processes as in gas turbines for aircrafts

Division of the two basic equations given above results in:

$$p_o/p_p = \Sigma \, m_o / \Sigma \, m_p \cdot T_o/T_p$$

and

$$T_p = \Sigma \, m_o / \Sigma \, m_p \cdot p_p/p_o \cdot T_o$$

With volume ratio $\mu = \Sigma \, m_o / \Sigma \, m_p$
and pressure ratio $\pi = p_p/p_o$
this will result in $T_p = \mu \cdot \pi \cdot T_o$.
For many gases μ is close to 1 (see Table 1.2), and so $T_p = \pi T_o$.

Table 6.34 summarizes the flame temperatures T_p for selected gases.
   Due to the high combustion temperature, the gas is ionized in parts, the grade of ionization is determined by temperature T and ionization energy E (Saha's equation):

$$N_i/N_o = \text{constant} \times (2\pi mkT)^{3/2}/h^3 \cdot e^{-E}/kT$$

N_i = number of ionized atoms
N_o = number of neutral atoms   (in basis state)
m = electron mass   (=$9.109 \cdot 10^{-31}$ kg)
k = Boltzmann's constant   (=$1.38 \cdot 10^{-23}$ J/K)
h = Planck's constant   (=$6.6256 \cdot 10^{-34}$ Js)
E = ionization energy, in text often given in eV (electron volts), with
      $1 \text{eV} = 1.602 \cdot 10^{-19}$ J

The main term in Saha's equation is e^{-E}/kT.
   Compared with this, $(2\pi mkT)^{3/2}/h^3$ is of minor interest, because T is in the range between 2400 K and 3140 K, resulting in $T^{3/2}$ variations with a ratio of 1.4965. The variations of e^{-E/kT_p} are much greater, see Table 6.35.

Table 6.35 Ionization values for combustible gases (according to [55])

Atom, molecule, radical	E Ionization eV	Energy $\times 10^{-18}$ J	T_p K	E/T_p $\times 10^{-22}$ J/K	e^{-E/kT_p}
H_2	15.4	2.467	2720	9.07	$2.93 \cdot 10^{-29}$
H	13.6	2.178	2720	8.01	
CH_4	13.0	2.083	2400	8.68	$4.68 \cdot 10^{-28}$
CH_3^+	13.1	2.098	2400	8.74	
CH_3	9.8	1.569	2400	6.54	
CH_2	11.8	1.890	2720^1	6.95	
C_2H_2	11.4	1.826	3140	5.82	$4.85 \cdot 10^{-19}$
C_2H_4	10.5	1.682	2720	6.18	$3.57 \cdot 10^{-20}$
C_2H_3	9.4	1.506	2720	5.54	
C_2H_6	11.8	1.890	2590	7.29	$1.15 \cdot 10^{-23}$
C_2H_5	8.8	1.409	2590	5.44	
C_3H_8	11.5	1.842	2650	6.95	$1.35 \cdot 10^{-22}$
C_3H_7	7.9	1.266	2650	4.78	
O_2	12.2	1.954	3140^2	6.22	$2.67 \cdot 10^{-20}$
O	13.6	2.178	3140	6.94	
N_2	15.6	2.499	3140^2	7.96	$8.91 \cdot 10^{-26}$
N	14.5	2.323	3140	7.40	
Ar	15.8	2.531	3140^2	8.06	$4.29 \cdot 10^{-26}$
CO	14.1	2.259	3140^2	7.19	$2.36 \cdot 10^{-23}$
CO_2	13.8	2.211	3140	7.04	$7.02 \cdot 10^{-23}$
H_2O	12.6	2.018	3140^2	6.43	$5.84 \cdot 10^{-21}$
OH	13.2	2.115	3140^2	6.74	

1 Combustion as C_2H_4
2 Combustion with C_2H_2 in a gas–air mixture

So, a 'ranking' of gases according to increasing ionization grade can be made as follows, referring to the e^{-E/kT_p} values:

$$\begin{array}{ll} H_2 & 2.93 \cdot 10^{-29} \\ CH_4 & 4.68 \cdot 10^{-28} \\ C_2H_6 & 1.15 \cdot 10^{-23} \\ C_3H_8 & 1.35 \cdot 10^{-22} \\ C_2H_4 & 3.57 \cdot 10^{-20} \\ C_2H_2 & 4.85 \cdot 10^{-19} \end{array}$$

An acetylene flame shows the highest ionization degree.

So, an electrical field with a reacting gas–air mixture inside will show a current due to ionization, see Figs 6.147 and 6.148. In Figs 6.149 and 6.150 the

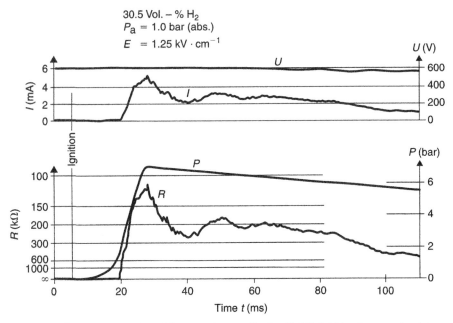

Figure 6.147   *Ion current* I *and flame resistance* R *of a 30.5% (v/v) hydrogen–air mixture.* p = *explosion pressure; DC field strength 1.25 kV/cm* (U = 600 V).

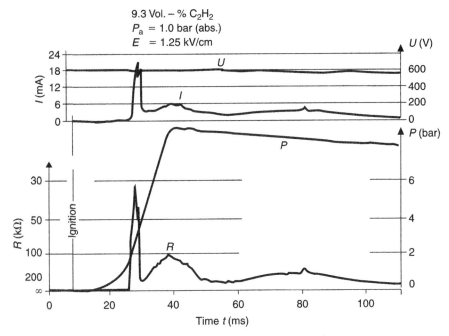

Figure 6.148   *Ion current* I *and flame resistance* R *of a 9.3% (v/v) acetylene–air mixture.* p = *explosion pressure; DC field strength 1.25 kV/cm* (U = 600 V).

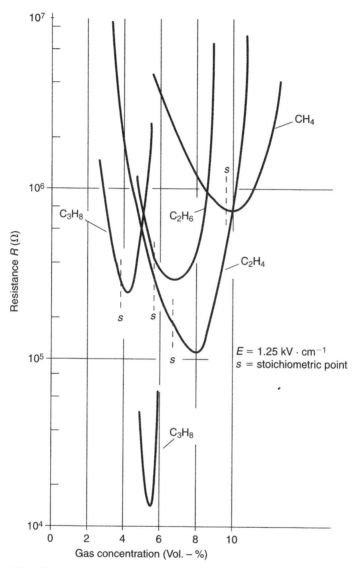

Figure 6.149 *Flame resistance R of different hydrocarbon–air mixtures versus gas concentration given in % (v/v). DC field strength 1.25 kV/cm.*

resistance of reacting hydrocarbon–air mixtures is plotted versus gas concentration. The stoichiometric points (s) are indicated. Obviously, the minimal values of resistance occur at 'rich' gas mixtures above the stoichiometric point due to the dominant role of the hydrocarbons.

Variations of the electric field strength result in variations of electrical conductance (Figs 6.151 and 6.152). With increasing field strength, there is a hyperbolical decrease of conductance and – with continuously increasing field strength – a sudden breakover (Fig. 6.152) at a field strength of only 10 … 15% of the 'normal' breakdown field strength in air.

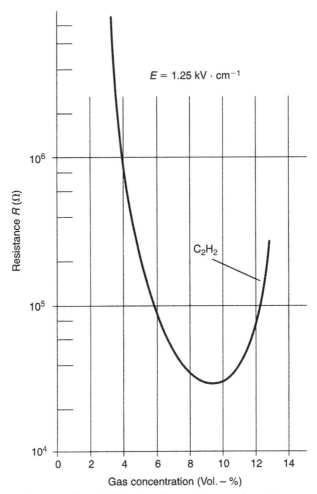

Figure 6.150 *Same as Fig. 6.149, given for acetylene–air mixtures.*

Generally speaking, the explosion of a gas–air mixture in a flameproof enclosure initiates a current between bare live parts due to ionization and, with increasing electrical field strength, a sudden breakdown in the clearance: short-circuit arcing in flameproof enclosures is not only a question of adequate planning of electrical equipment or conscientious assembly and maintenance of electrical apparatus, but may happen due to basic physics. To demonstrate this, a 1 AC busbar system was installed in a gas-tight and pressure resistant explosion vessel made of steel (a modified testing vessel for type tests of flameproof enclosures). The following combinations – applied 50 cps voltage–clearance – have been investigated with exploding C_2H_2–air mixtures:

Voltage (1 AC, 50 cps)	Clearance
500 V rms	4.5 mm
500 V rms	6.5 mm

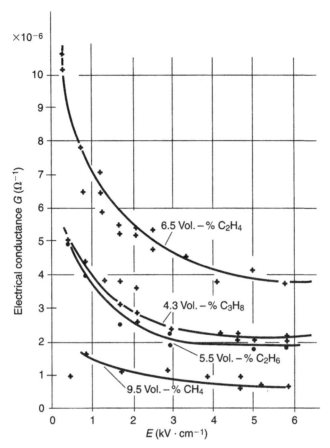

Figure 6.151 *Electrical conductance G of different hydrocarbon–air mixtures versus electrical field strength E (DC).*

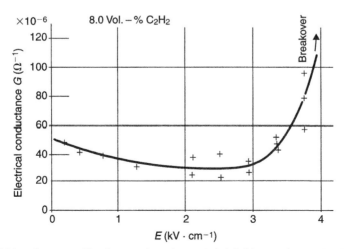

Figure 6.152 *Same as Fig. 6.151, given for an 8.0% (v/v) acetylene–air mixture.*

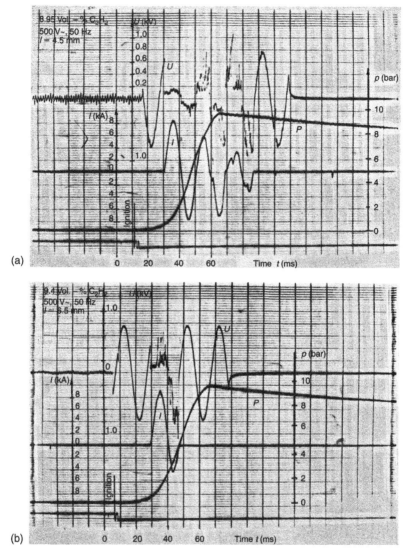

Figure 6.153   *Flame-induced short-circuit arcing over a 1 AC busbar system. Voltage U, arc current I and explosion pressure p versus time.*
(a)   8.95% (v/v) acetylene–air mixture; $U = 500\,V$ rms; clearance 4.5 mm
(b)   9.4% (v/v) acetylene–air mixture; $U = 500\,V$ rms; clearance 6.5 mm.

Voltage (1 AC, 50 cps)	Clearance
500 V rms	9 mm
1000 V rms	14 mm

The latter two comply with 'increased safety – e –', see Table 6.16. The results obtained hereby are given in Figs 6.153 and 6.154.

For all combinations, the flame – ignited by a low energy spark plug – of the gas–air mixture (or better the reaction zone) causes a breakdown with

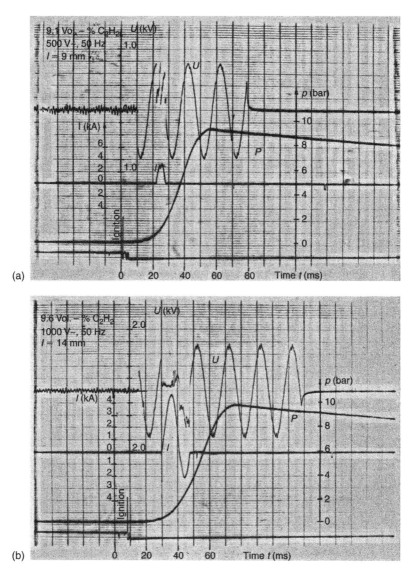

Figure 6.154 *Flame-induced short-circuit arcing over a 1 AC busbar system. Voltage U, arc current I and explosion pressure p versus time.*
(a) 9.1% (v/v) acetylene–air mixture; *U* = 500 V rms; clearance 9 mm
(b) 9.6% (v/v) acetylene–air mixture; *U* = 1000 V rms; clearance 14 mm.

subsequent short-circuit arcing over a time period of one to six half-periods. Even the introduction of clearances according to 'increased safety – e –' standards into flameproof standards would not solve this problem!

What happens in a flameproof enclosure with a short-circuit arc inside? First of all, the power and energy converted in an arc into radiation, heat transfer by convection and melting or vaporizing heat for metals shall be calculated.

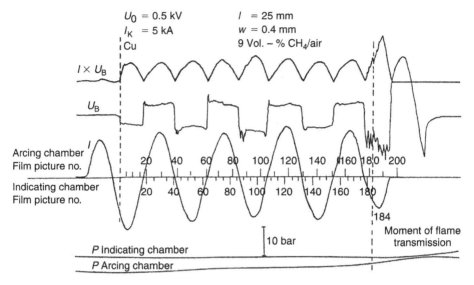

Figure 6.155   *Typical record of short-circuit arcing and the flame transmission initiated hereby. Arc voltage* U_B, *arc current* I *and arc power* I $\times$ U_B *versus time, as well
as explosion pressure* p *(results attained in the apparatus according to Fig. 6.157).
The moment of flame transmission (determined by evaluation of film records of
camera no.1) corresponds to the maximum arc power* I $\times$ U_B.
9% (v/v) CH_4–air mixture; Gap: $w = 0.4$ mm; Joint width: $l = 25$ mm; Voltage: 500 V
rms; Short-circuit current: 5 kA rms; Copper electrodes.

A sinusoidal current:

$$I = I_o \cdot \sin \omega t$$
$$\omega = 2\pi f$$
$$f = \text{AC frequency}$$

forming an arc between metallic electrodes remaining at a fixed clearance
(and not showing a time-dependent movement like the contacts of a
switchgear) is accompanied by an almost rectangular-shaped arc voltage U_B
symmetric to zero voltage (Fig. 6.155). The arc is an ohmic but not time constant load. The power in the arc is:

$$P = U_B \cdot I_o \cdot \sin \omega t$$

and the energy per half-period

$$W = \int_0^{T/2} P \, dt = U_B \cdot I_o \int_0^{T/2} \sin \omega t \, dt$$

with period duration $T = 1/f$.

The integration results in

$$W = U_B \cdot I_o[-1/2\pi f \times \cos 2\pi ft]_0^{T/2}$$
$$= U_B I_o[-1/2\pi f \times \cos 2\pi f \cdot T/2 + 1/2\pi f \times \cos O]$$
$$= U_B I_o[+1/2\pi f + 1/2\pi f]$$
$$= U_B I_o \cdot T/\pi$$

Table 6.36 Caloric values of enclosure and busbar materials

Material	Mean specific heat J/gK	Melting point K	Melting heat J/g	Heat content, liquid kJ/g*	Mean specific heat, liquid J/gK	Boiling point K	Vaporizing heat kJ/g	Heat content, vapour kJ/g*
Aluminium	0.904	931	370	0.95	1.09	2473	10.8	13.43
Iron, steel	0.460	1803	205	0.90	0.75	3300	6.28	8.30
Copper	0.39	1356	180	0.59	0.494	2583	4.77	5.97

*Values referring to +20°C (=293 K)

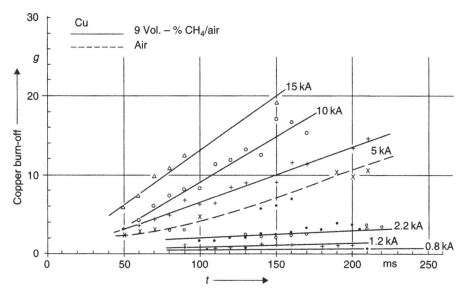

Figure 6.156 *Copper electrode burn-off versus arcing time* t *for short-circuit currents in the range 0.8 kA ... 15 kA rms.*
Straight lines: arc in 9% (v/v) CH_4–air mixture; Dashed line: arc in air.

As an example, in a low voltage AC system with $f = 50$ cps corresponding to $T = 20$ ms, with $U_B = 300$ V and $I_o = 30$ kA, W results in $W = 57.3$ kJ (per half-period!). After 100 ms, the energy release accumulates to 573 kJ.

To give an impression of the significance of these numbers in terms of quantities of melted or vaporized metals, Table 6.36 summarizes specific values for metals used for enclosure and busbar materials. Obviously, the energy release of a half-period arc can melt some 10 grams or vaporize some grams of steel.

For copper, the electrode burn-off versus arcing time has been measured for different short-circuit currents up to 15 kA (root mean square value), see Fig. 6.156.

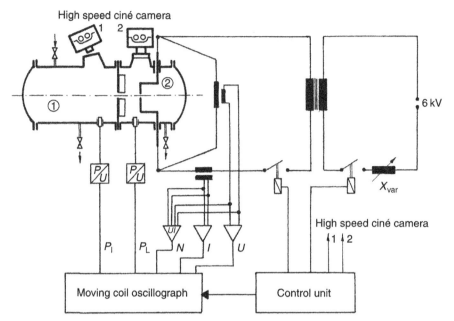

Figure 6.157   *Block diagram of the apparatus for investigations on arc-induced flame transmission according to [56].*
1: indicating chamber; 2: arcing chamber; *P/U*: pressure transducer with signal convertor.

So, one effect of short-circuit arcs inside a flameproof enclosure may be heating or even melting the enclosure material and thus exceeding the surface temperature limitations or burning a hole in the enclosure – far away from the basic safety rules of 'flameproof enclosure – d –'.

A quite different effect is an arc-induced flame transmission through the joints of a flameproof enclosure. To avoid any misunderstanding: these joints do comply with the d-standards and they have been shown to be safe by type testing for the corresponding explosion subgroup and test gas–air mixture. Investigations covering this field have been published in [35] and [56].

According to [56], two cylindrical vessels (no. 1, indicating chamber, 0.2 m³ volume, and no. 2, arcing chamber, 0.1 m³ volume) are flanged together (Fig. 6.157). The flange plate has been fitted with a variable joint (gap, joint width). The electrodes in the arcing chamber have been adjusted at a distance of 100 mm to the joint. Two high speed ciné cameras record the reactions in the chambers, especially camera no. 1 in order to detect a flame transmission into the indicating chamber. (The light emission in the visible part of the spectrum of a burning gas–air mixture is very weak (especially for a hydrogen flame), so that high speed records (some 10^3 pictures/second) are prevented by very short exposure times. However, the experiments with copper electrodes and the burn-off attained hereby dyes the flame intensive green due to the copper emission line at $\lambda = 521.8$ nm.) All relevant electrical values and the explosion pressures have been recorded.

Figure 6.155 represents a typical record. The ignition of the gas–air mixture in the indicating chamber is determined by evaluation of the films of camera no. 1. A number of parameters have been varied, e.g.:

- combustible gas
- electrode material
- short-circuit current
- joint dimensions.

One remarkable result obtained hereby shows that the major part of the energy released from the arc has been transferred by radiation and thermal convection to the surrounding gas–air mixture (or to air). Only a small percentage of the energy has been used for the electrode burn-off (see Table 6.37). Even including the energy obtained by oxidation of the copper burn-off according to

$$Cu + \tfrac{1}{2}O_2 \rightarrow CuO + 155\,kJ/mole$$

($=2.44\,kJ/g$) in the energy balance does not change the picture as a whole. Thus, the main part of the arc energy is transferred to the surrounding gas atmosphere. Some results are summarized in Figs 6.158 and 6.159.

Arc-induced flame transmission through a 'safe' joint (i.e. complying with and type tested according to the d-standards) occurs at gaps far below the values for the maximum allowable gap (see Tables 6.27, 6.28 and 6.29), e.g. for gaps down to 0.3 mm for CH_4–air mixtures and 25 mm joint width. Even arcing in pure air (without any combustibles) initiates a flame transmission to a surrounding CH_4–air mixture (see Figs 6.158(b) and 6.159(b)) at a gap down to 0.35 mm. The borderline:flame transmission to the non-transmitting area is correlated to the peak power $U_B \times I_o$ of the electrical arc at the moment of the flame transmission (defined by evaluation of camera no. 1 records). These records show that Cu particles enter for a certain time (passing the joint) the indicating chamber without any ignition of the gas–air mixture: the same is true for graphite electrodes.

A very different result has been obtained using aluminium electrodes (Fig. 6.160). There is no correlation: safe gap versus arc peak power, but a 'chaotic' distribution of flame transmissions down to very small gap values. This indicates a real 'particle-induced flame transmission' and has been proven by evaluating the films of camera no. 1: the aluminium particles entering the indication chamber act as a very effective ignition source.

The results can be summarized as follows:

- aluminium (or light alloys based upon aluminium) shall not be used as conductor material in flameproof enclosures
- arcing with short-circuit currents causes an ignition of the environmental gas–air mixtures at gap values smaller than the maximum allowed values given in the standards EN 50018 or IEC 60079-1. It seems that the very high power density in the surroundings of the electrical arc initiates the flame transmission. A power of 10^0 to 10^1 MW is concentrated on a 'small' volume estimated at $10 \dots 10^2\,cm^2$, thus forming power densities of $10^5 \dots 10^6\,W/cm^3$.

Table 6.37 Energy distribution for short-circuit arcs (copper electrodes) (according to [56])

RMS-short circuit current kA	Arc energy per half period W kJ*	Gas-air mixture	Electrode burn-off g/half-period	Energy content of electrode burn-off				Oxidation heat of electrode burn-off	
				Liquid		Vaporized		kJ/ half-period	% of W
				kJ	% of W	kJ	% of W		
15.0	33.68	9% (v/v)	1.35	0.796	2.36	8.06	23.9	3.29	9.76
10.0	22.45	CH$_4$ in air	1.01	0.596	2.65	6.03	26.9	2.46	10.96
5.0	11.22	(for all short	0.69	0.407	3.63	4.12	36.7	1.68	14.97
2.2	4.94	circuit	0.15	0.088	1.78	0.89	18.0	0.37	7.49
1.2	2.69	current	0.075	0.044	1.63	0.45	16.7	0.18	6.69
0.8	1.80	values)	0.035	0.021	1.17	0.21	11.7	0.08	4.44

* Arcing voltage $U_B = 250$ V

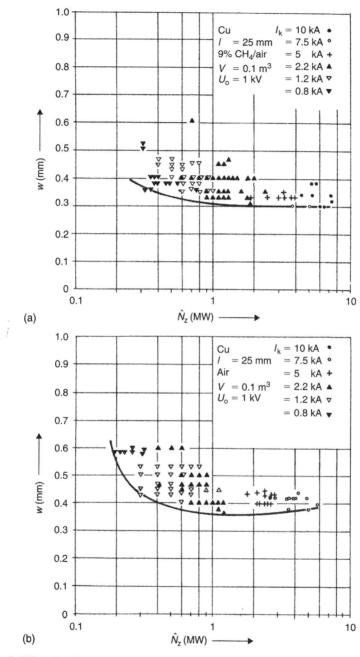

Figure 6.158 *Arc-induced flame transmissions. Gap* w *versus arc peak power* $\hat{N}_z$
at the moment of flame transmission.
Copper electrodes:
Joint width: l = 25 mm; Arcing chamber volume: 0.1 m³; Voltage: 1000 V rms; Short-
circuit currents: 0.8 ... 10 kA rms; (a) Arc burning in 9% (v/v) CH₄–air mixture,
(b) Arc burning in pure air. Indicating chamber filled with 9% (v/v) CH₄–air
mixture.

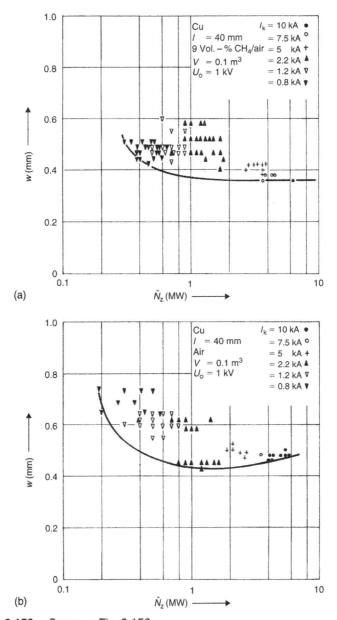

Figure 6.159 *Same as Fig. 6.158.*
Joint width: I = 40 mm, (a) Arc burning in 9% (v/v) CH_4–air mixture, (b) Arc burning in pure air. Indicating chamber filled with 9% (v/v) CH_4–air mixture.

Compared with the values for reacting gas–air mixtures (see Fig. 6.77, Table 6.26) this may explain the flame transmission through joints proven as safe for gas–air reactions.

To decide what provisions should be taken, first of all the risk of such an event – short-circuit arcing inside a flameproof enclosure and, at the same

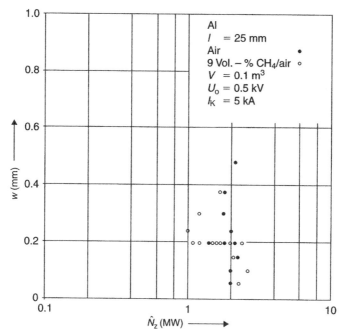

Figure 6.160 *Particle-induced flame transmission, caused by aluminium electrode burn-off. Gap w versus arc peak power $\hat{N}_z$ at the moment of flame transmission.*
Aluminium electrodes:
Joint width: $l = 25$ mm; Arcing chamber volume: 0.1 m^3; Voltage: 500 V rms; Short-circuit current: 5 kA rms. Arc burning in pure air (filled dots) or in 9% (v/v) CH$_4$–air mixtures (blank dots). Indicating chamber filled with 9% (v/v) CH$_4$–air mixture.

time, a combustible gas–air atmosphere outside – can be estimated as low. In spite of risk estimations, technical measures can be taken to avoid arc-induced flame transmissions:

- by reduction of gap values of the enclosure
- by installation of quick sensing and acting protection devices, as described in [6], [27], [57] and Chapter 12.

6.9 Intrinsic safety*

Standards: EN 50020
                 IEC 60079-11
                 (Intrinsically safe electrical systems: EN 50039)
Key letter: i

---

*Sections 6.9.2 to 6.9.6 by Dipl.-Ing. Michael Hagen, Product Manager Instrumentation, R. STAHL Schaltgeräte GmbH, Waldenburg/Germany.

6.9.1   Basic principles of intrinsic safety

To start a chemical reaction between a combustible substance (e.g. a gas such as hydrogen, methane or propane) and oxygen (within the scope of explosion protection as a constituent of air) caused by electrical sparking, a certain amount of spark energy is required. Values are given in Table 1.7 (for high voltage capacitor discharges) and in Table 1.9 (for low voltage circuits). With a sufficient low energy dissipation in electrical sparks a combustible gas–air mixture cannot be ignited. This fact covers intentionally generated sparks in relays and switches, in plug-and-socket connectors when plugging or unplugging an electric circuit or unintentionally generated by short-circuits or wire breaks in electric circuits.

The energy dissipation in sparks at make or break is given by circuit voltage U and current I, by energy storage elements such as capacitors (with an energy $W = 0.5CU^2$) and inductances (with $W = 0.5LI^2$), by the 'power delivery capability' of the power source in the circuit, i.e. by the voltage or current regulator design or the internal impedance of the power source, by the circuit frequency and by damping elements (resistances) in the circuit.

Another main factor which determines the ignition/no ignition limit of a gas–air mixture can be summarized as 'electrical contact design', including the choice of contact materials and the contact velocities during make and break. In long trials over some decades [24], [43], [51], an 'interrupter' (in the terminology of explosion protection: spark test apparatus) with counter-rotating tungsten wires and a cadmium disc has been proven to show the 'lowest' electrical values (voltage, current etc.) to ignite a gas–air mixture. This 'interrupter' (which is used to simulate all types of sparks at make or break in an electric circuit) has been overtaken as a part of all i-standards (IEC 60079-11, EN 50020) and is described in the more recent editions of these standards. Prior editions of IEC 60079-11 made a reference to IEC 60079-3, a special standard defining this apparatus in detail. In this book, the description is given in Section 8.3.

The third main factor determining ignition/no-ignition limits of a gas–air mixture is given by the physical values of this gas–air mixture (gas concentration, temperature, pressure or density). Remembering the strong correlation between MESG (maximum experimental safe gap) values and minimum ignition energy values as shown in Fig. 4.1, decreasing MESG values are correlated to decreasing minimum ignition energy values, and, in consequence, increasing pressure of a gas–air mixture (before ignition!) with decreasing MESG values (see Fig. 1.7) will result in lowered minimum ignition energy values and, indicating the identical physical behaviour, in lowered MIC (minimum ignition current) values according to:

$$\text{MESG (mm)} = 0.022 \times \text{MIC}^{0.87} \text{ (mA)}$$

as reported in [44].

For MIC, it follows:

$$\text{MIC (mA)} = (\text{MESG (mm)}/0.022)^{1.1494}$$

In logarithmic representation, it is:

$$\lg \text{MESG (mm)} = \lg 0.022 + 0.87 \cdot \lg \text{MIC (mA)}$$

or

$$\lg \text{MIC (mA)} = 1.1494 \cdot \lg (\text{MESG (mm)}/0.022)$$

Operating the 'interrupter' in an electric circuit, the ignition values (voltage, current) for a given gas–air mixture show a 'statistical' behaviour, i.e. for a circuit with well-defined electric values there will be no 'sharp' boundary between ignition/no-ignition areas, but a certain ignition probability, defined as the ratio number of ignitable sparks to total number of generated sparks. In other words, by no means will each single spark ignite the gas–air mixture. The reason has been found in the very complex processes during make and break. A typical diagram for the behaviour ignition probability versus current (with fixed values for all other parameters) is given in Fig. 6.161. In order to limit time and financial expenditures when operating the spark test apparatus, the ignition probability has been fixed to 10^{-3} and this is the general reference when determining the 'minimum ignition curves' in electrical circuits.

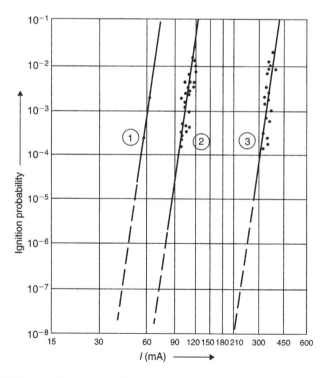

Figure 6.161 *Ignition probability versus current (according to [51]). Methane–air mixture, inductive circuits with:*
curve 1: $L = 300$ mH; curve 2: $L = 95$ mH; curve 3: $L = 9.25$ mH (air core coils).

Table 6.38 Reference gas–air mixtures to be used in the spark test apparatus (at $T = 293\,\text{K}$ and $p = 1.013 \cdot 10^5\,\text{Pa}$)

For Group	Gas	Gas concentration % (v/v) in air
I	Methane, CH_4	8.3 ± 0.3
IIA	Propane, C_3H_8	5.25 ± 0.25
IIB	Ethylene, C_2H_4	7.8 ± 0.5
IIC	Hydrogen, H_2	21.0 ± 2

As a reference the gas–air mixtures shown in Table 6.38 shall be used in the spark test apparatus to determine the minimum ignition curves for Group I and Group IIA/IIB and IIC circuits. The minimum ignition curves obtained hereby are shown in Figs 6.162–6.167, where the area *above* the curves indicates ignition probabilities higher than 10^{-3}, and areas *below* the curves indicate ignition probabilities lower than 10^{-3}. These diagrams may give an impression of the physical limits of ignitability of electric circuits with respect to voltage and current, capacitance and inductance.

In the test circuits, only 'linear' components are used: resistance R, capacitance C and inductance L (as an air core coil), combined with a DC voltage source. Its internal resistance R_i combined with the external elements act as a current limitation. Electronically stabilized voltage (or current) sources show a very different behaviour in the test circuit (Fig. 6.168). With linear elements in the circuit, the voltage–current diagram is a straight line, starting with the no-load voltage U_{NL} at $I = 0$ down to $U = 0$ in the short-circuit condition $I = I_{sc}$ according to $U = U_{NL} - U_{NL}/I_{sc} \cdot I$, with $R_i + R = U_{NL}/I_{sc}$ (left side diagram). The no-load voltage U_{NL} and the short-circuit current I_{sc} are the relevant parameters for the intrinsic safety of this circuit. The maximum output power of this circuit is $P = 0.25 \cdot U_{NL} \cdot I_{sc}$ at the resistance matching condition:

$$R_{Load} = R_i + R \sim R, \quad \text{with} \quad U = 0.5 \cdot U_{NL} \quad \text{and} \quad I = 0.5 \cdot I_{sc}$$

An electronically stabilized power source may show an idealized voltage–current diagram according to the right side curve. The output voltage remains constant $U = U_{NL}$ at all load conditions $I < I_{sc}$, and in the short-circuit-condition $I = I_{sc}$, U drops down to zero. At point A, the output power is $P \sim U_{NL} \cdot I_{sc}$, four times the maximum output power of a linear circuit at identical values for U_{NL} and I_{sc}.

The output power $P = 0.25 \cdot U_{NL} \cdot I_{sc}$ can be achieved by electronic stabilization (dashed line in the left side diagram) with $U = 0.5 \cdot U_{NL}$ and $I = 0.5 \cdot I_{sc}$, defining the intrinsic safety of the circuit (the minimum ignition curves cannot be applied as a reference in this case, however, due to the non-linearity of a stabilized power source).

In resistive circuits, the diagram I versus U at constant power, $P = U \cdot I$, $I = P/U$, should be expected to be hyperbolically shaped, or, in double-logarithmic representation, $\lg I = \lg P - \lg U$, to be a straight line. Obviously,

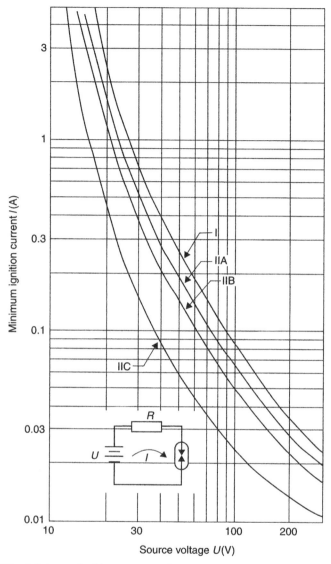

Figure 6.162   *Minimum igniting current* I *(in A) versus source voltage* U *(in V) for Groups I, IIA/B/C. Resistive circuits (according to IEC 60079-11 and EN 50020).*

in Fig. 6.162 there is a considerable deviation, indicating that P is not a constant value. For all groups, $P = U \cdot I$ decreases with increasing voltage, e.g. from $P = 45\,W$ at $U = 20\,V$ to $P = 9\,W$ at $U = 100\,V$ for Group I, and from $P = 9\,W$ at $U = 20\,V$ to $P = 2.5\,W$ at $U = 100\,V$ for Group IIC.

As explained in Fig. 6.168, the 'maximum available output power' fed by a linear voltage source (i.e. a constant internal resistance determines the U–I behaviour) into an external resistive load at the resistance matching condition equals 25% of the power values given above.

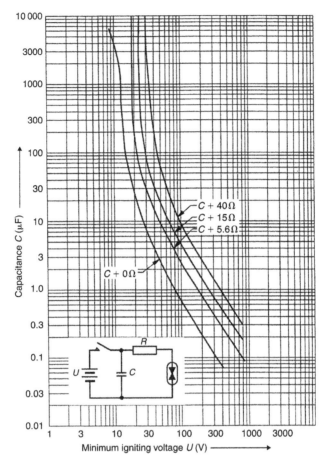

Figure 6.163   *Capacitance* C *(in μF) versus minimum igniting voltage* U *(in V) for Group I. Capacitive circuits (according to IEC 60079-11 and EN 50020). The curves correspond to values of current-limiting resistance as indicated.*

In the same way, capacitive circuits (see Figs 6.163 and 6.164) show curves C versus U deviating from those expected for a constant storage energy:

$$W = 0.5C \cdot U^2 \quad \text{or} \quad C = 2W/U^2 \text{ respectively}$$

or, in double-logarithmic representation:

$$\lg C = \lg 2W - 2 \lg U$$

which describes a straight line. The energy stored in the capacitor decreases from $W = 5 \cdot 10^{-3}$J at $U = 20$ V and $C = 25\,\mu\text{F}$ to $W = 4 \cdot 10^{-3}$J at $U = 200$ V and $C = 0.2\,\mu\text{F}$ for Group I, and, in the same way, from $W = 1 \cdot 10^{-3}$J at $U = 10$ V and $C = 20\,\mu\text{F}$ to $W = 5 \cdot 10^{-5}$J at $U = 100$ V and $C = 10$ nF.

A very opposite behaviour is shown in inductive circuits. Assuming a constant energy $W = 0.5LI^2$ stored in an air core inductance, the curve $L = 2W/I^2$

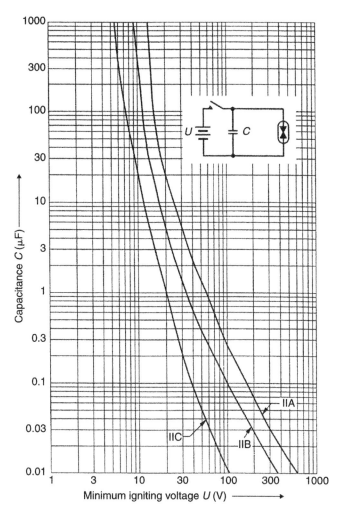

Figure 6.164　*Capacitance C (in μF) versus minimum igniting voltage U (in V) for Groups IIA/B/C. Capacitive circuits (according to IEC 60079-11 and EN 50020).*

is expected to be hyperbolically shaped, or in double-logarithmic representation, $\lg L = \lg 2W - 2\lg I$, to be a straight line. This is exactly what has been measured (see Figs 6.165 and 6.166). The diagrams L versus I show (in the low current region) a straight line, corresponding to a constant energy:

$W = 525\,\mu J$　for Group I
$W = 320\,\mu J$　for Group IIA
$W = 160\,\mu J$　for Group IIB
and $W = 40\,\mu J$　for Group IIC

stored in the inductance.

This is the physical background of the values of minimum ignition energy given in Table 1.9!

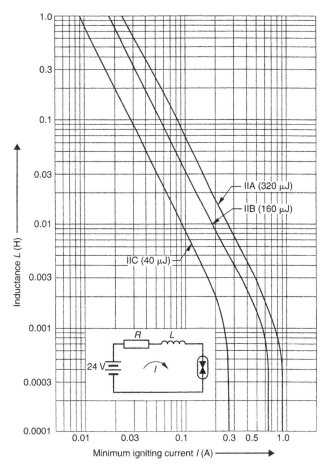

Figure 6.165 *Inductance L (in H) versus minimum igniting current I (in A) for Groups IIA/B/C. Inductive circuits, circuit test voltage 24 V DC (according to IEC 60079-11 and EN 50020). The energy levels indicated refer to the constant energy portion of the curve.*

At higher currents ($>1\,$A for Group I, $>0.1\,$A for Groups IIA/B/C), the curves L versus I are parallel to the L-axis, and so the dependence of I on L disappears completely.

In addition, the curves split up in different 'branches' depending on voltage in the low inductance region (see Figs 6.166 and 6.167). This indicates a transition to pure resistive circuits. The time constants L/R for the transition to pure resistive circuits are in the range $L/R = 2 \cdot 10^{-5}$s (for Group I) to $L/R = 1 \cdot 10^{-5}$s (for Group IIC) or even lower.

Though Figs 6.162–6.167 have been used to give an imagination of physical limiting electric values in electric circuits (in this case, for DC circuits), it should be emphasized that the main objective of these diagrams is that of reference curves in type testing i-circuits (see Section 8.3) and that in practice the allowable electrical values are reduced by safety factors, e.g. 1.5.

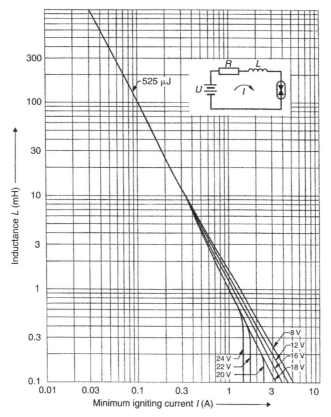

Figure 6.166 *Inductance* L *(in mH) versus minimum igniting current* I *(in A) for Group I. Inductive circuits, circuit DC voltages as indicated (according to IEC 60079-11 and EN 50020). The energy level of 525 μJ refers to the constant energy portion of the curve.*

The reduction of, e.g., current values to $\times 1.5^{-1} = 2/3$ lowers the ignition probability (10^{-3} in the diagrams) by some orders of magnitude (see Fig. 6.161).

A short survey of voltage and current limitation in resistive circuits and for restriction of capacitances and inductances is given in Table 6.39.

So the first main feature of intrinsic safety is the limitation of voltage and current in the circuits, as well as the limitation of stored energy in capacitances or inductances.

A very generalized principle of an intrinsically safe electric circuit is shown in Fig. 6.169. Power source, voltage and current limitation are located in a safe area or shall be explosion protected (e.g. in a flameproof enclosure) if located in a hazardous environment. The electric circuit entering the hazardous area as an intrinsically safe circuit is not capable of producing ignitable sparks at make or break.

In addition, surface temperatures shall be limited to prevent any ignition of the potentially explosive atmosphere. All components (including the wiring)

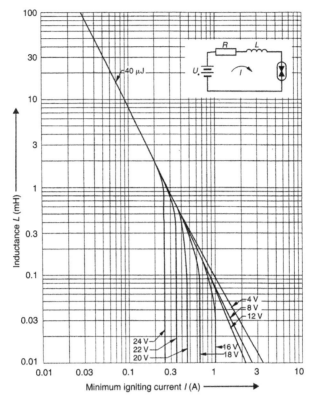

Figure 6.167   *Inductance L (in mH) versus minimum igniting current I (in A) for Group IIC. Inductive circuits, circuit DC voltages as indicated (according to IEC 60079-11 and EN 50020). The energy level of 40 µJ refers to the constant energy portion of the curve.*

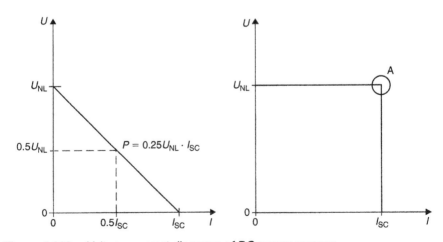

Figure 6.168   *Voltage–current diagrams of DC power sources.*
Left side: unstabilized power source with resistance $R = U_{NL}/I_{SC}$, no-load voltage U_{NL}, short-circuit current I_{SC}.
Right side: electronically stabilized power source with constant output voltage U_{NL} and current limitation at I_{SC}, maximum output power at point A.

Table 6.39 Examples for limitation of electrical values in intrinsically safe circuits for Group IIB and IIC (including a safety factor of 1.5) (by courtesy of R. STAHL Schaltgeräte GmbH, Waldenburg/Germany)

Resistive circuits			*Capacitive circuits*			*Inductive circuits*		
U V	*I mA* IIC	*I mA* IIB	*U V*	*C nF* IIC	*C nF* IIB	*I mA* (at 24 V)	*L mH* IIC	*L mH* IIB
16	687	1700	16	460	2750	80	6	22
18	440	1110	18	309	1780	90	5	18
20	309	785	20	220	1410	100	4	15
22	224	575	22	165	1140	120	2.5	10
24	174	433	24	125	930	140	1.6	8.0
26	143	357	26	99	770	170	0.8	5.5
28	120	299	28	83	650	200	0.5	4.0
30	101	253	30	66	560	300	0.2	1.8
32	87.8	216	32	56	475	400	0.15	0.8
34	78	187	34	48	406	500	0.1	0.2

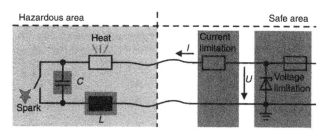

Figure 6.169 *Basic principle of an intrinsically safe electrical circuit.*

of intrinsically safe electrical circuits shall be assumed to be exposed to a hazardous environmental atmosphere and so the temperature limitations according to the temperature class rating are to be met. So, well-defined relations wire diameter cross-section to maximum permissible continuous currents are given for the T rating of (copper) wiring (Table 6.40) and for printed board wiring (Table 6.41). Compared to the current densities (current per cross-sectional area) common in electrical power engineering – some 10^0 A/mm^2 up to 10^1 A/mm^2 – the values in Table 6.40 (550 A/mm^2 to 34 A/mm^2) and in Table 6.41 (230 A/mm^2 to 90 A/mm^2) seem to be surprisingly high. The 'solution' is an increasing ratio of cooling conductor surface to conductor volume (which causes ohmic losses) for 'thin' conductors.

For a cylindrical conductor with length l and radius r, the cross-sectional area is πr^2 and the total volume $V = \pi r^2 \cdot l$. The surface S is $2\pi r l$, and so the ratio S/V follows with $2/r$. Diameters from 0.035 mm to 0.5 mm, i.e. a ratio 0.07:1, will be loaded with inverse current density ratios, in this example 16, 17, corresponding to 1:0.062.

Table 6.40 Temperature class rating of copper wiring (referring to an ambient temperature of +40°C), according to EN 50020 and IEC 60079-11

Diameter mm	Cross-sectional area mm^2	Maximum permissible continuous current* for T class... A		
		T1–T4, Group I	T5	T6
0.035	$9.62 \cdot 10^{-4}$	0.53	0.48	0.43
0.05	$1.96 \cdot 10^{-3}$	1.04	0.93	0.84
0.1	$7.85 \cdot 10^{-3}$	2.1	1.9	1.7
0.2	$3.14 \cdot 10^{-2}$	3.7	3.3	3.0
0.35	$9.62 \cdot 10^{-2}$	6.4	5.6	5.0
0.5	$1.96 \cdot 10^{-1}$	7.7	6.9	6.7

* For AC currents, the rms value shall be referred to

Table 6.41 Temperature class rating of printed board wiring (referring to an ambient temperature of +40°C), according to EN 50020 and IEC 60079-11 (Cu single layers of 35 μm thickness, thickness of printed board ⩾1.6 mm)*

Minimum track width mm	Maximum permissible continuous current** for T class... A		
	T1–T4, Group I	T5	T6
0.15	1.2	1.0	0.9
0.2	1.8	1.45	1.3
0.3	2.8	2.25	1.95
0.4	3.6	2.9	2.5
0.5	4.4	3.5	3.0
0.7	5.7	4.6	4.1
1.0	7.5	6.05	5.4
1.5	9.8	8.1	6.9
2.0	12.0	9.7	8.4
2.5	13.5	11.5	9.6
3.0	16.1	13.1	11.5
4.0	19.5	16.1	14.3
5.0	22.7	18.9	16.6
6.0	25.8	21.8	18.9

* Correction factors for the continuous current values are specified in case of:
- lower board thickness
- conducting tracks on both sides
- multilayer boards
- deviating track thickness
- tracks passing under components dissipating 0.25 W or more.

Track lengths ⩽10 mm shall be disregarded for T class rating
** For AC currents, the rms value shall be referred to

Printed board wiring shows a rectangular-shaped conductor cross-section. With layer thickness d and track width w, the cross-sectional area is $w \cdot d$ and the surface of a track with length l follows with:

$$S = 2 \cdot w \cdot l + 2 \cdot d \cdot l$$
$$S = 2l(w + d)$$

In most cases, $w \gg d$, and so the surface is $S = 2lw$. A cylindrically shaped conductor with radius r and length l shows a cross-sectional area πr^2 and a surface $S = 2\pi r l$. For a cross-sectional area equivalent to the printed board wiring, the radius r can be calculated as follows:

$$w \cdot d = \pi r^2$$
$$r = (w \cdot d / \pi)^{0.5}$$

The surface is $S = 2\pi l (w \cdot d / \pi)^{0.5}$.

Then the surface ratio: surface of printed board wiring (rectangular) to surface of cylindrical conductor (wire), S_{print}/S_{wire}, can be calculated:

$$S_{print}/S_{wire} = 2wi \cdot (2\pi l)^{-1} \cdot (w \cdot d / \pi)^{-0.5}$$
$$= w/\pi \cdot \pi^{0.5} \cdot (wd)^{-0.5}$$
$$S_{print}/S_{wire} = (w/\pi \cdot d)^{0.5}, \text{ so}$$

S_{print}/S_{wire} is proportional to $(w/d)^{0.5}$ at constant cross-sectional area.

So, broad tracks as printed board wiring show increased permissible current densities: a wire 0.5 mm in diameter (cross-sectional area: 0.196 mm^2) is limited to 7.7 A or 39.3 A/mm^2 respectively (for T1...T4 rating), whereas a 5×0.035 mm^2 layer (0.175 mm^2) is limited to 22.7 A or 129.7 A/mm^2 respectively, and a 6×0.035 mm^2 layer (0.210 mm^2) is limited to 25.8 A or 122.9 A/mm^2 respectively for the same T rating.

The T rating of explosion protected apparatus is based upon the classification according to Table 4.1, referring to experimentally determined ignition temperatures of combustible gas (vapour, mist)–air mixtures (see Table 1.3). The maximum surface temperature is limited to a certain value, e.g. +135°C for class T4, in general below the ignition temperature of the gas–air mixture forming the environmental atmosphere at the place of duty of the electrical apparatus. A significant 'margin of safety' is obtained by the method of test for ignition temperature (given in IEC 60079-4), especially caused by the large surface exposed to the mixture to be tested. Thus, 'small' surfaces (i.e. smaller than 10 cm^2) show increased ignition temperatures [49]. And consequently, 'small' surfaces may exceed the T classification limitations (this is a general rule, given in EN 50014 and IEC 60079-0, but its major field of application has been found in electronic components as parts of intrinsically safe electrical circuits), see Table 6.42.

As a second main feature of intrinsic safety, current densities in 'thin' internal wires and printed board wiring may increase up to some 10^2 A/mm^2, and the surface temperature of 'small' components may exceed the limits of T classification by far.

Due to these limitations, the fields of application of intrinsic safety – i – are mainly:

- data transmission and signalling, including wireless communication (Figs 6.170–6.172, showing components of a remote control and monitoring system (including communication) for coal mines (underground) – 6.170(a)–(j) stationary installations, 6.171(a) and (b) vehicle-mounted components, 6.172 portable transmitter)

Table 6.42 T4 classification of 'small' components* (according to EN 50020 and IEC 60079-11)

Total surface areas (excluding lead wires)	T4, Group I classification requirements
$S < 20\,\text{mm}^2$	Surface temperature $\leqslant +275°C$
$S \geqslant 20\,\text{mm}^2$	Power dissipation $\leqslant 1.3\,\text{W}^{**}$
$20\,\text{mm}^2 \leqslant S < 10\,\text{cm}^2$	Surface temperature $\leqslant +200°C$

* For T5 classification, the surface temperature shall not exceed +150°C for
  $S < 10\,\text{cm}^2$
** Referring to an ambient temperature of +40°C

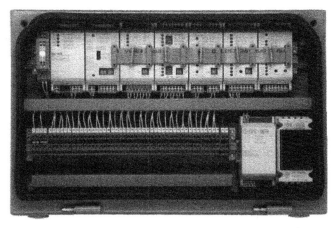

Figure 6.170(a) *Process data station. A housing (up to height H/width W/depth D H = 700 × W = 700 × D = 210 mm³) contains:*
- CPU (32 kbyte program memory, 8 kbyte data memory)
- I/O modules (8 fold, optocouplers or relays)
- Diagnostic module
- Interface modules (level conversion TTL/RS 232, TTL/TTY
  U/I conversion:
  0–1 V/0–1 mA, 0–20 mA, 4–20 mA, 0–10 V)
- Power supply:
  Fieldbus: 2 wire; Max. length: approx. 20 km; Transmission rate: up to 19 200 bit/s;
  Frequencies: 1.6/2.4 kcps, 4.8/6.4 kcps, 6.4/9.6 kcps, 25.4/38.4 kcps; Max.: 256
  stations; Type of protection: EEx ia I/EEx ib I; Certificate: BVS 92.C.1059 X.

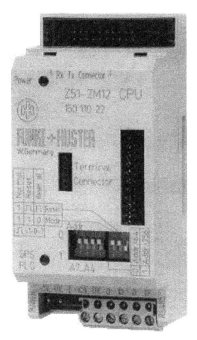

Figure 6.170(b)   *Central module (CPU), to be clipped on a 35 mm standard rail. With a microprocessor, RAM and EPROM/OTP, it coordinates, controls and monitors all the functions of a station.*

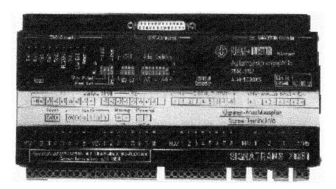

Figure 6.170(c)   *Compact automation unit for:*
- switch point control systems
- packing material supply systems
- shaft signal systems
- floor conveyor control systems.

The instruction list of the programmable controller is stored in an EEPROM: 3800 instructions, with:
- 1024 flag bits
- 256 flag words
- 64 timers
- 32 count values.

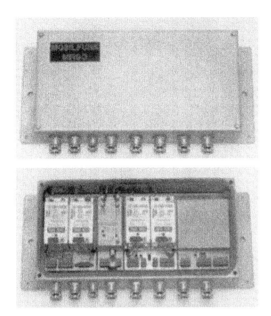

Figure 6.170(d) *Fixed station of a multichannel voice and data radio system.*
Type of protection: SYST EEx ia I/SYST EEx ib I; Certificate: BVS 92.C.1212 X;
Transmissions bands: between 27 Mcps and 35 Mcps; Modulation: Frequency
modulation, 21 channels for transmission band, 21 channels for receiving band;
Frequency distance between channels: 50 kcps; 27 Mcps band: 26.50–27.50 Mcps;
35 Mcps band: 34.50–35.50 Mcps; Dimension: Height × width × depth, 220 × 510 ×
90 mm³.

Figure 6.170(e) *Pressure chamber loudspeaker.*
Rated continuous power: 20 W; Frequency range: 0.6–10 kcps; Type of protection:
EEx ia I; Certificate: BVS 92.C.1130; Dimensions: Height × width × depth,
129 × 110 × 96 mm³; Weight: approx. 2.3 kg.

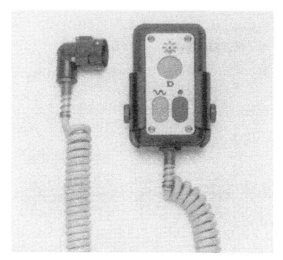

Figure 6.170(f)   *Microphone, electret type.*
Operational voltage: 1.5–10 V; Operation current: 300 μA at 1.5 V; Frequency range:
100 cps–5 kcps; Typical output voltage: 100 mV; Type of protection: EEx ia I;
Certificate: BVS 92.C.1120; Dimensions: Height × width × depth,
116 × 60 × 30 mm³; Weight: 0.55 kg including retaining clip.

Figure 6.170(g)   *Amplifier station for an active or passive radiating cable (leaky
feeder) branching. Optional equipment:*
• amplifier module:
  Amplification: +20 dB; Power input, rated voltage: 12 V DC; Operational current:
  35 mA; Type of protection: EEx ia I; Certificate: BVS 93.C.1136 U
• attenuation element:
  Frequency range: DC...100 Mcps; Dissipated power: <1 W; Impedance: 50 ohms;
  Attenuation: −3/6/9 dB.

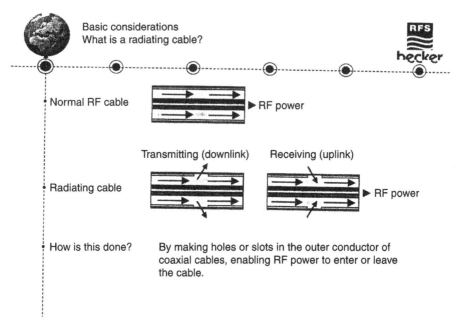

Figure 6.170(h)   *Operational principle of a radiating cable (leaky feeder).*

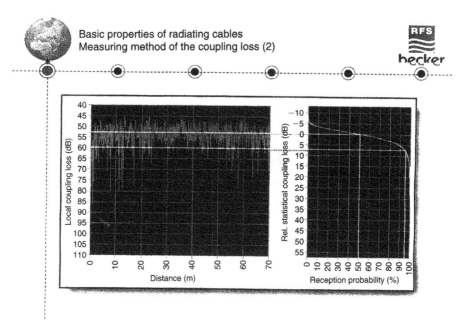

Figure 6.170(i)   *Basic properties of radiating cables (leaky feeders).*

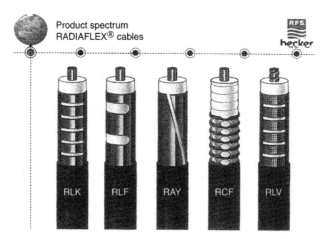

Figure 6.170(j) *Types of radiating cables (leaky feeders).*
Technical data:
Type RLK:
Outer diameter range: ½″–1⅝″; Frequency range: wide band: 30–1950 Mcps; ultra wide band: … 2600 Mcps
Applications:
• broadband;
• low coupling loss at 900/1800/2200 Mcps.
Type RLF:
Outer diameter range: ⅝″–1⅝″; Frequency range: standard: 75–1000 Mcps; wide band: 75–2000 Mcps
Applications:
• robust, low loss
• broadband.
Type RAY:
Outer diameter range: ⅝″–1⅝″; Frequency range: standard: 30–1000 Mcps; ultra wide band: … 2400 Mcps
Applications:
– very low coupling loss
Type RCF:*
Outer diameter range: ⅜″–1⅝″; Outer conductor: milled corrugated copper (quasi continuous slot); Frequency range: 30–2400 Mcps
Type RLV:
Outer diameter: 1¼″, 1⅝″; Outer conductor: Slot groups of increasing density, decreasing coupling loss compensates rising longitudinal loss at 450/900 Mcps; Frequency range: standard: 75–1000 Mcps; wide band: 450–2060 Mcps; ultra wide band: 900–2400 Mcps; Applications: system loss hardly varies along the cable.

* The radiating cable (leaky feeder) of the remote control and monitoring system described in Fig. 6.170 shows the following technical data:
Type of cable: RCF; Natural impedance: 50 ohms
Resistance:
• inner conductor 1.48 ohms/km
• outer conductor 1.90 ohms/km.
Attenuation (at 30 Mcps): 11.5 dB/km; Outer diameter: 16 mm; Weight: approx. 0.26 kg/m.

Fig. 6.171(a) *Mobile station (e.g. on a locomotive) containing a transmitter module, a receiver module and an antenna adaption module.*
Transmitter module:
Frequency range: 26.5–27.5 Mcps; 34.5–35.5 Mcps; Number of channels: 21; HF output: +8 dBm; Low frequency input: −6 dBm/600 ohms; Low frequency range: 300–3400 cps; Type of protection: EEx ia I; Certificate: BVS 92.C.1207 U
Receiver module:
Frequency range: 34.5–35.5 Mcps; 26.5–27.5 Mcps; Number of channels: 21; Bandwidth/6 dB: approx. 14 kcps; Low frequency output: −6 dBm/600 ohms; Low frequency bandwidth: 0.3–3.4 kcps; Type of protection: EEx ia I; Certificate: BVS 92.C.1208 U
Antenna adaption module:
Transmitting frequency range: 26.5–27.5 Mcps; Receiving frequency range: 34.5–35.5 Mcps; Rated output power: 50 mW/channel; Max. output power: 500 mW; Rated input voltage: up to 2.23 mV (−40 dBm); Max. input voltage: 7.07 mV (−30 dBm); Power supply: 12–15 V DC; Rated current: 100 mA; Type of protection: EEx ia I; Certificate: BVS 92.C.1206 U
Enclosure:
Dimensions: Height × width × depth, 220 × 201 × 90 mm^3; Weight: approx. 5.5 kg (without internal components).

Figure 6.171(b) *Antenna of the mobile station.*
Type: $\lambda/4$, shortened; Frequency range: 26.5–27.5 Mcps and 34.5–35.5 Mcps; Wave impedance: 50 ohms; Operational voltage: >10 V; Power: 1 W; Type of protection: EEx ia I; Certificate: BVS 92.C.1193; Dimensions: overall length: 450 mm; Weight: approx. 0.7 kg.

Figure 6.172   *Portable transceiver.*

Frequency range	27/35 Mcps
Number of channels	21
Modulation	frequency modulation
Transmitting power	50 mW
Sensitivity (receiver)	1 μV
Type of protection	EEx ia I
Certificate	BVS 94.C.1102
Weight	approx. 1.5 kg (including accumulator).

- remote control and monitoring (in many cases, i-circuits form part of an extended data network), see Figs 6.7 and 6.173–6.179 (some sensors described in these figures operate in liquids with a relative dielectric constant ε_r deviating considerably from that of air or water – a survey is given in Table 6.43)
- power supply of small sensors (Figs 6.180–6.182 – showing inductive sensors, where, during undisturbed operation, an internal oscillator picks up a defined current: when a metallic part enters the sensitive area, the magnetic field of the oscillator is damped hereby, the oscillations break off and the supply current decreases – and Figs 6.183–6.187)
- lighting with low demands for illuminance and/or luminous flux, e.g. for escape routes and emergency exits (see Figs 6.8 and 6.188)
- hand-held instrumentation (Figs 6.189 and 6.190).

Conductive parts of an intrinsically safe circuit shall be separated from those of non-intrinsically safe circuits or from other i-circuits in order to strictly prevent any voltage transfer or current displacement. Separation elements are clearances (in air), distances through a casting compound or a

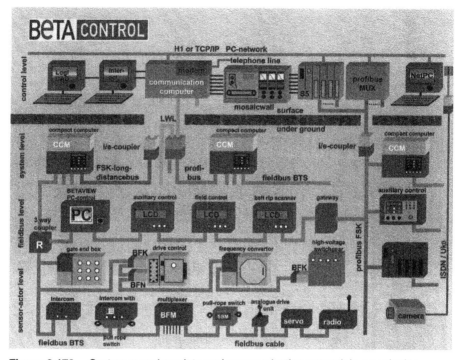

Figure 6.173 *System overview: data and communication network in a coal mine.*
Group I intrinsically safe circuits are located at the sensor–actuator level (multiplexers, pull-rope switches, analogue drive units, wireless communication) and at the fieldbus level (low and high voltage switchgear, drive controls and frequency convertors, control panels, belt rip detector/scanner). At the system level (underground) and at surface installations (control level), the i-circuits are linked (in parts) to non-i-circuits via so-called i/e-couplers or to fibre-optic data lines (LWLs). The relative high voltage and power level of Group I i-circuits (see Figs 6.162–6.167) enables long transmission distances.

solid insulation, creepage distances in air or under coating and partitions. Figure 6.191 gives examples for the separation of conductive parts.

A defined separation shall be ensured between:

- primary and secondary windings of a transformer
- earthed metallic parts and parts of an unearthed i-circuit
- the terminals of components relevant to current limitation, e.g. a resistor
- parts of different supply circuits (not in itself intrinsically safe)
- parts of different i-circuits.

In Table 6.44, separation elements are listed depending on the voltage between parts to be separated.

For galvanically separated circuits within the apparatus, the value of voltage to be considered shall be the highest voltage that can appear across the separation when the two circuits are connected together at any one point, derived from:

- the rated voltages of the circuits or

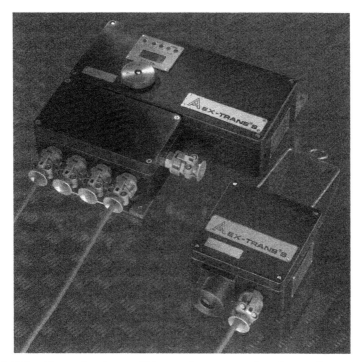

Figure 6.174   *Group I, category M1 – methane transmitter for continuous monitoring of the CH$_4$ concentration in coal mines. Catalytic oxidation. The heat generated hereby causes a temperature rise in a branch of a resistor bridge.*
Type of protection: EEx ia I; Marking: Ex I M1 EEx ia I; Certificate: DMT 99 ATEX E 009; Measuring range: 0–3% (v/v) CH$_4$; Ambient temperature range:
$-20°C \leqslant T_{amb} \leqslant +55°C$; Power supply: 9–14 V, 380 mA
Dimensions ($H \times W \times D$):
• sensor head $250 \times 160 \times 90\,mm^3$
• transmitter $355 \times 360 \times 90\,mm^3$.
Weight:
• sensor head   1.9 kg
• transmitter   4.0 kg.
Signal outputs (corresponding to 0–3% (v/v) methane):
• analogous   10–100 mV
• frequency shift 6–15 cps.

• the maximum voltages (specified by the manufacturer) which may safely be supplied to the circuits or
• any voltages generated within the same apparatus.

For sinusoidal voltages the peak voltage is $\sqrt{2}\,\times$ rms value of the rated voltage.

Between parts of a circuit, the value of voltage to be considered shall be the maximum peak value of the voltage that can occur in either part of that circuit. This may be the sum of the voltages of different sources connected to that circuit. (One of the voltages may be ignored if it is less than 20 per cent of the other.)

Separations conforming to Table 6.44 are considered not to be subject to failure to a lower insulation resistance.

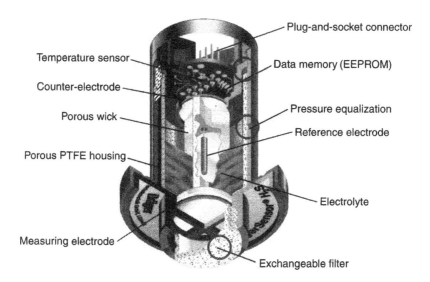

Figure 6.175   *Electrochemical sensor head for oxidizable or reducible gases. Electrons generated at a measuring electrode by an electrochemical reaction form a current indicating the gas concentration.*
Type of protection: EEx ia I; EEx ia IIC T4/6; Certificates: INERIS 97.D.7020 X for Group I; BVS 94.C.2032 for Group II
Gases and measuring ranges (short survey):
NH_3: 0–100 ppm; 0–1000 ppm; CO: 0–100 ppm; 0–1000 ppm; HCN: 0–20 ppm; H_2S: 0–20 ppm; 0–500 ppm; O_2: 0–25% (v/v)
Ambient temperature range:
$-40° \leq T_{amb} \leq +40°C$ for T6 rating
$-40° \leq T_{amb} \leq +65°C$ for T4 rating
Power supply: 16.5 V–30 V DC; Dimensions ($H \times W \times D$) (including transmitter): $130 \times 210 \times 92$ mm^3; Weight (including transmitter): 1.8 kg;
Signal outputs:
- analogous   4 mA–20 mA
- digital   HART.

Following the philosophy of separation, terminals for intrinsically safe circuits shall be separated from terminals for non-i-circuits, e.g.:

- the clearance between i- and non-i-terminals shall be at least 50 mm (see Fig. 6.192) or
- partitions used to separate i- and non-i-terminals shall extend to within 1.5 mm of the enclosure walls or
- they shall provide a minimum distance of 50 mm between these terminals in any direction around the partition (see Fig. 6.193) or
- metal partitions (at least 0.45 mm thick) between the terminals shall be earthed or
- non-metallic insulating partitions (at least 0.90 mm thick) shall be used.

Plugs and sockets used for connection of external i-circuits shall be separate from and non-interchangeable with those for non-i-circuits.

(a)

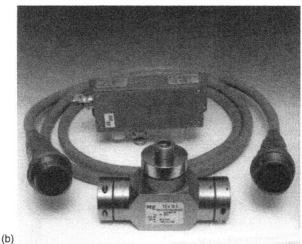

(b)

Figure 6.176 *(a) Detection system for ferromagnetic parts. The antenna is a closed cable loop (total loop length 8 m) embracing the conveyor belt, fed with an AC voltage at 250–2500 cps. Ferromagnetic parts entering the antenna loop area cause a frequency shift and start an alarm.*
Type of protection: EEx ib I; Certificate: BVS 96.D.1146; Supply voltage: 10–12 V DC; Supply current: <100 mA; Ambient temperature range: $-20°C \leqslant T_{amb} \leqslant +70°C$
(b) Antenna with adapter and monitoring unit.

Whereas the need for separation of i-circuits from non-i-circuits is evident, separation between or interconnections of different i-circuits shall be explained in more detail (Fig. 6.194). In a simple model, two intrinsically safe DC power supplies with no-load voltages U_{10} and U_{20} and constant

Figure 6.177 *Two-wire transmitter for pH/redox or conductivity measurements.*
pH-measuring range: pH 2.00–16.00; Resolution: pH 0.01; Automatic temperature
compensating range: $-20°C–+150°C$; Redox measuring range: $-1500–+1500\,mV$;
Resolution: 1 mV; Conductivity measuring range: 0–2000 mS/cm;
Resolution: 0.1 μS/cm; Measuring frequency: 2 kcps;
Technical data:
Output signal: 4–20 mA; Load: max. 600 ohms;
HART signal output:
Auxiliary power: $+12–+30\,V\,DC$; Power consumption: max. 700 mW; Measured
value display: LCD; Ambient temperature range: $-10°C–+55°C$;
Power supply and signal circuit:
Type of protection: EEx ib IIC T4;
Sensor circuit:
Type of protection: EEx ia IIC T4; IP protection code: IP 65; Dimensions
$(H \times W \times D)$: 233 × 103 × 137 mm^3; Weight: approx. 1.25 kg.

internal resistances R_1 and R_2 shall be connected in series by faulty wiring.
The highest voltage at the output terminals will be $U_{tv} = U_{10} + U_{20}$, the
short-circuit current at the terminals:

$$I_{sc} = (U_{10} + U_{20})(R_1 + R_2)^{-1}$$

The parameters relevant to intrinsic safety, no-load voltage and short-circuit current of the stand-alone circuits are U_{10} and $I_{sc1} = U_{10}/R_1$ or U_{20} and $I_{sc2} = U_{20}/R_2$, respectively. In series connection, U_{tv} may exceed the voltage limitation of, e.g., a capacitive circuit. The short-circuit current I_{sc} can be divided into:

$$I_{sc} = U_{10}(R_1 + R_2)^{-1} + U_{20}(R_1 + R_2)^{-1}$$

with $\qquad I_1 = U_{10}(R_1 + R_2)^{-1}, I_2 = U_{20}(R_1 + R_2)^{-1}$

$$I_{sc} = I_1 + I_2$$

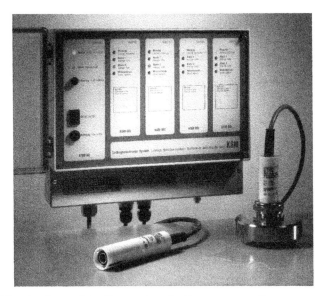

Figure 6.178 *Leakage detection system for flange connections, pump seals or container walls (max. four sensor heads). The capacitive sensor head (normally exposed to air) shows an increased capacitance when plunged in a liquid with a relative dielectric constant $\varepsilon_r > 1$. (For values of ε_r, see Table 6.43.) This causes a frequency shift in an oscillator circuit and hereby generates an alarm.*
Type of protection: EEx ib IIB T6; Certificate: PTB Ex-90.C.2145; Speed of response: <5 s; $U_o = 28$ V; $I_{sc} = 110$ mA; $P = 0.8$ W
Dimensions of sensor head:
Diameter: 25 mm; Length: 110 mm
(The monitoring unit shall be installed in a non-hazardous area).

following the relations:

$$I_1 = U_{10}(R_1 + R_2)^{-1} < U_{10}/R_1 = I_{sc1}(R_2 > 0)$$

and

$$I_2 = U_{20}(R_1 + R_2)^{-1} < U_{20}/R_2 = I_{sc2}(R_1 > 0)$$

resulting in

$$I_{sc} < I_{sc1} + I_{sc2}$$

or

$$I_{sc} = I_{sc1}(1 + R_2/R_1)^{-1} + I_{sc2}(1 + R_1/R_2)^{-1}$$

In the special case of $U_{10} = U_{20}$ and $R_1 = R_2$, it is $I_{sc} = I_{sc1} = I_{sc2}$, but in general, I_{sc} may exceed I_{sc1} or I_{sc2}. The limitation for I_{sc} can be calculated as follows:

Assuming

$$I_{sc1} \geq I_{sc2},$$
$$I_{sc2}/I_{sc1} \leq 1$$

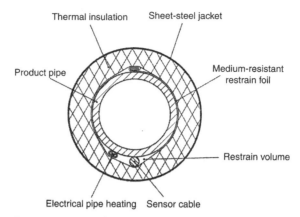

Figure 6.179   *Cross-sectional view of an integrated leakage detection system for product pipelines (electrical pipe heating with thermal insulation and sensor cable).* Medium: organic/inorganic liquids

A sensor cable as a 'stretched' capacitor normally exposed to air increases its capacitance when soaked with a liquid showing a relative dielectric constant $\varepsilon_r > 1$. (For values of ε_r, see Table 6.43.) This initiates a frequency shift in an oscillator circuit and hereby generates an alarm.

Type of protection for the monitoring unit, to be installed in a non-hazardous area: [EEx ib] IIB

Sensor cable: $U_o = 15.4\,V$; $I_o = 23.1\,mA$; Max. cable capacitance: $C_o = 2.1\,\mu F$; Max. external inductance: $L_o = 210\,mH$;

Max. sensor cable lengths (according to type): 10–600 m

Ambient temperature range: $-40°C \leqslant T_{amb} \leqslant +80°C$; Cable impedance: $Z = 50$ ohms; Cable capacitance: $\sim 84\,pF/m$ (in air); Cable diameter: 8–10 mm.

Table 6.43   Relative dielectric constant ε_r (for DC, temperature range $+20°C$–$+25°C$) – for vacuum, $\varepsilon_r = 1.0$ (exactly)

Substance	Chemical formula	ε_r
Air (at $1.013\,25 \cdot 10^5\,Pa$)	–	1.000 59
Acetaldehyde	CH_3CHO	14.8
Acetone	CH_3COCH_3	21.5
Ammonia	NH_3	16.9
Benzaldehyde	C_6H_5CHO	17.5
Benzene	C_6H_6	2.28
Benzine/Petrol/Fuel oil	–	2.20
Diethyl ether	$(C_2H_5)_2O$	4.34
Ethanol	C_2H_5OH	25
Ethylbenzene	$C_6H_5C_2H_5$	2.4
Formic acid	$HCOOH$	57.9
Hexane	C_6H_{14}	1.89
Hydrocyanic acid	HCN	116
Methanol	CH_3OH	33.7
Toluene	$C_6H_5CH_3$	2.38
Water	H_2O	80.3

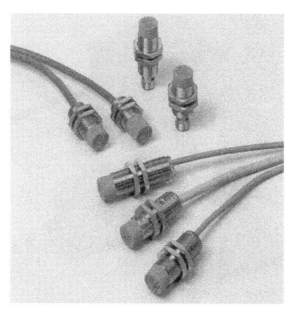

Figure 6.180 *Cylindrical-type inductive sensor.*
Supply voltage: 8 V DC; Current: 3 mA (object not detected); 1 mA (object detected);
Switching frequency: 200–1000 cps; Self-inductance: 44–50 μH; Self-capacitance:
65–89 nF; IP protection code: IP 67 or 68; Ambient temperature range:
$-25°C \leqslant T_{amb} \leqslant +100°C$.

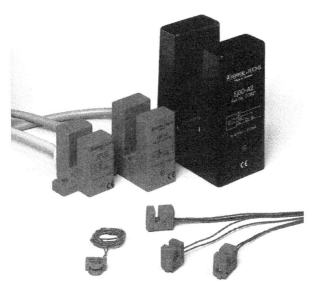

Figure 6.181 *Slot-type inductive sensor.*
Supply voltage: 8 V DC; Current: 3 mA (object not detected); 1 mA (object detected);
Switching frequency: 150–5000 cps; Self-inductance: 100–1500 μH;
Self-capacitance: 30–150 nF; IP protection code: IP 67; Ambient temperature
range: $-25°C \leqslant T_{amb} \leqslant +70°C$ or $+100°C$.

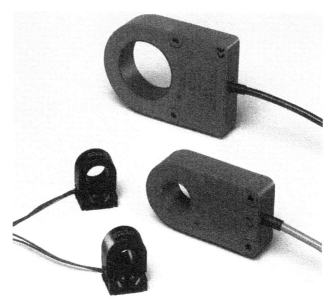

Figure 6.182   *Annular-type inductive sensor.*
Supply voltage: 8 V DC; Current: 3 mA (object not detected); 1 mA (object detected); Switching frequency: 500–2000 cps; Self-inductance: 10–45 μH; Self-capacitance: 25–30 nF; IP protection code: IP 67; Ambient temperature range: $-25°C \leqslant T_{amb} \leqslant +70°C$.

Figure 6.183   *One-way light barrier (transmitter/receiver separated).*
Type of protection: EEx ia IIC T6; Certificate: PTB Ex-97.D.2264; Maximum optical path length: 30 m; Light wavelength: 660 nm (red)
Transmitter:
Supply voltage: 6–16 V DC
Receiver:
Supply voltage: 8 V DC; Current: $\geqslant 2.2$ mA (object not detected); $\leqslant 1$ mA (object detected); Switching frequency: 100 cps; IP protection code: IP 64; Ambient temperature range: $-25°C \leqslant T_{amb} \leqslant +70°C$.

Figure 6.184 *Reflection light barrier (transmitter/receiver integrated, folded optical path).*
Type of protection: EEx ia IIC T6; Certificate: PTB Ex-93.C.2118; Sensitive range: 2 m; Light type: wavelength 660 nm (red); Supply voltage: 6–20 V DC; Current: $\geqslant 2.7$ mA (object not detected); $\leqslant 1$ mA (object detected); Switching frequency: 100 cps; IP protection code: IP 67; Ambient temperature range: $-25°C \leqslant T_{amb} \leqslant +70°C$.

Figure 6.185 *Incremental encoder. A shaft-mounted code disc with copper segments or transparent zones is sampled by slot-type inductive or optical sensors. Per shaft rotation a defined number of pulses are generated.*

Sensor type	Inductive	Optical
Type of protection	EEx ia IIC T6	EEx ib IIC T6
Pulse rate per rotation	25	100
Max. pulse frequency	5 kcps	1 kcps
Supply voltage	8 V DC	8 V DC
Internal resistance	1 kohm	1 kohm
Self-inductance	30 µH	–
Self-capacitance	20 nF	177 nF
Max. speed	3000 min^{-1}	3000 min^{-1}
Torque	<5 N cm	<5 N cm
IP protection code	IP 65	IP 65.

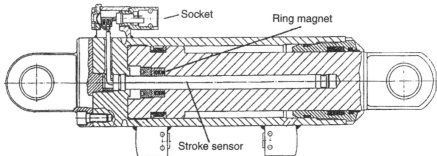

Figure 6.186(a) *Hydraulic cylinder with integrated piston stroke sensor. The stroke sensor fixed at the cylinder bottom, a tube made of stainless steel, contains a chain of resistors and, connected to these, an adequate number of Reed contacts activated by a piston-integrated annular-shaped permanent magnet. A magnetically activated Reed contact works like the tap of a potentiometer.*

Rated pressure	320 bar (3.2 · 10^7 Pa)
Max. stroke	3150 mm
Diameter	18–24 mm
Resolution	<4 mm
Ring magnet	annular-shaped permanent magnet

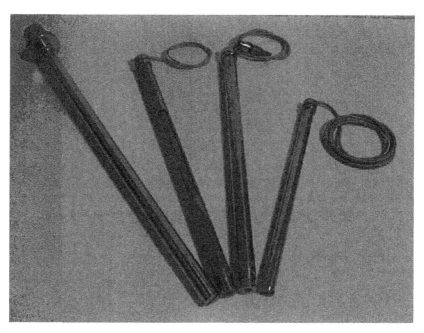

Figure 6.186(b) *Stroke sensors.*

Type of protection	EEx ia I
Certificate	BVS 91.C.1095
Supply voltage	12 V DC
Output voltage	0.5–4.5 V corresponding to stroke.

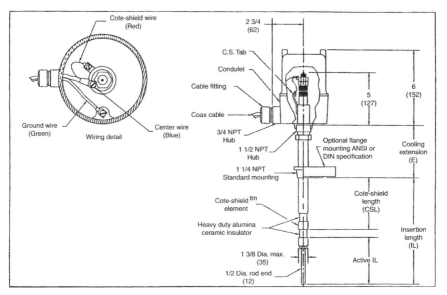

Figure 6.187   *Sensor of a continuous level transmitter for liquids. A metallic rod in a tank forms a capacitance together with the (metallic) tank walls. A liquid with a relative dielectric constant $\varepsilon_r > 1$ (for values of ε_r see Table 6.43) increases this capacitance proportional to the liquid level. The transmitter may be either integrated into the sensor head or located separately.*
Type of protection: EEx ia IIC T...; Supply voltage: 15–28 V DC; Power consumption: 1 W; Frequency of the sensor circuit: 110 kcps; Capacitance of sensor: 0–9 pF or 0–90 pF; Output (analogue): 4–20 mA; IP protection code: IP 65; Ambient temperature range: $-20°C \leqslant T_{amb} \leqslant +60°C$.

Figure 6.188   *Escape route lighting system. An electroluminescence or, as an option, an LED lighting strip in an i-circuit is fed via a power supply in 'encapsulation – m –' and 'increased safety – e –'.*
Type of protection/certificates:
Power supply: EEx em [ia] IIB T4; BASEEFA Ex 92 C 3397; Back-up battery enclosure: EEx emd IIB T4; BASEEFA Ex 92 C 3559 X; Lighting strip: EEx ia IIC T4; BASEEFA Ex 92 C 2569 X; Lighting strip circuits: 8 output channels; Output voltage: 37 V; Output current (per channel): typically 50 mA; Frequency: 1 kcps.

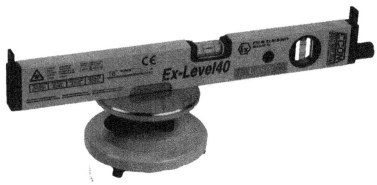

Figure 6.189 *Laser spirit level.*
Type of protection: EEx ia IIC T4; Certificate: PTB Ex-97.D.2019;
Laser:
Semiconductor type: wavelength 650 nm (red); Optical output power: <1 mW;
Spirit level:
Horizontal bubble, accuracy: ±0.014° = ±0.25 mm/m; Vertical bubble, accuracy:
±0.043°; Dimensions: 400 × 50 × 21 mm³.

It is:

$$I_{sc}/I_{sc1} = (1 + R_2/R_1)^{-1} + I_{sc2}/I_{sc1} \times (1 + R_1/R_2)^{-1}$$
$$\leqslant (1 + R_2/R_1)^{-1} + (1 + R_1/R_2)^{-1} = 1$$
$$I_{sc} \leqslant I_{sc1}$$

An analogous calculation can be made with:

$$I_{sc2} \geqslant I_{sc1}$$
$$I_{sc1}/I_{sc2} \leqslant 1$$

resulting in

$$I_{sc} \leqslant I_{sc2}$$

Thus, additional to an increased output voltage, an increased short-circuit current may exceed the limitations of intrinsic safety.

When connected in parallel, the two power supplies are assumed – as a first step – to show identical no-load voltages $U_{10} = U_{20} = U_0$. With internal resistances R_1 and R_2, the individual short-circuit currents are:

$$I_{sc1} = U_0/R_1$$
$$I_{sc2} = U_0/R_2$$

resulting in a total value of:

$$I_{sc} = I_{sc1} + I_{sc2}$$
$$I_{sc} = U_0(R_1 + R_2) \cdot R_1^{-1} \cdot R_2^{-1}$$

In this case, with $I_{sc} > I_{sc1}$ and $I_{sc} > I_{sc2}$ at a constant no-load voltage U_0, the limitations of intrinsic safety may be exceeded.

In the operational mode, e.g. with voltage U_L at the terminals common to both voltage sources, the output current is:

$$I = I_1 + I_2$$

Figure 6.190 *Datalogger for voltage/current/temperature measurements.*
Marking: Ex II 2G EEx ia IIC T4; Type of protection: EEx ia IIC T4; Certificate:
TÜV 98 ATEX 1318 X

Voltage range	0–2.300 V
• Resolution	1 mV
• Accuracy	±0.1%, ±1 digit
Current range	0–25.00 mA
• Resolution	1 μA
• Accuracy	±0.1%, ±1 digit
Temperature range	(Pt 100)
	−100°C–+500°C
• Resolution	0.1°C
• Accuracy	±0.3°C, ±1 digit
Memory storage capacity	200 000 samples

A/D convertor:
Resolution: 16 bit; Number of channels: 1 or 2 or 3; Sample interval: 1 s–24 h;
Dimensions: $210 \times 90 \times 40\,\text{mm}^3$; Weight: 0.77 kg.

and, with

$$U_0 - I_1 \cdot R_1 = U_L$$
$$U_0 - I_2 \cdot R_2 = U_L$$

the ratio I_1/I_2 follows:

$$I_1/I_2 = R_2/R_1$$

As a second step, the more general case $U_{10} \neq U_{20}$ shall be considered. The voltage at the common terminals, U_{tv}, is identical for both 'branches' due to a balancing current I' and so:

$$U_{tv} = U_{10} - R_1 \cdot I'$$
$$U_{tv} = U_{20} + R_2 \cdot I'$$
$$U_{10} - R_1 \cdot I' = U_{20} + R_2 \cdot I'$$

is valid for the no-load condition.

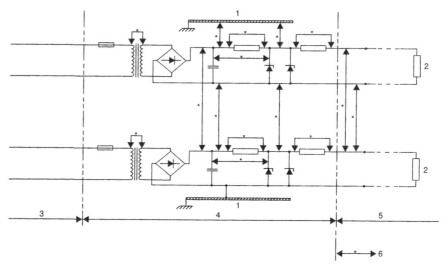

Figure 6.191   *Separation of conducting parts (according to EN 50020 and IEC 60079-11).*
1 Chassis
2 Load
3 Non-intrinsically safe circuit
4 Part of intrinsically safe circuit, not in itself intrinsically safe
5 Intrinsically safe circuit
6 $\longleftarrow \longrightarrow$ dimensions to which Table 6.44 is applicable.

I' results in:

$$I' = (U_{10} - U_{20})(R_1 + R_2)^{-1}$$

In the no-load case, a balancing current $I' = I_1 = I_2$ is present in both branches. The no-load terminal voltage U_{tv} will follow with:

$$U_{tv} = U_{10} - R_1(U_{10} - U_{20})(R_1 + R_2)^{-1}$$
$$U_{tv} = (U_{10} \cdot R_2 + U_{20} \cdot R_1)(R_1 + R_2)^{-1}$$

or

$$U_{tv} = U_{20} + R_2(U_{10} - U_{20})(R_1 + R_2)^{-1}$$

with the same value:

$$U_{tv} = (U_{10} \cdot R_2 + U_{20} \cdot R_1)(R_1 + R_2)^{-1}$$

In the special case $R_1 = R_2 = R$, U_{tv} is:

$$U_{tv} = 0.5(U_{10} + U_{20})$$

with a balancing current:

$$I' = (U_{10} - U_{20}) \cdot (2R)^{-1}$$

Table 6.44 Clearances, creepage distances and separations (according to EN 50020 and IEC 60079-11)

Voltage V (peak values)	10	30	60	90	190	375	550	750	1.0 kV	1.3 kV	1.575 kV	3.3 kV	4.7 kV	9.5 kV	15.6 kV
Clearance mm	1.5	2.0	3.0	4.0	5.0	6.0	7.0	8.0	10.0	14.0	16.0	*			
Separation distance															
• through casting compound mm	0.5	0.7	1.0	1.3	1.5	2.0	2.4	2.7	3.3	4.6	5.3	9.0	12.0	20.0	33.0
• through solid insulation mm	0.5	0.5	0.5	0.7	0.8	1.0	1.2	1.4	1.7	2.3	2.7	4.5	6.0	10.0	16.5
Creepage distance															
• in air mm	1.5	2.0	3.0	4.0	8.0	10.0	15.0	18.0	25.0	36.0	49.0	*			
• under coating mm	0.5	0.7	1.0	1.3	2.6	3.3	5.0	6.0	8.3	12.0	16.3	*			
Comparative tracking index CTI ** ***															
• for ia	–	100	100	100	175	175	275	275	275	275	275	*			
• for ib	–	100	100	100	175	175	175	175	175	175	175	*			

* Except for separation distances no values for voltages higher than 1.575 kV are proposed at present (1999–02)
** At voltages up to 10 V, the CTI of insulating materials is not required to be specified. The CTI testing method is described in IEC 60112
*** Table 6.17 gives a survey of insulating materials and their CTI classification

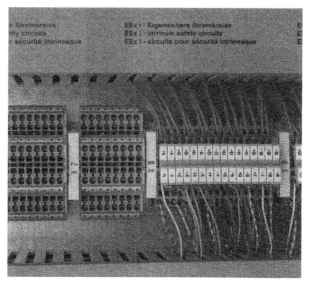

Figure 6.192 *Separation of i-terminals from non-i-terminals.*

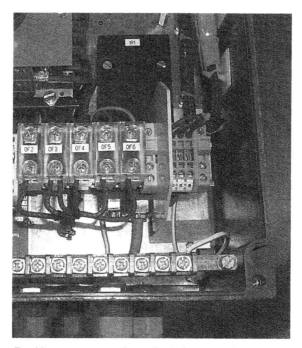

Figure 6.193 *Partition to separate i-terminals from non-i-terminals.*

Generally, U_{tv} follows with:

$$U_{tv} = (U_{10}R_2 + U_{20}R_1)(R_1 + R_2)^{-1}$$
$$U_{tv} = U_{10}(R_1/R_2 + 1)^{-1} + U_{20}(R_2/R_1 + 1)^{-1}$$

which indicates that U_{tv} can never exceed U_{10} or U_{20}, whichever is the highest.

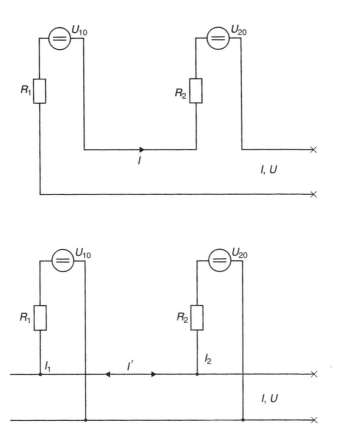

Figure 6.194 *Interconnections of two different intrinsically safe circuits. Top: series connection; Bottom: parallel connection.*

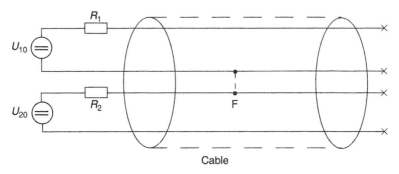

Cable

Figure 6.195 *Interconnection of two different intrinsically safe circuits caused by a cable fault (at point F).*

Based upon these considerations, Fig. 6.195 will demonstrate that faults in cables, e.g. a conductive path between two different intrinsically safe circuits, may initiate increased voltage or current values exceeding the limitations of intrinsic safety.

So, the third feature of intrinsic safety is a strict electrical separation between

- **i- and non-i-circuits**
- **different i-circuits (or, as an alternative, a careful assessment shall prove that interconnections will not adversely affect this type of protection).**

In addition, external cables with different i-circuits are included in the 'separation assessment'.

Apart from separation considerations, Fig. 6.191 shows a very general principle in safety technology: the principle of redundancy, in this case the doubling of components relevant to current and voltage limitation (resistors connected in series, Zener diodes in parallel).

6.9.2 Classification and marking

Classification and marking of intrinsically safe apparatus concerning explosion groups (I, IIA, IIB, IIC) and temperature classes (T1–T6) are organized in the same way as for, e.g., 'flameproof enclosure – d –' (see Tables 4.1 and 4.2, Chapter 4, and Chapter 5 for marking). Some special features, however, which are of importance for type of protection 'intrinsic safety – i –' only, will be explained in the following.

6.9.2.1   Intrinsically safe and associated apparatus

In principle, 'intrinsic safety – i –' is a characteristic of an electric circuit. It may run within an autonomous apparatus, e.g. within a hand lamp. The electric circuit, however, may be formed just as well by two (or more) apparatus, e.g. a transmitter which is connected to a limiting stage (in the following described as 'Ex i-isolator').

As a rule, this Ex i-isolator is installed in a safe area and connected to additional apparatus not ex-protected, e.g. to a single loop controller or to a PLC (Programmable Logic Controller). Therefore, in the field of intrinsic safety two types of apparatus are distinguished:

- *intrinsically safe apparatus:* an electrical apparatus in which all the circuits are intrinsically safe
- *associated apparatus:* an electrical apparatus which contains both intrinsically safe and non-intrinsically safe circuits and is constructed so that the non-intrinsically safe circuits cannot adversely affect the intrinsically safe circuits.

Note:

These are the definitions according to the i-standards EN 50020 and IEC 60079-11.

Intrinsically safe apparatus is designed for use or installation in a potentially explosive gas atmosphere.

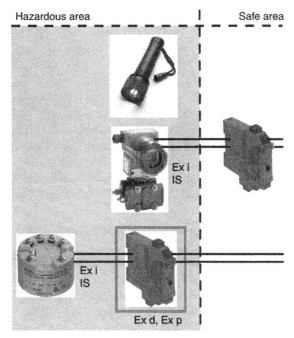

Figure 6.196 *Associated and intrinsically safe apparatus.*

Associated apparatus is commonly installed in a safe area. Many applications of 'intrinsic safety' in remote control and monitoring instrumentation are assembled in such a way that an intrinsically safe apparatus, e.g. a sensor or actuator in the hazardous area, is connected with an associated apparatus, e.g. a safety barrier or an Ex i-isolator in the safe area (see Fig. 6.196). With that, the associated apparatus takes over the function to safely limit current and voltage in the intrinsically safe circuit to permissible values.

Indeed, it is a requirement of associated apparatus even if an Ex i-isolator designated for a safe area is installed in a potentially explosive gas atmosphere and guarded by an additional type of protection, e.g. 'flameproof enclosure – d –'.

6.9.2.2 Categories ia and ib

Another special feature of 'intrinsic safety – i –' also results from its appropriateness for zone 0 applications. Hazardous areas classified as zone 0 are characteristic of the presence of an explosive gas atmosphere at a high probability, e.g. in the interior of a ventilated tank. Thus, special safety precautions are required.

Corresponding to these facts, there is a differentiation in the so-called categories ia and ib for zone 0 and zone 1 applications. To comply with the special high degree safety requirements imperative for zone 0 areas, apparatus or electrical circuits are subject to a two-fault analysis. The i-standards define a fault as any defect of any component, separation, insulation or connection between components, not defined as infallible*, upon which the intrinsic safety of a circuit depends.

Note:

After enactment of Directive 94/9/EC (ATEX 100a), see Section 3.3, the term 'category' describes additional, but different, facts, namely, the term 'apparatus category', see Table 2.3, which is not of relevance for 'intrinsic safety' only. This 'category' classifies apparatus according to its appropriateness for the different zones, e.g. category 2G for zone 1 application. It is planned that in future the term 'category' classifying intrinsically safe apparatus will be replaced by the term 'level of protection' in order to avoid confusion.

For the minor degree of exposure to danger in zone 1, a single-fault analysis is considered adequate.

The difference between category ia and ib apparatus will be pointed out in an example (Fig. 6.197). In Ex i-isolators Zener diodes are commonly used for voltage limitation. Semiconductors are considered as fallible components, however. Following the fault analysis one has to proceed from a high resistive Zener diode which no longer performs its protective function. So, two Zener diodes shall be applied for category ib apparatus. The protective function will be ensured even in the case of a single fault. For category ia apparatus, three Zener diodes shall be applied accordingly (assuming two faulty diodes).

In the United States the discrimination between categories ia and ib is necessary only if the IEC-based classification of hazardous areas has been adopted. According to the differentiation into Divisions 1 and 2 (still predominant), one single type of 'intrinsic safety (IS)' has been defined, requiring a two-fault assessment.

This is based upon the definition of Division 1, which may be considered with a certain simplification as a combination of zones 0 and 1 (see Tables 2.1 and 2.2).

6.9.2.3   Marking

The classification of intrinsically safe apparatus as described above according to:

- explosion group
- temperature class

---

* EN 50020 and IEC 60079-11 specify:

- infallible component or infallible assembly of components: component or assembly of components that is considered not subject to certain fault modes as specified (in these standards). The probability of such fault modes occurring in service or storage is considered to be so low that they are not to be taken into account
- infallible separation or insulation: separation or insulation between electrically conductive parts that is considered not subject to short-circuits. The probability of such fault modes occurring in service or storage is considered to be so low that they are not to be taken into account

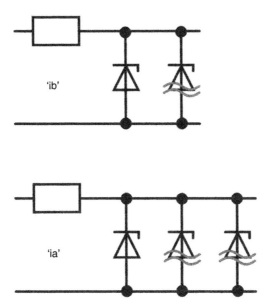

Figure 6.197 *One- and two-fault analysis applied to a voltage-limiting circuit.*

- categories (ATEX 100a categories, and the 'i' categories ia, ib)
- intrinsically safe and associated apparatus.

and the possibilities of operation arising from it for different applications can be understood completely from the marking of the apparatus (or from equivalent data given in the certificate). Some examples are given in Table 6.45.

Generally, the brackets [ ] indicate an associated apparatus (see examples 3 and 4 in Table 6.45) or electrical circuits deviating from the ATEX 100a category of the apparatus (see example 2 in Table 6.45). To some extent, the brackets may be read as 'The apparatus contains additional electrical circuits which comply with the category marked in brackets'.

Moreover, the marking of intrinsically safe apparatus contains further information as follows:

- manufacturer and type of the apparatus
- number of certificate
- safety-related maximum values for voltage U, current I, power P, inductance L and capacity C
- permissible ambient temperature range, if deviating from $-20°C–+40°C$
- for associated apparatus only: voltage U_m, defined as the maximum permissible voltage which may be applied to terminals of non-intrinsically safe circuits of associated apparatus without adversely affecting the type of protection 'intrinsic safety'. Commonly, U_m is rated for the mains voltage ($U_m = 250\,V$).

Since the space available for marking is rather limited on small apparatus, the descriptive documentation and/or the certificate should be referred to for a detailed Ex-protection specification.

Table 6.45   Marking and application of intrinsically safe and associated apparatus

Marking	Description of apparatus	Examples
Part A Marking according to Directive 94/9/EC (ATEX 100a) and EN 50014–50020, 50028		
Ex II 2G EEx ib IIC T6	intrinsically safe apparatus, zone 1 application	transmitter, positioner, solenoid valve, measuring device
Ex II 2 (1) G EEx ib [ia] IIC T4	intrinsically safe apparatus, installed in zone 1, circuit(s) suitable for zone 0	transmitter for zone 1 with sensor in zone 0
Ex II (1) G [EEx ia] IIC	associated apparatus, installed in a safe area, circuit suitable for zone 0	safety barrier, Ex i-isolator
Ex II 2G EEx de [ib] IIB T4	associated apparatus, installation in zone 1 circuit suitable for zone 1	power supply for an intrinsically safe apparatus (e.g. sensor), fitted into an EEx d enclosure with an EEx e terminal compartment
Part B Marking as required by IS* Standard ANSI/UL 913 – 1999 (FM Class 3610 – 1999, UL 913 – 1999)		
IS* class I Division 1 Groups A, B, C, D	intrinsically safe apparatus, installed in Division 1	transmitter (according to Division concept, NEC Article 500)**
IS* class I zone 1 AEx ib IIB T4	intrinsically safe apparatus, installed in zone 1	transmitter (according to zone concept, NEC Article 505)**
IS* circuit class I Division 1 Groups A, B, C, D	associated apparatus, installation in a safe area, circuit suitable for Division 1	safety barrier, Ex i-isolator (according to Division concept, NEC Article 500)**
IS* circuit class 1 zone 0 [AEx ia] IIC	associated apparatus, installation in a safe area, circuit suitable for zone 0	safety barrier, Ex i-isolator (according to zone concept, NEC Article 505)**

* IS = Intrinsically safe
** NEC = National Electrical Code (of USA)

Within the European Community, CE marking is mandatory. It indicates that the apparatus complies with all relevant EC directives in their entirety. In the example of Fig. 6.198, directives 94/9/EC (ATEX 100a) and 89/336/EEC (EMC) are referred to.

Made in Germany, www.stahl.de

BINARY OUTPUT

TYP 9651/40-12-10

$\text{II(1)G [EEx ia] IIC/IIB}$
DMT 98 ATEX E 024 X
0102 ISM Group 1 Class B

$T_a = -20°C{-}+65°C \text{ or } -20°C{-}+45°C$
$U_m = 250 \text{ V}$
$U_o = 11.2 \text{ V} \quad C_i = 1 \text{ nF}$
$I_o = 75 \text{ mA} \quad L_i = 0$
$P_o = 210 \text{ mW}$

[EEx ia]	IIC	IIB
$L_o/\text{mH} \leq$	6.7	25
$C_o/\text{nF} \leq$	1840	12600

	⊖→ [EEx ia]	⊖→ 0 V = OFF, 24 V = ON	TEST	SUPPLY
1	d26(+) . z26	d8 . b8		$-d2(+)\overline{}$
2	d28(+) . z28	d10 . b10	d20	d4 —— 24 V-
3	d30(+) . z30	d12 . b12		(18–35 V-)
4	d32(+) . z32	d14 . b14		z4(−) ———

Figure 6.198 *Example for marking of an associated apparatus (binary output to drive a solenoid valve).*

6.9.3 Apparatus for intrinsically safe electrical circuits

As mentioned above, the main domain of application of 'intrinsic safety' is remote control and monitoring instrumentation. More than two million intrinsically safe circuits for sensors and actuators alone are installed anew or modernized year by year.

Some examples are given in Fig. 6.199:

- intrinsically safe apparatus without an electrical connection to the environment (hand lamp)
- intrinsically safe apparatus (transmitter, terminal) connected to an associated apparatus (safety barrier, Ex i-isolator)
- intrinsically safe apparatus (cables, terminal box)
- associated apparatus, explosion protected (remote I/O) with intrinsically safe and non-intrinsically safe electrical circuits
- intrinsically safe fieldbus.

In the following, intrinsically safe apparatus are described, without any claim for completeness.

6.9.3.1 Simple intrinsically safe apparatus

The so-called 'simple' electrical apparatus plays a special part. It is defined as follows:

An electrical component or combination of components of simple construction with well-defined electrical parameters which is compatible with the intrinsic safety of the circuit in which it is used

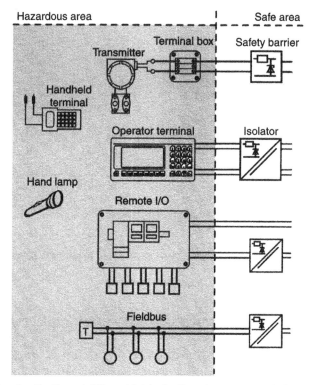

Figure 6.199   *Application of different intrinsically safe apparatus in hazardous areas.*

Simple intrinsically safe apparatus are not subject to certification (see Section 6.9.5).

Simple intrinsically safe apparatus	Examples
Passive components *without* an energy storage	terminal boxes, plug-and-socket connectors, switches, Pt 100 resistance transmitter sensor
Passive components *with* an energy storage with well-defined parameters (L, C) which shall be considered when determining the overall safety of the system	indicators (moving-coil instrument), loudspeakers, cables
Active components (sources of generated energy) which do not generate more than 1.5 V, 100 mA and 25 mW	thermocouples, photocells

In 'classic' intrinsically safe instrumentation multicore cables lead from the control room into the explosion protected system. Several sensors/actuators are wired to a terminal box and connected to the multicore cable.

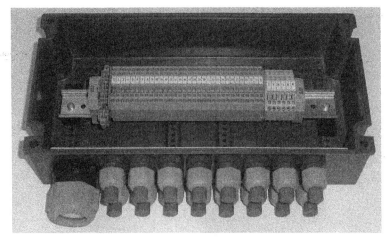

Figure 6.200   *Terminal box for intrinsically safe circuits (cover removed).*

The terminal box for intrinsically safe electrical circuits (Fig. 6.200) does not need to be certified, but it has to comply with constructional requirements, however. These are:

- for enclosures made of plastic materials electrostatic charging shall be avoided (see Section 6.1)
- clearances, creepage distances and separation distances between terminals and metallic parts of the enclosure shall comply with the relevant requirements (e.g. see Table 6.44).

6.9.3.2   Complex intrinsically safe apparatus

As a rule, such apparatus consists of an electronic circuit arrangement, whose electrical and thermal reaction – especially under fault conditions – can be assessed no more without any difficulty. So, in this case, a certificate is required (see Section 6.9.5).

Examples for complex intrinsically safe apparatus are as follows (see also Figs 6.201 and 6.202):

Apparatus	*Examples*
Apparatus without individual voltage supply	sensors/actuators in 2-wire technique: transmitter, positioner, proximity sensor, solenoid valve
Apparatus with individual voltage supply	4-wire transmitter, mobile telephone, handheld terminal, operator terminal

Figures 6.201 and 6.202   *Examples of intrinsically safe temperature transmitters.*

Transmitters are used to record different physical quantities, e.g. temperature, presssure, flow, level, pH value and so on. The mechanical design
corresponds to requirements at the place of duty. Due to the small power
demand, type of protection 'intrinsic safety – i –' is an obvious choice in the
field of explosion protection in most cases.

Frequently, the electronic components are assembled on a so-called 'round
card' and may be fitted into different round-shaped housings.

The operator terminal according to Fig. 6.203 is intrinsically safe in total.
An individual electrical circuit connected to a suitable Ex i-isolator is used
for power supply and data transmission respectively, the latter attached to an
automation device.

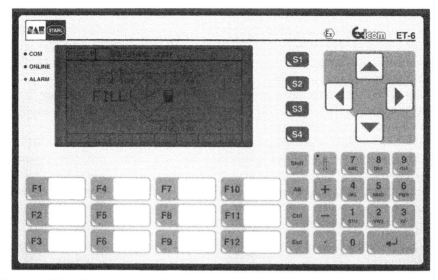

Figure 6.203 *Intrinsically safe operator terminal.*

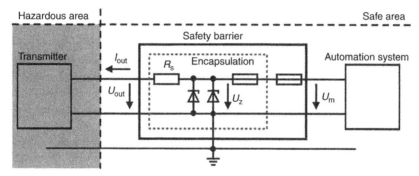

Figure 6.204 *Schematic circuit diagram of a safety barrier.*

6.9.3.3 Safety barriers/Zener barriers

Safety barriers (or 'Zener barriers') form an essential technique in the field of intrinsic safety.

As a rule, they are installed in the safe area as associated apparatus. They are used to limit the electrical values for sensors, actuators etc. located in the hazardous area to levels permissible for intrinsic safety. A certificate is required each time.

Due to the galvanic isolation lacking between intrinsically safe and non-intrinsically safe electrical circuits the power-limiting characteristic of components commonly used for it (e.g. small transformers, optocouplers, relays) is lacking as well. So, a 'robust' safety barrier shall be constructed.

For voltage limitation, Zener diodes connected in parallel are generally used (see Fig. 6.204). They are protected against thermal overload by means of a fuse (in case of long-term overvoltage at the non-intrinsically safe side of the circuit). According to requirements for category ia or ib, two or three diodes are connected in parallel. An infallible resistor (see Section 6.9.4) limits the current in the intrinsically safe circuit. Alternatively, electronic current limiters (with a rectangular-shaped output characteristic) may be used, but restricted to category ib only.

Safety barriers commonly used are rated for a maximum voltage U_m of 250 V (DC or AC rms) so that they may be connected to all apparatus supplied with 'normal' mains voltage.

Due to its simple construction and the lack of galvanic isolation, a safety barrier shall be connected to the equipotential bonding system, which is stipulated imperatively in hazardous areas. As a rule, a minimum conductor cross-section of 4 mm^2 (for copper) or an earthing resistance lower than 1 ohm shall be used.

An enormous variety of different safety barriers is available for the user. The process of selection of a suitable safety barrier is as follows:

- polarity with respect to the equipotential bonding system (in most cases: positive)
- rated voltage (at which the Zener diodes are still non-conducting in normal operation). Typical rated voltages are in the range 3 to 24 V DC
- series resistance of or voltage drop across the safety barrier in case of electronic current limitation.

The safety-related parameters U, I, P, C and L can be matched to the corresponding values of the relevant intrinsically safe apparatus (sensors, actuators etc.) by variations of rated voltage and series resistance. Intrinsic safety may be engineered in a very flexible way using safety barriers. After all, even 'hopeless' cases may be put into practice frequently. Additional advice for engineering is given in Section 6.9.5.

A protection against polarity reversal cannot be realized for the non-intrinsically safe side of a safety barrier without increasing its series resistance which is considered to be frequently disturbed. One solution consists of the implementation of a second fuse so that it can be replaced if the safety barrier has been operated at reversal polarity or supplied with overvoltage (see Fig. 6.205). The 'internal' fuse protecting the Zener diodes as well as the other components shall be guarded against external access. So, safety barriers are encapsulated for the most part.

The advantages and disadvantages of safety barriers may be compared as follows:

Advantages	Disadvantages
Small size	Shall be connected to the equipotential bonding system imperatively
Inexpensive	Safety is based on the 'quality' of the equipotential bonding system
Simple construction, high availability	Susceptible to, e.g., polarity reversal, overvoltage
No need for auxiliary power	Series resistance may be disturbed

Figure 6.205 *Example of a two-channel safety barrier with replaceable fuses.*

6.9.3.4 IS interface with galvanic isolation (Ex i-isolators)

If a high quality equipotential bonding system cannot be put into practice or a steady supply voltage cannot be guaranteed, a galvanic isolation between intrinsically safe and non-intrinsically safe electrical circuits is recommended. Such apparatus, e.g. power supplies for transmitters or switching repeaters, are equivalent to safety barriers with respect to the philosophy of intrinsic safety. The additional feature may be seen in the galvanic isolation and functional characteristics which may be practicable.

Where an intrinsically safe circuit is earthed by construction at the place of the field device (e.g. this is frequently the case using thermocouples), an Ex i-isolator with galvanic isolation shall be inserted. The reason is that intrinsically safe circuits shall be earthed at one single point only.

The 'classic' Ex i-isolator is an associated apparatus for installation in a safe area. In this case, too, Zener diodes and resistors are used for voltage and current limitation. However, the components may be rated lower, since the components inserted for galvanic isolation are able to transfer only a limited (e.g. transformers) or absolutely no electric power (e.g. optocouplers). For the maximum voltage U_m, the above-mentioned is valid concerning safety barriers.

In the block diagram (Fig. 6.206), the galvanic isolation running through all functional units is shown. This is an example of a power supply with a

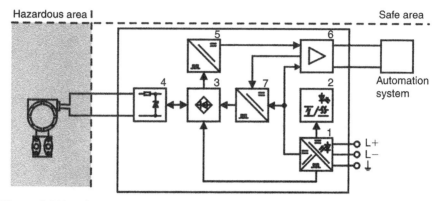

Figure 6.206   *Schematic circuit diagram of a transmitter power supply with galvanic isolation between input, output and auxiliary power.*

three-port isolation, i.e. all three external circuits (to the transmitter, to the automation system and to the auxiliary power supply) are galvanically isolated from each other. In addition to the basic function to transmit a 4 … 20 mA signal, a wire break and short-circuit control have been fitted.

In practice, these apparatus will be selected according to the type of the sensor or actuator desired. Manufacturers offer a broad range of varying functions. The safety-related data (U_o, I_o, P_o, C_o, L_o, see Section 6.9.5) are matched to the 'customary' values of the field devices. Sometimes, variants with different Ex i-parameters are available.

Externally, the only difference between this Ex i-isolator (Fig. 6.207) with galvanic isolation and a safety barrier is its increased size. The complex electronics and, of course, the components limiting current and voltage are hidden inside. In this example, a trip amplifier for DIN rail mounting is shown, which is suitable for operating temperature sensors in intrinsically safe circuits. The marking is:

Ex II (1)G [EEx ia] IIC

The advantages and disadvantages of Ex i-isolators (with galvanic isolation) may be summarized as follows:

Advantages	*Disadvantages*
No connection with equipotential bonding system required	More expensive than safety barriers
No disturbing series resistance	Increased size compared with safety barriers
High level interference suppression	As a rule, an additional power supply is required
Simple selection, since Ex i-isolators are optimized for quite different applications	

Figure 6.207 *Example of an Ex i-isolator with galvanic isolation (limit indicator).*

6.9.3.5 HART (Highway Addressable Remote Transducer) technology in intrinsically safe electrical circuits

As early as in the middle of the 1980s the Rosemount company introduced a technology outlined for transmitters. It combined the 'classic' 4 … 20 mA current loop with a digitized data transmission – on a single pair of wires. Basically, even several measuring devices can be operated on a single cable. This operational mode is called 'multi-drop'.

Above all, HART has succeeded in the field of process engineering, with feasibilities, e.g., to remotely change the meter range of an instrument or to interrogate diagnostic data. So, numerous products are available in an explosion protected design. A number of them are certified as intrinsically safe apparatus.

HART (see Fig. 6.208) superposes the 4 … 20 mA analogous measuring signal with a digital FSK (Frequency Shift Keying) current signal, which enables a bidirectional data transmission from/to the measuring instrument. The digital coding is as follows:

- '1' corresponds to a frequency of 1200 cps
- 'O' corresponds to a frequency of 2200 cps.

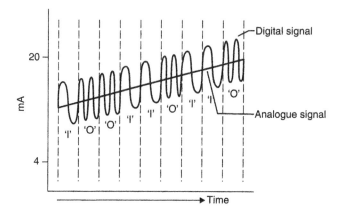

Figure 6.208　*HART signal as a superposition of an analogous (measuring) signal (4–20 mA) and a digital FSK signal.*

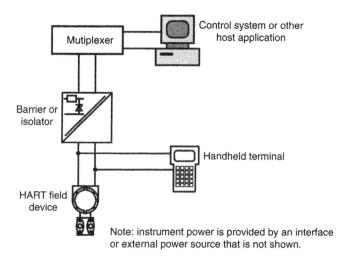

Note: instrument power is provided by an interface or external power source that is not shown.

Figure 6.209　*Communication with a HART field device via a handheld terminal or a PC (personal computer).*

The FSK signal is free of DC components. The original measuring signal can be restord easily by filtering.

There are two methods for communicating with a HART field device (Fig. 6.209):

- the use of a handheld terminal, which shall be suitable for duty 'on location' and which shall be certified as an intrinsically safe apparatus
- the use of a personal computer which enables communication with the field devices from a central control point. Several field devices require a multiplexer or a remote I/O.

When planning intrinsically safe electrical systems it shall be taken into account that the handheld terminal may be connected to the intrinsically safe circuit in addition to other apparatus. Thus, the safety-related parameters of the terminal show especially low levels. Besides, the safety barrier shall be suitable for the HART technology (impedance, FSK signal transmission, bidirectional communication).

Currently HART transmitters and valve positioners are 'state of the art'. They do not differ outwardly from their pure analogue predecessors. A wide range of varying functions and measuring principles is available with a HART interface. Numerous products are deliverable with a certificate for intrinsic safety, too. The same holds for safety barriers, multiplexers and remote I/O systems which are suitable for HART communication as well.

Advantages and disadvantages of HART field devices may be summarized as follows:

Advantages	Disadvantages
Parameters (e.g. measuring range) are changeable without intervention in the instrument	A new technology asks for instruction and experience in practice
Diagnostic data interrogatable, high level of diagnosis profundity. To some extent, self-diagnosis supports predictive maintenance	If necessary, additional multiplexers are required for communication
Central administration of measuring instruments in their entirety is feasible ('asset management')	–
Measured data can be read out with high precision in digitized form	–
Several measuring instruments can be operated on a twin-wire cable ('multi-drop capability')	–

6.9.3.6   Remote I/O systems for hazardous areas

In manufacturing industries, e.g. car assembly or beverage bottling, modern distributed control systems are well founded. For some years, these technologies have been available for planners and users in the field of process engineering for plants which are explosion protected. They may be illustrated by some catchwords:

- intelligent measuring instruments and actuators in HART technology, asset management
- remote I/O systems (distributed Input/Output)
- standardized fieldbus systems, e.g. Profibus, ControlNet, Foundation Fieldbus.

Currently, remote I/O systems of several manufacturers are available for operation in zone 1 or Division 1. To a large extent, these systems are based upon the technology of intrinsic safety. The power supply only is designed according to another type of protection, e.g. 'flameproof enclosure – d –'. A remote I/O suitable for hazardous areas may be an integration of input (I) and output (O) assembly units of automation systems and of Ex i-isolators of 'classic' design. The installation may be made in a hazardous area and supersede the classic field distribution box (terminal box). Standardized fieldbus systems may be used for data transmission from/to the automation system.

These remote I/O systems for hazardous areas are characterized as follows:

- installation in zone 1/Division 1
- associated apparatus
- in most cases combination of several types of protection
- auxiliary power supply usually in 'increased safety – e –' or in conduit technique ('flameproof enclosure – d –')
- intrinsically safe interfaces for sensors and actuators
- fieldbus intrinsically safe or according to 'increased safety – e –'
- 'hot swap' module replacement in the operational state, for non-intrinsically safe modules, too.

Remote I/O systems are also available which can be installed in a safe area or in zone 2/Division 2 only, although they are equipped with intrinsically safe inputs and outputs.

Typically, a remote I/O system is a modular assembly (see Fig. 6.210). In this figure, the power supply, the connection to the fieldbus, the internal bus (power and data line) are shown as well as input and output modules. Input and

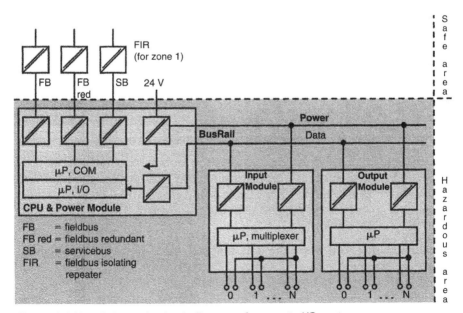

Figure 6.210 *Schematic circuit diagram of a remote I/O system.*

output circuits of the I/O modules, auxiliary power supply and fieldbus are galvanically isolated from each other. The fieldbus is operated intrinsically safe. A fieldbus isolating repeater is used as an Ex i-isolator (installed in the safe area).

Modules for various signal types (analogue IN (AI), analogue OUT (AO), digital IN (DI), digital OUT (DO), temperature, frequency etc.) are available covering all the systems which are known today. The signal circuits are intrinsically safe.

A typical installation 'on location' of a remote I/O system is shown in Fig. 6.211. At the right side/top in the housing, the power supply with an integrated fieldbus interface can be seen. The module complies with 'flameproof enclosure – d –' and is designed in such a way that it can be replaced under live voltage and explosion hazard ('hot swap'). The auxiliary power lines and terminals comply with 'increased safety – e –'. All I/O modules comply with 'intrinsic safety – i –'. Signal circuits and fieldbus are intrinsically safe as well (blue coloured cables and terminals).

In the following, advantages and disadvantages of remote I/O systems are summarized:

Advantages	Disadvantages
One single fieldbus instead of numerous individual cables	A new technology (fieldbus) asks for instruction and experience in practice
Ex i-isolators, marshalling cabinets, terminal boxes 'on location' are dropped	Not yet applicable for safety-related signals/functions
Extension in a simple way	–
Little space required within the control room	–
Less weight by fewer cables (offshore installations)	–
Time and money saving in planning, installation, start-up and maintenance	–
High level of diagnosis profundity (data terminal, field station, module, signal)	–

6.9.3.7   Fieldbus for intrinsically safe field devices

During international fieldbus standardization at the start of the 1990s a physical layer had been specified as IEC 61158-2 with an option for 'intrinsic safety'. In the meantime, two standardized fieldbus systems comply with said standard: Profibus PA and Foundation Fieldbus H1. These are the most important characteristics:

- data transmission rate 31.25 kbit/s
- modulation ±10 mA into 50 ohms (=1 V peak-to-peak)

Figure 6.211   *Example of a zone 1 field station with remote I/O.*

- Manchester II code
- cable length (trunk) up to 1900 m, branch line (spur) up to 120 m
- power supply (for field devices connected to the bus) and data transmission on a single pair of wires
- power supply for a field device 10 mA at 9 ... 32 V DC
- number of field devices in a safe area: up to 31
- number of field devices in a hazardous area 4 ... 6
- terminating impedance $R = 100$ ohms connected in series with $C = 1 \mu F$.

On the basis of these technical boundary conditions, such a fieldbus is suited (independently of the bus protocol applied here) for fully digitized sensors and actuators.

Meanwhile, an increasing number of transmitters and positioners etc. is available either with a Profibus PA or a Foundation Fieldbus H1 interface (or with both of them).

As a result of current – and voltage – limitations required, and in case of 'classic' Ex i-engineering (see Section 6.9.5), some restrictions are imposed upon the intrinsically safe variant. The number of field devices to be connected to the bus system is limited to approximately five, in particular in case of a current consumption exceeding the designated value of 10 mA. Besides, the length of cables shall be limited to some 100 m due to capacitance and inductance limitations. A way out is demonstrated below (FISCO model, see Section 6.9.6).

The schematic circuit diagram of an intrinsically safe fieldbus according to IEC 61158-2 (physical layer) is given in Fig. 6.212. The field devices are certified as intrinsically safe apparatus. Four-wire transmitters with an external power supply are explosion protected by an additional type of protection,

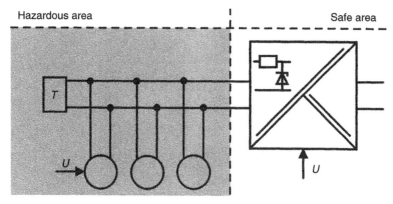

Figure 6.212 *Schematic circuit diagram of an intrinsically safe fieldbus according to IEC 61158-2 (physical layer).*

Figure 6.213 *Example of a fieldbus installation with Profibus PA.*

in many cases 'flameproof enclosure – d –'. A special Ex i-isolator (fieldbus power supply, segment coupler, fieldbus isolator) implements the galvanic isolation between the intrinsically safe and non-intrinsically safe bus segments. In addition, the current (approx. 80 mA) needed to supply the 2-wire field devices is delivered via the Ex i-isolator.

Deviating from conventional installations (in many cases characterized by several field devices connected to a common terminal box/junction box), the fieldbus cable with its intrinsically safe circuit passes from transmitter to transmitter (Fig. 6.213). Each transmitter is fitted with a branch box (T box) to connect it to the fieldbus (Fig. 6.214). At the 'physical' end of the bus segment, a terminating impedance shall be fitted.

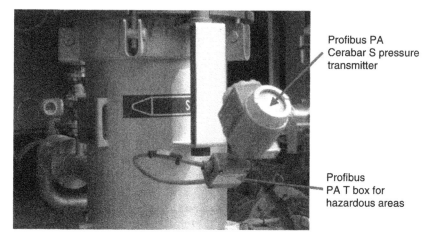

Profibus PA
Cerabar S pressure
transmitter

Profibus
PA T box for
hazardous areas

Figure 6.214 *Profibus PA pressure transmitter with T box.*

Advantages and disadvantages of fieldbus systems are given as follows:

Advantages	*Disadvantages*
One single fieldbus instead of many individual cables	A new technology (fieldbus) asks for instruction and experience in practice
One single Ex i-isolator only required for several field devices	Not yet applicable for safety-related signals/functions
Little space required within the control room	Limited number of field devices per bus segment (5 … 10)
Less weight by fewer cables (offshore installations!)	Field devices with a higher power demand cannot be fed via the bus (e.g. flow transmitters)
Time and money saving in planning, installation, start-up and maintenance	–
High level of diagnosis profundity	–

The technology of explosion protected (intrinsically safe) digital fieldbus systems in process engineering is still in its infancy (pilot plants). Nevertheless, it may already be recognized today that in parallel with the 'pure' intrinsically safe fieldbus there will be alternatives as well (Fig. 6.215). The main bus segment (trunk) can be designed in 'increased safety – e –' with the advantage that an increased current is available to supply the field devices. In this conception, the Ex i-isolator (multibarrier) between fieldbus and field device is fitted near to the field devices in the hazardous area and shall be protected accordingly (see Fig. 6.216).

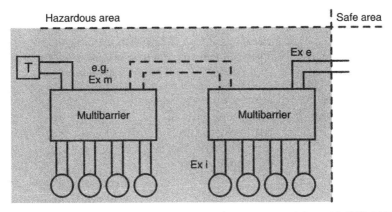

Figure 6.215 *Schematic circuit diagram of a fieldbus complying with IEC 61158-2. The main bus is designed according to 'increased safety – e –', the spurs are intrinsically safe.*

Figure 6.216 *Barrier as an intrinsically safe power supply of field devices for IEC 61158-2 fieldbus systems. 'Mixed' types of protection, marking: Ex II 2 (1)G EEx me [ia] IIC T4. Installation in zone 1.*

6.9.4 Constructional requirements for intrinsically safe apparatus

Constructional requirements for intrinsically safe and associated apparatus refer to all components and provisions related to 'intrinsic safety – i –'. A complete description of these constructional requirements would go beyond the scope of this chapter. So, only the requirements which are essential for a better understanding are mentioned and described in brief.

As mentioned above, fault analysis is of central significance: the number of (countable) faults determines the category of an apparatus. The correlation between faults and fallible or infallible components and assemblies of components is explained by some examples as follows:

A component, assembly of components, interconnection, separations is	
... infallible	*... fallible*
A fault is not to be taken into account	current limiting resistor complying with all requirements and imposed with ≤2/3 of its rated value
Countable fault*	Zener diodes for voltage limitation imposed with ≤2/3 of its rated value
Non-countable fault**	arbitrary interconnection/ interruption within an integrated circuit

*Countable fault = fault which occurs in parts of electrical apparatus conforming to the constructional requirement of this standard
**Non-countable fault = fault which occurs in parts of electrical apparatus not conforming to the constructional requirements of this standard
(These definitions are quoted from EN 50020 and IEC 60079-11)

6.9.4.1   Safe galvanic isolation and separation elements

For intrinsically safe circuits, a basic protection measure is a safe isolation from all other non-intrinsically safe circuits. With the exception of safety barriers a safe galvanic isolation is required. As a rule, this is guaranteed by observance of geometrical distances. It is differentiated between:

- clearance
- creepage distance in air
- creepage distance under coating
- separation distance through casting compound
- separation distance through solid insulation.

These distances for isolation (separation elements) depending on voltage are given in Table 6.44 according to EN 50020 and IEC 60079-11. Their valuation is summarized as follows:

Distances for isolation	*Valuation*
Distance ≥ EN/IEC value	Separation element is infallible
1/3 × EN/IEC value < distance < EN/IEC value	Separation element considered as subject to countable short-circuit faults
Distance < 1/3 × EN/IEC value	Separation element considered as subject to non-countable short-circuit faults if this impairs intrinsic safety

For a safe galvanic isolation, certain components are applied, e.g.:

- transformers
- optocouplers
- relays.

The distances for isolation for these components are effective inside as well as outside.

The terminals of all non-intrinsically safe electrical circuits shall be fused so that the rated values of the components cannot be exceeded (Zener diode with a fuse).

6.9.4.2   Current and voltage limitation

A safe current and voltage limitation plays a special part, of course. The most important techniques for limitation are summarized as follows:

Current limitation	*Examples*
Infallible resistors	<ul><li>film-type resistors</li><li>wire wound-type resistor with protection to prevent unwinding of the wire</li><li>printed resistors covered by a coating or encapsulated</li></ul>
Blocking capacitors	<ul><li>two capacitors connected in series</li><li>high reliability capacitors</li><li>solid dielectric-type capacitor</li></ul>
Blocking diodes	<ul><li>three diodes connected in series (category ia)</li></ul>
Other semiconductors e.g. transistors	<ul><li>two semiconductors connected in series (category ib only)</li></ul>
Voltage limitation Diodes, Zener diodes, thyristors	<ul><li>two semiconductors connected in parallel (category ib)</li><li>three semiconductors connected in parallel (category ia)</li><li>transient conditions/effects shall be taken into account</li></ul>
Shunt safety assemblies* Diodes, Zener diodes	<ul><li>diodes shall form at least two parallel paths</li><li>connections of the shunt components shall be infallible</li></ul>

* An assembly of components shall be considered as a shunt safety assembly when it ensures the intrinsic safety of a circuit by the utilization of shunt components

A so-called 'two-thirds rated value criterion' is valid for all limiting components: the component shall be imposed with not more than two-thirds of its rated value for current, voltage and power.

6.9.4.3 Casting by a compound

Casting by a compound can be used for intrinsically safe apparatus or components. However, this technique shall not be confused with type of protection 'encapsulation – m –' as described in Section 6.5. Here, casting is a part of 'intrinsic safety – i –' and not an autonomous type of protection. Casting, in most cases applied to certain parts of the wiring only, is used as:

- isolation of live parts (specified separation distances through casting compound)
- exclusion technique for hazardous atmospheres (i.e. the compound prevents an immediate access to an environmentral gas–air atmosphere to 'hot' or sparking components, e.g. fuses)
- 'equalizer' for dissipated heat (from small components)
- protection of components against access from outside (relevant for, e.g., safety barriers).

In principle, intrinsically safe apparatus do not need an enclosure since 'explosion protection' is inherent due to the design of the circuitry. In cases where 'intrinsic safety – i –' may be adversely affected by any access to live parts (e.g. infallible creepage distances in air), an enclosure ensuring an IP code of (at least) IP 20 (according to IEC 60529) shall be fitted.

For Group I apparatus (application in coal mines), a degree of protection of IP 54 is required in general.

6.9.5 Installation of intrinsically safe electrical circuits

A fundamentally different installation philosophy has to be recognized in the installation of intrinsically safe circuits. In comparison with all other types of installations, where care is taken to confine electrical energy to the installed system as designed so that a hazardous environment cannot be ignited, the integrity of an intrinsically safe circuit has to be protected from the intrusion of energy from other electrical sources *so that the safe energy limitation in the circuit is not exceeded even when breaking, shorting or earthing of the circuit occurs.*

As a consequence of this principle the aim of the installation rules for intrinsically safe circuits is to maintain separation from other circuits.

(quoted from EN 60079-14 and IEC 60079-14 respectively, clause 12.1)

In most cases, an intrinsically safe circuit is put into practice so that a safety barrier or an Ex i-isolator or an Ex i-input/output is connected to measuring devices or positioning elements (sensors, actuators) located in a hazardous area via a cable (see Fig. 6.217). Hence, explosion protection cannot be 'purchased ready for use' for the most part, but it has to be put into practice by the planner and electrician. Only in cases where apparatus and cables are

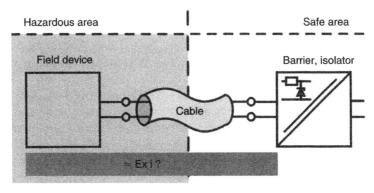

Figure 6.217 *The standard design: sensor/actuator + cable + Ex i-isolator.*

'well matched' and fitted correctly may the attribute 'intrinsically safe' be granted. In the following, planning is described in detail.

6.9.5.1 Selection of apparatus

The selection of apparatus according to the data of a gas/the gases (explosion Groups I, IIA/B/C, temperature classes T1–T6) does not differ from rules valid for e.g. 'flameproof enclosure – d –'.

For 'normal' zone 1 applications, category ib is sufficient. For circuits in zone 0 only, category ia shall be selected.

A guideline for ascertainment of apparatus categories according to their application in different zones is given in the following:

Criterion	Zone 0	Zone 1	Zone 2
Safety level	very high	high	normal
No ignition	in normal operation and with the application of two faults	in normal operation and with the application of one fault	in normal operation
Safety factor (relating to voltage or current)	1.5	1.5	1.0
Category of apparatus	ia	ib	nL*

* nL (limited energy apparatus) is an 'n' concept (IEC 60079-15, EN 50021), see Section 2.4

Thus, based on the knowledge of the gases occurred and specified zones, each apparatus can be defined completely (as far as explosion protection is concerned), e.g. Ex II 2G EEx ib IIC T4, obviously designed for zone 1 application.

6.9.5.2   An excursion: energy limitation for zone 2

Zone 2 applications and the (E) Ex nL technique (corresponding to 'intrinsic safety – i –') for limited energy apparatus and circuits are mentioned here to finalize this subject area.

The basic idea is to simplify techniques which have proved successful in the field of intrinsic safety in such a way that they will comply with the risks of zone 2. In comparison with 'intrinsic safety – i –', the fault assessment can be dropped and the reference curves (Figs 6.162–6.167) are applied with a safety factor of 1.0 (i.e. voltage and current values do not have to be reduced additionally in comparison with those given in the reference curves). Other requirements relevant for 'intrinsic safety – i –' are still applicable, e.g. the two-thirds rating of components on which the type of protection depends.

In practice, type of protection 'nL' scarcely plays a part until now. As a field of application, operationally sparking sensors in remote controlling and monitoring may be considered, e.g. electrical contacts or potentiometers.

In the USA, type of protection 'NI' ('Non-Incendive') is widely used for Division 2 applications. Operational principle and requirements are comparable with European regulations according to EN 50021. As opposed to European practice, 'nL' apparatus shall be certified by a testing body. In Europe, following Directive 94/9/EC ('ATEX 100a Directive'), a manufacturer's declaration and internal control of production is adequate for category 3G apparatus (see Section 3.3).

6.9.5.3   Certificates

Not all of the apparatus inserted in an intrinsically safe circuit need to have a certificate. This is based on the fact that, e.g., a terminal, a plug-and-socket connector or even a contact sparking in normal operation will not cause a risk if the electrical circuit is intrinsically safe.

So, the question arises which apparatus shall be certified. The following survey indicates that simple intrinsically safe apparatus do not need a certificate:

Apparatus	*Certificate for zone 1 required?*	*Examples*
Simple intrinsically safe apparatus		
Passive component without any energy storage (C, L)	no*	switch/contact, terminal, junction box, potentiometer, resistance thermometer, plug-and-socket connector
Component with a source of stored energy	no* – but capacitance and inductance shall be taken into account	coil, moving-coil instrument, capacitor

(continued)

Apparatus	Certificate for zone 1 required?	Examples
Component with a source of generated energy	yes – exception: none of the values: • 1.5 V • 100 mA • 25 mW shall be exceeded*	thermocouple, photocell
All other intrinsically safe apparatus	yes	transmitter, positioner, inductive proximity switch, handheld terminal, mobile telephone
All associated apparatus	yes	safety barrier, Ex i-isolator

* Constructional requirements given in the i-standards shall be met. On the whole, this relates to electrostatic discharge, clearances and creepage distances, separation distances between connection facilities. A special marking is not required

Certificates are issued not only for individual apparatus. In the same way, an 'intrinsically safe system' e.g. composed of a sensor, cable, power supply and meter amplifier, can be certified. This practice is customary in the Anglo-Saxon countries (United Kingdom, USA). Here, only those apparatus shall be planned and fitted which are listed or specified in the system certificate. So, a high level of safety results for the planner and user since the intrinsically safe system in its entirety (including all components) has been examined. When, however, another component has to be fitted, e.g. due to the non-availability of a specified device, the system shall be certified again (see Section 6.9.6).

In contrast to this safe but not very flexible practice the so-called 'entity concept' has been established. Here, intrinsically safe and associated apparatus may be interconnected in an arbitrary combination as long as certain 'rules' are observed. However, the high level of flexibility of this practice shall be verified by an arithmetical proof of intrinsic safety for the most part. In the following, this practice will be explained in detail.

6.9.5.4 Simple interconnection of an intrinsically safe circuit with an associated apparatus

From a 'functional' point of view, in an intrinsically safe circuit signals or data are transmitted from one device to another.

These signals or data may be:

- 4 ... 20 mA or a thermoelectric voltage in the mV range
- the voltage-coded signals of an RS 485 interface or
- simply the supply voltage.

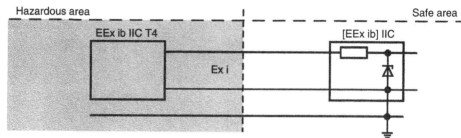

Figure 6.218 *Scheme for the assessment of 'intrinsic safety' for simple intrinsically safe standard circuits.*

The electrical parameters are current, voltage, internal resistance etc. which are specified as rated data for normal operation. These data shall not be confused with the safety-related maximum values which are the result of a fault analysis (theoretical maximum values) and which cannot occur in practice. These safety-related maximum values form the basis of planning in the field of 'intrinsic safety – i –'. They may be concluded from the certificate.

For each interconnection of an intrinsically safe circuit with an associated apparatus a proof of 'intrinsic safety – i –' shall be performed on the basis of the safety-related maximum values. As far as the simple standard design given in this example (Fig. 6.218) is concerned, it is assumed that the associated apparatus only – the Ex i-isolator – can feed electrical power into the intrinsically safe circuit. Therefore the safety–related parameters are marked as out-parameters with the subscript 'o'. On the other side, the intrinsically safe apparatus may consume electrical power only. Its values are marked as in-parameters, subscript 'i'.

Now it shall be stated whether the maximum permissible input values U_i, I_i and P_i of the intrinsically safe apparatus are higher than or equal to the maximum output values U_o, I_o or P_o of the associated apparatus. The internal capacitance and inductance C_i and L_i of the intrinsically safe apparatus shall be accordingly lower than or equal to the maximum permissible external capacitance and inductance C_o and L_o as established for the associated apparatus.

It is:

$$U_i \geqslant U_o$$
$$I_i \geqslant I_o$$
$$P_i \geqslant P_o$$
$$C_i + C_{cable} \leqslant C_o$$
$$L_i + L_{cable} \leqslant L_o$$

An apparatus showing data as given in Fig. 6.219 complies with explosion Group IIC. For IIB applications, however, the increased maximum values C_o and L_o (as indicated for IIB) may be used, resulting from the reference curves for explosion Group IIB. The maximum values for voltage and current, U_o and I_o, are laid down by design, so that the increased values permissible for IIB in principle cannot be applied here.

Digital output type 9651/*0-19-10
(terminals d26-z26; d28-z28; d30-z30; d32-z32)

Max. values:	Voltage	$U_o = 27.6\,\text{V}$
	Current	$I_o = 93\,\text{mA}$
	Power	$P_o = 642\,\text{mW}$
	Capacitance	$C_i = 1\,\text{nF}$
	Inductance	$L_i = \text{negligible}$

The respective maximum permissible value of inductance L_o or capacitance C_o is to be taken from the following table:

	IIC	IIB
C_o (nF)	85	667
L_o (mH)	4.7	17

Figure 6.219 *Abstract of a type test certificate of an associated apparatus, establishing the maximum output parameters.*

In this procedure, the cable is considered simply as a concentrated capacitance and inductance according to its length. With this simplification, however, things are at the safe side each time. The manufacturers may be asked for the cable parameters. As a 'rule of thumb' 200 nF/km and 1 mH/km are usually adequate. In practice, real cable parameters are generally lower than recommended values.

6.9.5.5   *Some peculiarities with interconnections*

Only in the case of two apparatus identical with respect to category and explosion group will their interconnection show the same properties. If this does not apply, it is:

- 'ia' with 'ib' results in 'ib'
- 'IIC' with 'IIB' results in 'IIB'.

To a certain extent, the interconnection 'inherits' the poorest properties of both apparatus. If during the design of a category ia circuit capacitance C and inductance L are examined *simultaneously* (for zone 0 application), then various combinations (C_o, L_o) are given in the certificate (Fig. 6.220) for the most part. As a rule, these values are smaller than those derived from the reference curves.

Trapezoidal characteristics, e.g. of an output for solenoid valves may result in two different Ex i-specifications for a single pair of terminals:

- category ia: in this case, only the current limiting resistor is considered

Electrical data:

Input circuits

Type 9460/12-08-11 equipped with 2-wire transmitter:

Terminals (+/−):   with intrinsically safe type of protection EEx ia IIC/IIB

1/2, 3/4, 5/6, 7/8, 9/10, 11/12, 13/14, 15/16   for connection to passive intrinsically
                                                safe circuits only
                                                Maximum values:

U_o = 26.2 V
I_o = 86 mA
P_o = 561 mW

Linear characteristic curve

The maximum permissible L_o and C_o can
be found in the following table (C_i and L_i
included):

IIC		IIB	
L (mH)	C (nF)	L (mH)	C (nF)
2.5	42	18.0	320
2.0	46	2.0	340
1.0	59	1.0	390
0.5	74	0.5	470
≤0.2	97	0.2	590
		0.1	700
		≤0.05	750

Figure 6.220   *Abstract of a type test certificate, establishing various maximum permissible L_o/C_o-combinations.*

- category ib: *in addition*, an electronic current limiting device is to be taken into account. The current I_o is reduced accordingly. In most cases, C_o and L_o are smaller than specified for category ia. For applications which do not ask for category ia, both specifications may be used alternatively (Fig. 6.221).

6.9.5.6   Interconnection of several i-apparatus with an associated apparatus

If several intrinsically safe apparatus are interconnected in an intrinsically safe circuit (e.g. a transmitter and an indicator are connected in series in a 4 ... 20 mA current loop), then a comparison between the OUT parameters and the IN parameters of each single intrinsically safe apparatus shall be made. Internal capacitances and inductances of both apparatus shall be added for this assessment.

Electrical data:

Type 9475/12-04-11

Output circuit terminals (+/−):   with intrinsically safe type of protection EEx ia IIC/IIB, or EEx ib IIC/IIB, for connection to passive intrinsically safe circuits only

1/2, 5/6, 9/10, 13/14

Maximum value category ia:

$U_o = 19.9\,V$
$I_o = 150\,mA$
$P_o = 742\,mW$
Linear characteristic curve

Maximum value category ib:

$U_o = 19.9\,V$
$I_o = 60\,mA$
$P_o = 714\,mW$

Trapezoidal characteristic curve

The maximum permissible external inductances and capacitances can be found in the following table (C_i and L_i included):

IIC				IIB			
ia		ib		ia		ib	
L_o (mH)	C_o (nF)	L_o (mH)	C_o (nF)	L_o (mH)	C_o (nF)	L_o (mH)	C_o (nF)
				5.0	830	5.0	840
1.1	100	2.0	100	2.0	840	2.0	840
1.0	100	1.0	100	1.0	840	1.0	840
0.5	120	0.5	120	0.5	850	0.5	860
0.2	150	0.2	150	0.2	1000	0.2	1000
0.1	190	0.1	190	0.1	1200	0.1	1200
≤0.05	220	≤0.05	220	≤0.05	1417	≤0.05	1417

Figure 6.221 *Abstract of a type test certificate, establishing various maximum permissible L_o/C_o combinations for category ia as well as for category ib.*

A model for proving type of protection 'intrinsic safety – i –' with two interconnected intrinsically safe apparatus is given as follows:

Hazardous area		Cable	Safe area	
i-apparatus 1	U_{i1}		$\geqslant U_o$	associated
	I_{i1}		$\geqslant I_o$	apparatus
(C_{i1}, L_{i1})	P_{i1}		$\geqslant P_o$	(C_o, L_o)
i-apparatus 2	U_{i2}		$\geqslant U_o$	
	I_{i2}		$\geqslant I_o$	
(C_{i2}, L_{i2})	P_{i2}		$\geqslant P_o$	
	$C_{i1} + C_{i2}$	$+C_{cable}$	$\leqslant C_o$	
	$L_{i1} + L_{i2}$	$+L_{cable}$	$\leqslant L_o$	

6.9.5.7   *Interconnections with several associated apparatus: EN 60079-14, IEC 60079-14*

Deviating from the method described above these are cases with several associated apparatus interconnected in one single circuit. It may then happen that two or more apparatus are 'active in the sense of intrinsic safety'. This indicates that more than one individual apparatus may feed power into the intrinsically safe circuit.

In these cases it shall be determined whether one (or more) of the following may happen:

- addition of currents (always true for parallel connection, the most frequent case)
- addition of voltages (always true for series connection) or
- addition of currents and voltages.

Therefore, a detailed knowledge of the interconnection of all apparatus inserted in this circuit is required.

In the case of an addition of currents, 'intrinsic safety – i –' will be proved step by step as follows:

- add the currents of the associated apparatus $I_o = I_{o1} + I_{o2}$
- determine the maximum voltage $U_o = U_{o1}$ or $U_o = U_{o2}$ whichever is the highest
- examine whether the 'new' combination (U_o, I_o) complies with the reference curves (or not), pay attention to the safety factor
- calculate the permissible C_o with respect to the capacitive reference curve for voltage U_o, pay attention to the safety factor
- calculate the permissible L_o with respect to the inductive reference curve for current I_o, pay attention to the safety factor
- compare the 'new' parameter set (U_o, I_o, C_o, L_o) with the INPUT parameters $(U_i, I_i$ etc.) of the intrinsically safe apparatus as described above, pay attention to the cable parameters C_{cable} and L_{cable}.

Caused by addition of currents and increasing values (U_o, I_o) produced thereby, the explosion group of the interconnected apparatus may change. Both associated apparatus are assumed to be specified for IIC. Their interconnection, however, may exceed the values permissible for IIC, so that the total assembly complies with IIB only.

Besides, the maximum power P_o is increased due to the interconnection for the most part. Thus, the temperature class of the intrinsically safe apparatus shall be investigated.

In the case of an addition of voltages (associated apparatus connected in series) the calculation shall be made accordingly.

In each case, the interconnection is classified as category ib, even though all associated apparatus comply with category ia. This 'downgrade' of category makes allowance for the fact that the assessment is performed without any test by calculation only. The procedure is described in detail in an annex to EN 60079-14 and IEC 60079-14. For documentation, a description of the system is required.

Generally, non-optimized values are obtained following this procedure since the resulting characteristic shows a deviating shape as a matter of fact. However, this approximation is at the safe side each time.

Finally, it should be stated that the procedure described here shall be applied for linear characteristics only. For deviating characteristics publication PTB-ThEx-10 should be referred to. In addition, rear feed from the circuit into the associated apparatus shall be taken into account (the rated values of current and voltage limiting components shall not be exceeded). As far as safety barriers are concerned, rear feed is uncritical due to their 'robust' design.

6.9.5.8 Interconnections according to EN 60079-14/IEC 60079-14: an example of how to proceed

An intrinsically safe solenoid valve may show the following data:

- $U = 12\,V$, $I_{min} = 35\,mA$ (switch-on condition)
- EEx ia IIC T6
- $U_i = 28\,V$, $I_i = 105\,mA$
- $P_i = 650\,mW$ (for T6)
- $P_i = 800\,mW$ (for T4)
- $L_i \sim 0$, $C_i = 20\,nF$.

This solenoid valve shall be controlled by an intrinsically safe digital output. Two identical channels show data as follows:

- no-load voltage $U = 25\,V$, internal resistance $R_i = 630\,ohms$, functional current limitation to 25 mA
- EEx ia IIC T6
- $U_o = 28\,V$, $I_o = 49\,mA$
- $P_o = 345\,mW$
- $L_o = 15\,mH$, $C_o = 83\,nF$
  (both values for IIC)

Functional planning follows the diagram given in Fig. 6.222.

Since the output current of the digital output is not sufficient to ensure a safe switch-on of the solenoid valve two channels shall be operated in parallel. This assembly can feed approx. 40 mA at 12 V so that there will be a certain reserve to cover, e.g. cable resistances.

Safety-related planning is explained according to Fig. 6.223.
Calculate the values of the assembly:

- $I_o = I_{o1} + I_{o2} = 2 \times 49\,mA = 98\,mA$
- U_o = maximum value of $(U_{o1}, U_{o2}) = 28\,V$
- (U_o, I_o) complies with IIC (see Table 6.39), safety factor 1.5 is included

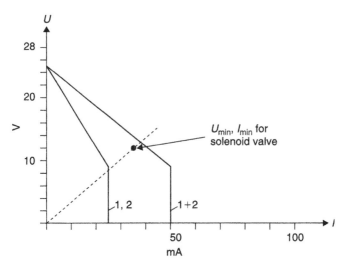

Figure 6.222 *Functional assessment of two outputs connected in parallel.*

- C_o results in 83 nF (for IIC, 28 V, see Table 6.39), safety factor 1.5 is included
- L_o results in 4 mH (for IIC, 98 mA, see Table 6.39), safety factor 1.5 is included.

Compare INPUT parameters with OUTPUT parameters (see Fig. 6.224):

- $U_o = 28\,\text{V} \leqslant U_i = 28\,\text{V}$ (OK)
- $I_o = 98\,\text{mA} \leqslant I_i = 105\,\text{mA}$ (OK)
- $P_o = 2 \times 345\,\text{mW} = 690\,\text{mW} \leqslant P_i = 800\,\text{mW}$ (OK for T4 only!)
- $C_o = 83\,\text{nF} \geqslant C_i + C_{\text{cable}}$
  $83\,\text{nF} - C_i \geqslant C_{\text{cable}}$
  $C_{\text{cable}} \leqslant 63\,\text{nF}$
- $L_o = 4\,\text{mH} > L_i + L_{\text{cable}}$
- $L_i \sim 0$
  $L_{\text{cable}} \leqslant 4\,\text{mH}$.

Assuming a 2-wire cable with 200 nF/km and 1 mH/km per wire then there is a length limitation due to capacitance:
$$l_C = 63/200\,\text{km}$$
$$l_C = 0.315\,\text{km}$$
and a length limitation due to inductance:
$$l_L = 1/2\ (\text{wires}) \times 4/1\,\text{km}$$
$$l_L = 2\,\text{km}$$
Since $l_C < l_L$, the cable length shall not exceed 315 m.
Finally, the interconnection shows properties as follows:

- EEx ib (as mentioned above, the interconnection of several associated apparatus generally results in category ib)

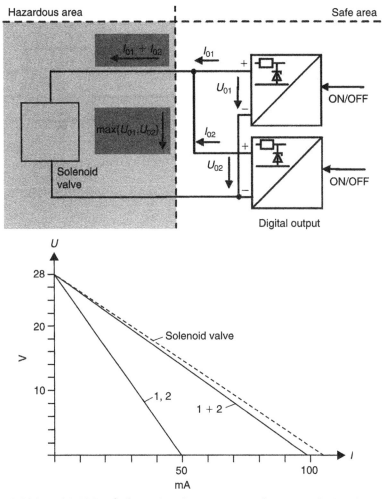

Figures 6.223 and 6.224 *Safety-related assessment of current addition (two outputs connected in parallel).*

- IIC (all values of the interconnection comply with IIC)
- T4 (due to $P_o = 690\,\text{mW}$)
- cable length $\leqslant 315\,\text{m}$.

6.9.5.9 Verification of intrinsic safety for the H1 fieldbus

In principle, the intrinsically safe fieldbus forms an interconnection of several intrinsically safe and, perhaps, of several associated apparatus.

The comments given above demonstrate that – without any further restrictions – the assessment of intrinsic safety could be realized with an insupportably high expenditure only. For such a fieldbus, the constellation

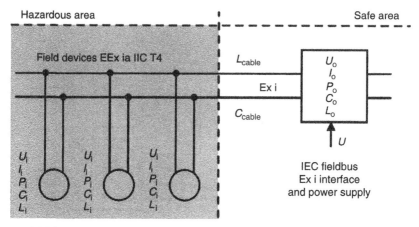

Figure 6.225 *Safety-related parameters of fieldbus components.*

according to Fig. 6.225 is typical: several field devices (intrinsically safe or associated apparatus) are connected to the fieldbus power supply (an associated apparatus) via a single cable only. This fieldbus power supply contains the barrier to isolate the non-intrinsically safe bus and ensures the DC supply of the field devices. The assessment of intrinsic safety for such a fieldbus may be arbitrarily long winded and arduous especially in the case of field devices with an individual power supply (4-wire transmitters), which shall be considered as associated apparatus with their OUTPUT parameters.

The Fieldbus Foundation made the following recommendation to simplify matters:

- the safety-related maximum values for field devices are specified 'severely'
- from this, the parameters of the power supply can be deduced. So, field devices and power supply are matched (theoretically) each time
- in addition, field devices which are not supplied via the bus are 'passive in the meaning of safety technology', i.e. $U_o = 0$ and $I_o = 0$.

So, there is a need only to verify whether the parameters of each single field device will harmonize with the parameters of the power supply (or not). The capacitances and inductances of the field devices and the cable shall be added as usual. The specifications of the Fieldbus Foundation for safety-related parameters of field devices are summarized in Table 6.46.

Due to these restrictions the 'intrinsic safety' of a fieldbus will be established in a simple way. However, practice has shown that a rather small amount (4 … 6) only of field devices depending on cable length can be operated via a single bus following this proceeding. The reason is the limitation of capacitance C and inductance L. In fact, a power supply with high output current provides the current for many field devices (10 mA each). However, the values for C_o and L_o still permissible are accordingly low. In contrast, a

Table 6.46   Safety-related parameters of
field devices as specified by Fieldbus
Foundation

$U_i \geqslant 24\,V$	
$I_i \geqslant 250\,mA$	
$P_i \geqslant 1.2\,W$	$U_o \leqslant 24\,V$
$C_i \leqslant 5\,nF$	$I_o \leqslant 250\,mA$
$L_i \leqslant 20\,\mu H$	$P_o \leqslant 1.2\,W$
$U_o = 0$	
$I_o = 0$	

Note: C_o and L_o are deduced from the values
of U_o and I_o as put into practice

low current power supply with high values for C_o and L_o can provide a few
field devices only with the current required. Remedial measures are taken by
the so-called FISCO model (see Section 6.9.6).

6.9.5.10   Verification of intrinsic safety according to PTB-Report ThEx-10

The method to calculate interconnections according to EN 60079-14 or IEC
60079-14 shows certain 'drawbacks':

- the reference curves for inductive circuits are applicable for voltages not
  higher than 24 V only
- the reference curves are applicable for pure ohmic, capacitive or inductive
  circuits only. Mixed circuits with combined capacitances C and inductances L are not in the scope
- the reference curves are not applicable to non-linear sources.

To prove intrinsic safety with more complex interconnections by way of calculation, it is recommended to follow PTB-Report ThEx-10* in this matter.
    The core of this procedure is a graphic summation of the output character-
istics of all sources involved. The summation characteristic (output) is to be
plotted in a suitable diagram (Fig. 6.226) which indicates the compliance of
the circuit with 'intrinsic safety' even in the case of a simultaneous inductive
and capacitive wiring.
    As an essential advantage of this method, all safety-related information
and boundary conditions can be deduced from one diagram. The safety factor 1.5 has been incorporated into the diagram already.
    The proof of 'intrinsic safety' by way of calculation ascertains that the
apparatus and their interconnection comply with the 'rules' of intrinsic

---

* Johannsmeyer, U. and Kraemer, M.: Interconnection of non-linear and linear intrinsically safe circuits, PTB-Report ThEx-10, Physikalisch-Technische Bundesanstalt,
Braunschweig and Berlin/Germany, 2000

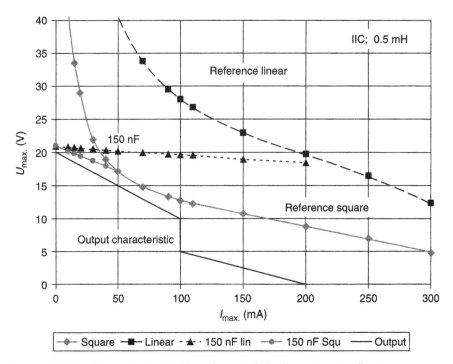

Figure 6.226 *Example of a limiting characteristic diagram according to PTB-Report PTB-ThEx-10. The output current/voltage diagram is indicated ('output characteristic').*

safety. In addition, attention shall be paid to some essential points during planning and installation of intrinsically safe circuits in order to guarantee a safe operation. Most of the provisions serve to isolate an intrinsically safe circuit from other electrical circuits adequately:

- requirements for cables
- marking of intrinsically safe circuits
- isolation of intrinsically safe from non-intrinsically safe circuits
- isolation of intrinsically safe circuits from each other
- as far as required, connection to the equipotential bonding system.

6.9.5.11 Cables for intrinsically safe circuits

Generally, cables for intrinsically safe circuits are subject to voltage tests. Cables intended for installation in chemical plants (Group II application) shall be tested with at least 500 Vrms AC between (Fig. 6.227(a)):

- conductor and earth
- conductor and screening
- screening and earth.

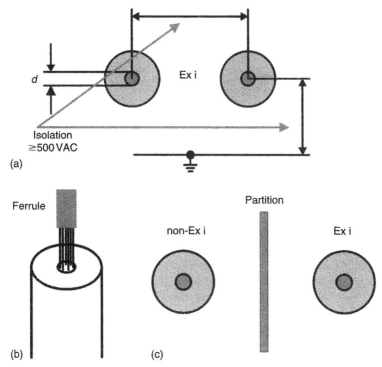

Figure 6.227(a) to (c)   *Requirements for cables in intrinsically safe circuits (schematic representation).*

In addition, the diameter of individual conductors within the area subject to explosion hazards shall not be less than 0.1 mm. This applies also to the individual wires of a finely stranded conductor.

These requirements are given in EN 60079-14 and IEC 60079-14 for Group II installations.

For Group I installations, requirements are given in the directives listed in Table 3.6, Part A. Directive 82/130/EEC asks for a voltage test for cables (after manufacturing) with:

- 1000 V rms AC between the entirety of all screenings or armours (which shall be connected together) and the bundle of all conductors (which shall be connected together)
- 2000 V rms AC between each single conductor of an n-wire cable and the residual n-1 conductors of this cable (which shall be connected together). These test requirements are given in supplement B, annex 3 of said directive, and, for the second 'generation' of EN 50020, in Directive 98/65/EC, supplement II, annex 3.

The ends of multistrand or finely stranded conductors shall be protected against splicing, e.g. by use of a ferrule (Fig. 6.227(b)). Suitable terminals may provide the same protection by design. A screening – if required – shall be earthed at one single point only, usually in the safe area. Exceptions may be made according to EN 60079-14 and IEC 60079-14.

Type of protection 'intrinsic safety – i –' shall not be adversely affected by external electric or magnetic fields. Precautions may be taken as follows:

- screening (of conductors)
- twisted conductors (twisted pair)
- distance from non-intrinsically safe sources.

The conductors of intrinsically safe and non-intrinsically safe circuits shall not pass the same cable. In cable ducts, an insulating layer or an earthed metallic partition (see Fig. 6.227(c)) is required.

Several intrinsically safe circuits may run in a single multicore cable without taking cable faults into account (short-circuits between conductors or conductor break), if:

- the conductor insulation will be capable of withstanding an rms test voltage of twice the nominal voltage of the intrinsically safe circuit with a minimum of 500 V
- a voltage test with at least 500 Vrms AC applied between any armouring and/or screening(s) joined together and all the cores joined together will be passed successfully
- a voltage test with at least 1000 Vrms AC applied between a conductor bundle comprising one-half of the cable cores joined together and a bundle comprising the other half of the cores joined together will be passed successfully. This test is not applicable for multicore cables with conducting screenings for individual circuits.

These requirements are part of EN 60079-14 and IEC 60079-14 for Group II applications and are additional to those explained above:

- the cable complies with all these testing requirements and shows conducting screenings providing individual protection for intrinsically safe circuits in order to prevent such circuits becoming connected to one another (so-called cable type A)
- the cable complies with all these testing requirements and is fixed, effectively protected against damage, and, in addition, no circuit within the cable shows a maximum voltage U_o exceeding 60 V (so-called cable type B).

Similar requirements for Group I installations are given in Directives 82/130/EEC and 98/65/EC.

If the cable does not comply with these requirements, a certain amount of faults shall be assumed. Generally, in industrial practice (one pair of wires for each single arbitrary intrinsically safe circuit) a fault assessment can hardly be performed. In these cases, cables of type A or B shall be used. A fault assessment is reserved for special cases.

6.9.5.12   *Marking of intrinsically safe electrical circuits*

As distinguished from other types of protection, 'intrinsic safety – i –' permits working at live parts, i.e. during continuous operation, if required.

Sparks generated at break or unintentional short-circuiting of an intrinsically safe circuit are not perilous. Confusions with non-intrinsically safe circuits which do not have this characteristic shall be strictly excluded. Therefore, intrinsically safe circuits shall be marked as such.

Standards do not ask for a defined way of marking. If a colour is used, however, it shall be light blue. This is the marking used for the most part in practice: light blue coloured cables, terminals, markings of connections of apparatus etc.

6.9.5.13   Equipotential bonding

Intrinsically safe circuits may be isolated from earth or connected to the equipotential bonding system at one single point, if this system covers the total installation area of the intrinsically safe circuits.

Safety barriers (without galvanic isolation) ensure 'intrinsic safety' only, when connected to the equipotential bonding system.

The requirements according to EN 60079-14 and IEC 60079-14 are either:

- at least two separate conductors each rated to carry the maximum possible current which can flow continuously, each with a minimum of $1.5\,mm^2$ (copper) or
- at least one conductor with a minimum of $4\,mm^2$ (copper) or
- an impedance less than 1 ohm between the point of connection and the earth point.

If equipotential bonding is lacking, there will be a voltage $U = \varphi_2 - \varphi_1$ between the intrinsically safe circuit and earth (see Fig. 6.228). An unintentional connection of both points with each other would generate a current limited by the resistance of the cable only. Equipotential bonding forces $\varphi_1 \sim \varphi_2$ and thus prevents voltages and currents worth mentioning.

The advantage of galvanic isolation is obvious: even at different electric potentials, a current cannot occur.

6.9.6   Advanced methods of intrinsic safety

Again and again, testing houses, manufacturers and users of intrinsically safe products have put the question, how could the narrow borderlines of classic 'intrinsic safety' concerning available power be passed over? These borderlines are determined by reference curves as well as by installation rules. The following approaches are dealt with in more detail below:

- several intrinsically safe circuits isolated from each other; intrinsically safe systems
- increased values for current and voltage are permissible for explosion Group IIB than for Group IIC
- for voltages lower than approx. 10 V, the reference curves permit currents of some amperes. In this field the effective power can be increased

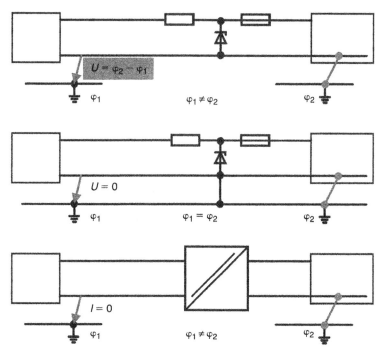

Figure 6.228   *Equipotential bonding of safety barriers – a must.*

- the so-called FISCO model eliminates restrictions resulting from, e.g., cable parameters (capacitance, inductance) for fieldbus systems
- more recent research (for the example of fieldbus systems, too) takes advantage of the fact that systems operated with alternating current permit an increased power level than those operated with direct current
- an increased power level may be handled in such a way that high power component assemblies are explosion protected according to a type of protection deviating from 'intrinsic safety – i –', in other words, a combination of several types of protection is recommended.

6.9.6.1   Intrinsically safe systems according to EN 50039

In certification practice, the so-called intrinsically safe systems are known as well. Deviating from a (single) apparatus which is certified individually, an interconnection of several apparatus is under consideration for an intrinsically safe system, including the connecting cables for the most part. Several intrinsically safe circuits and different categories of these circuits as a part of an intrinsically safe system are permissible.

The standard EN 50039 permits uncertified systems, too. In this case, type of protection 'intrinsic safety' shall be established by knowledge of the data of all apparatus including those not certified and the cables.

This proof is subject to experts in most cases. Documentation as a 'system's description' is imperative.

In addition, it may be mentioned as a special feature of intrinsically safe systems that a fault analysis (as an assessment for individual apparatus) (two faults are assumed for category ia, one fault for category ib) may be applied to the system as a total and not to each individual apparatus.

A separate chapter of standard EN 50039 is dedicated to 'cables' in an intrinsically safe system. Special requirements for multicore cables with several intrinsically safe circuits are listed as well as the faults to be assumed between intrinsically safe circuits (wire breaks and short-circuits) if these requirements are not met. In the meantime, these requirements form a part of standards EN 60079-14 and IEC 60079-14 respectively (see Section 6.9.5).

An example for a 'system certificate' is given in Fig. 6.229. In addition to the usual marking, the letters, 'SYST' are placed in front.

With, e.g., an intrinsically safe RS 485 bus application, the potentialities of this method will be demonstrated in brief.

In this case (see Fig. 6.230), the resulting total current $I_{osum} > 2\,A$, caused by addition of currents of the apparatus connected to the bus. A 'classic' consideration results in very short cable lengths with inductances correspondingly low. During extensive trials using the spark test apparatus it could be ensured that even 'long' cables may be installed without any risk as long as the relevant cable parameters L' (inductance per unit length) and C' (capacitance per unit length) do not exceed defined limiting values. The result has been documented in a SYST certificate.

6.9.6.2   Increased values for explosion Group IIB

If the variety of combustible gases in a chemical plant can be restricted to gases with an increased MIC (**minimum ignition current**), i.e. if gases classified as IIC can be excluded, then the reference curves, e.g. for Group IIB, may be applied: for the same voltage, increased currents are permissible. The difference is a considerable one and is shown as follows:

	Parameter	*Explosion Group*	
		IIC	*IIB*
Table 6.39, including a safety	U	28 V	28 V
factor of 1.5, resistive circuits	I	120 mA	299 mA
Effective rated voltage	U_N	24 V	24 V
Current limiting resistance approx.	R_i	250 ohms	110 ohms
Short-circuit current	$I_{max.} = U_N/R_i$	96 mA	218 mA
Power matching	$P_{max.} = 0.25 \times U_N \times I_{max.}$	0.58 W	1.31 W

Referring to a rated voltage of 24 V, the maximum available power is approx. 1.3 W for Group IIB. For Group IIC, the corresponding value is 0.58 W.

Uncertified Translation

Physikalisch-Technische Bundesanstalt
Braunschweig und Berlin

(1) **C E R T I F I C A T E   O F   C O N F O R M I T Y**

(2) **PTB No. Ex-96.D.2193 X**

(3) This certificate is issued for the electrical apparatus

Intrinsically Safe Field Bus System RS 485

(4) of R. STAHL SCHLATGERÄTE GMBH
D-74653 Künzelsau

(5) Type of construction of this electrical apparatus as well as the various permissible
versions are specified in the annex to this certificate.

(6) Physikalisch-Technische Bundesanstalt as certification body approves in accordance
with Article 14 of the Council Directive of the EC of 18 December 1975 (76/117/EEC),
the conformity of this electrical apparatus with the harmonized European Standards

Electrical apparatus for potentially explosive atmospheres

EN 50 014:1977 + A1...A5 (VDE 0170/0171 Part 1/1.87)   General Requirements
EN 50 039:1980   (VDE 0170/0171 Part10/4.82) Intrinsically Safe Electrical
                                                        Systems "i"

after having successfully tested the system in regard to type of construction. The
results of this test are specified in a confidential test record.

(7) The apparatus is to be marked as follows:

SYST EEx ib IIC T6 or SYST EEx ib IIB T6

(8) The manufacturer has the responsibility to ensure that each apparatus marked as
above is in conformity with the test documentation listed in the annex to this certificate
and that the required routine tests have been carried out successfully.

(9) The electrical apparatus may be marked with the mark of conformity specified in
Annex II of the Council Directive of 6 February 1979 (79/196/EEC).

On behalf of PTB   Braunschweig, 24.01.1997
(signature)
Dr.-Ing. Johannsmeyer
Oberregierungsrat

Figure 6.229   *Example of a certificate according to EN 50039.*

Applications are given in the field of, e.g., power supply of intrinsically
safe apparatus whose power demand can be satisfied with a Group IIB circuit only, i.e. reference is made to the IIB curves.

An interesting application arises out of the fieldbus technique, too (see
Fig. 6.231). In this example, the main bus segment is designed to comply with

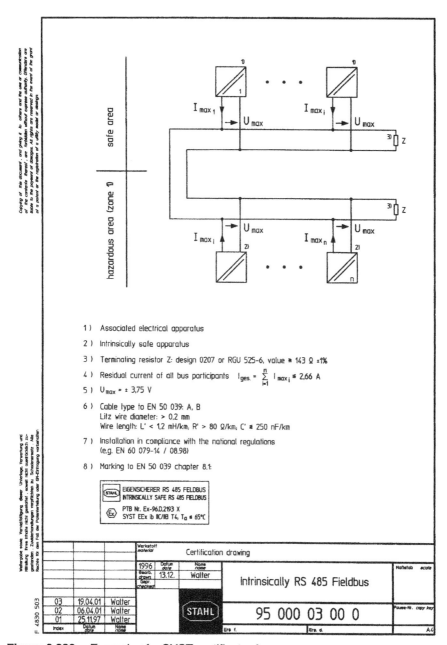

1) Associated electrical apparatus

2) Intrinsically safe apparatus

3) Terminating resistor Z: design 0207 or RGU 525-6, value ≈ 143 Ω ±1%

4) Residual current of all bus participants $I_{ges.} = \sum_{i=1}^{n} I_{max_i} ≈ 2.66$ A

5) $U_{max} = ± 3.75$ V

6) Cable type to EN 50 039: A, B
Litz wire diameter: > 0.2 mm
Wire length: L' < 1.2 mH/km, R' > 80 Ω/km, C' ≤ 250 nF/km

7) Installation in compliance with the national regulations
(e.g. EN 60 079-14 / 08.98)

8) Marking to EN 50 039 chapter 8.1:

> EIGENSICHERER RS 485 FELDBUS
> INTRINSICALLY SAFE RS 485 FELDBUS
> PTB Nr. Ex-96.D.2193 X
> SYST EEx ib IIC/IIB T4, Ta ≤ 65°C

			Werkstoff material	Certification drawing			Haftetab acate
			1996	Datum date	Name name		
			Bearb. drawn	13.12.	Walter	**Intrinsically RS 485 Fieldbus**	
			Gepr. checked				
03	19.04.01	Walter		**STAHL**		**95 000 03 00 0**	Pause-Nr. copy key
02	06.04.01	Walter					
01	25.11.97	Walter					
Index	Datum date	Name name		Ers. f.		Ers. d.	A4

F 4830 503

Figure 6.230 *Example of a SYST certificate: Annex.*

explosion Group IIB, thus enabling increased supply currents. The branch connections to the field devices are equipped with an additional current limiting resistor each. So, the field devices are operated according to explosion Group IIC in a local area.

Of course, the thermal behaviour of the additional resistors shall be taken into account accordingly.

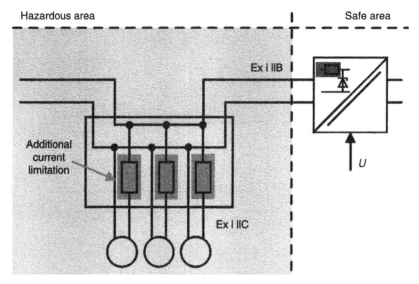

Figure 6.231   *A fieldbus is designed according to explosion Group IIB. To extend it with field devices into a hazardous area endangered by Group IIC gases, an additional current limitation is installed in the terminal box.*

6.9.6.3   Reference curves at low voltages

Overlooking the reference curves for resistive circuits (Fig. 6.162) it is obvious that relatively high currents are permissible for low voltages. More detailed information is given in Tables A.1 (resistive circuits, Groups IIA/B/C) and A.2 (capacitive circuits, Groups IIA/B/C) of EN 50020 or IEC 60079-11 respectively. (Table 6.39 is an abstract of Tables A.1 and A.2 as far as resistive and capacitive circuits are concerned.)

To give an example, Table A.1 indicates a minimum ignition current of 3.33 A (safety factor of 1.5 is included) for 12.1 V. Physical effects result in the fact that spark ignition does not occur for voltages below approx. 9–11 V. This behaviour may be used to increase power in intrinsically safe circuits to values exceeding 0.5–1 W as available in general.

The effective 'intrinsically safe voltage' is in the range of approx. 8–10 V. So, for many applications that require an increased voltage level a DC/DC convertor has to be inserted. Of course, this DC/DC convertor in an intrinsically safe circuit shall be explosion protected itself accordingly. This increases the technical expenditure required.

For currents exceeding approx. 500 mA it shall be taken into account that inadmissible high temperatures may occur for inadequate small conductor cross-sections (cable 'near to breakoff') or for poor contact making (terminal tightened insufficiently). This effect is well known as ignition by incandescence.

In particular, for explosion Group IIC the inductance values permissible for higher currents are considerably lower than 0.1 mH. So, leakage inductances arising everywhere shall be considered during planning and installation.

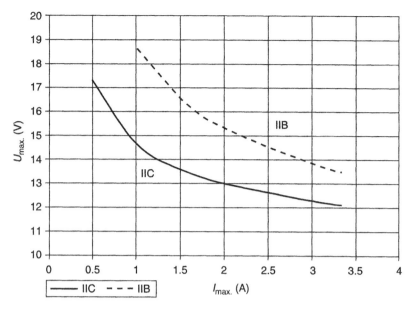

Figure 6.232 *Detailed representation of IIB and IIC reference curves for low voltages (resistive circuits). The safety factor of ×1.5 is incorporated.*

A summary may be given as follows:

- for voltages below approx. 9–11 V, no spark ignition will occur
- theoretically, currents in the ampere range can be achieved (see Fig. 6.232)
- there are restrictions caused by leakage inductances and ignition by incandescence.

In the example of Fig. 6.233 a power supply for installation in zone 1 is protected according to 'flameproof enclosure – d –'.

The outputs (max. four) to supply the other system modules comply with 'intrinsic safety – i –' with maximum values as follows:

- 9.5 V
- 1 A
- 10 μH.

The power available per individual output is approx. 8.5 W.

So, it can supply one or more system modules according to their current demand. The DC/DC convertor mentioned above is fitted to the individual modules as required. In this example, cable lengths do not exceed some metres, so that, in practice, the small inductance permissible does not impose a restriction.

6.9.6.4 The FISCO model (fieldbus intrinsically safe concept)

Each proof of intrinsic safety achieved by evaluations or in practice is an approximation, i.e. the work is done with simplifications. In order not to

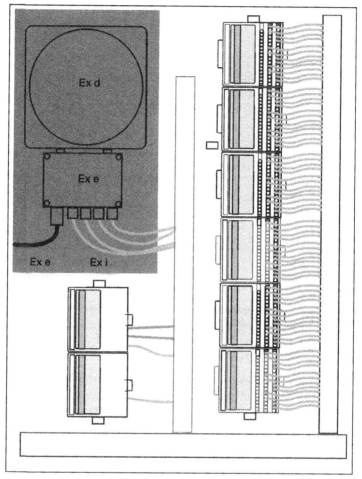

Figure 6.233   *Intrinsically safe power supply, installed in a flampeproof enclosure. Four outputs of 8.5 W each.*

jeopardize safety, simplifications shall come 'to the safe side'. As a rule, cables are considered as non-resistive and as a concentrated capacitance and inductance. Of course, this does not go along with physical reality but results in an approximation still useful in most cases.

Increasing work in the intrinsically safe fieldbus domain has raised problems which are closely related to the so-called cable parameters. The problems:

- the fieldbus shall be intrinsically safe
- the devices connected to the bus shall be supplied additionally with auxiliary power via the bus (two wires only)
- this results in relatively high safety-related maximum values U_o and I_o for the Ex i-isolator feeding the bus
- this results in relatively low values for maximum permissible capacitance C_o and inductance L_o

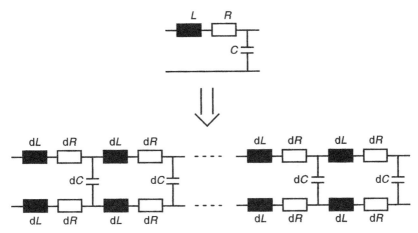

Figure 6.234 *Cable models with concentrated and differential cable parameters.*

- and results in short cable lengths only if the cable is considered as a concentrated capacitance and inductance.

The solutions:

- the behaviour of a cable can be explained very well by a model with differential resistances, capacitances and inductances (dR, dC and dL) distributed over the cable length (see Fig. 6.234)
- (concentrated) energy storage elements which are discharged rapidly are critical for spark ignition
- resistances dR are current limiting elements within the meaning of 'intrinsic safety' and act on spark ignition as an impediment.

So, in a research project, PTB Physikalisch-Technische Bundesanstalt, Germany, has investigated the influence of real cables on spark ignition theoretically and in practical trials using the spark test apparatus.*

The results of these investigations (known as 'FISCO') have confirmed the assumption that – with a certain simplification – a cable does not support but impedes spark ignition. So, a fieldbus with a 'long' cable ignites less efficiently than a fieldbus with a 'short' cable under different fault assumptions and test conditions.

The boundary conditions for the FISCO model are summarized as follows (see Fig. 6.235):

- only the power supply is 'active' with respect to safety (U_o, I_o are defined)
- all field devices are passive with respect to safety ($U_o = 0$, $I_o = 0$)

---

* Results have been published in: Johannsmeyer, U.: Investigations into the intrinsic safety of fieldbus systems, Report PTB-W 53, Physikalisch-Technische Bundesanstalt, Braunschweig and Berlin/Germany, 1994

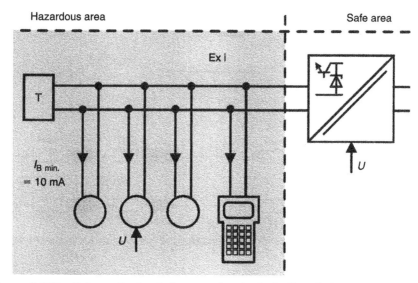

Figure 6.235   *Schematic circuit diagram of an intrinsically safe fieldbus with power supply, field devices and cable termination (T).*

- in addition, this applies to field devices which are supplied by an external voltage source (4-wire transmitters). A galvanic isolation to the auxiliary power source is required
- the energy storage elements which are effective at the terminals of the fieldbus devices are limited to $L_i < 10\,\mu H$ and $C_i < 5\,nF$.

Some proposals for a power supply have been worked out:

- e.g. rectangular-shaped output characteristic with
- $U_o = 15\,V\,DC$
- $I_o = 130\,mA$
- rated values (DC): 14 V, 110 mA

and for the cable termination (T) according to IEC 61158-2:

- 100 ohms connected in series with $1\,\mu F$
- at both ends of the bus
- the capacitance and the potential heating of the resistor shall be taken into account.

Different testing circuits have been put into practice (Fig. 6.236) to investigate the spark ignition characteristic of a fieldbus power supply with and without a cable and/or cable termination.
The results may be summarized as follows:

- the circuit *without cable* and *without cable terminations* ignites *best!*
- so, long cables are practicable.

These results can be generalized, of course. In practice, today they are applicable to Profibus PA and Foundation Fieldbus H1 only, because the apparatus

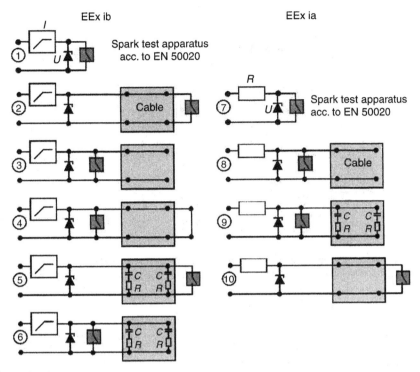

Figure 6.236 *Different test circuits for the investigation of the ignitability of a field-bus power supply with and without cable and cable termination.*

to be installed shall be certified according to the boundary conditions mentioned above.

The further development of the FISCO model will be incorporated into international standardization as IEC 60079-27 'Electrical apparatus for explosive gas atmospheres, FISCO-Concept for H1 intrinsically safe Fieldbus systems'. In this standard, the safety-related data will be specified in detail:

- fieldbus power supply with a linear or rectangular-shaped output characteristic
- field devices as 'passive' apparatus. Galvanic isolation is required if not fed via the bus
- bus termination and
- cable data (R_{cable}, C_{cable}, L_{cable}), maximum permissible length for bus cable and branches.

A system specified to such a degree is intrinsically safe with a maximum of 32 field devices and two terminations. The 'proof of intrinsic safety' is thereby reduced to a list of the apparatus installed and their certificate. The marking includes 'FISCO' and the type of apparatus, e.g. power supply, field device, termination.

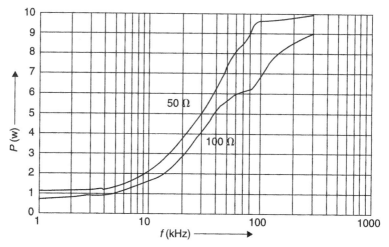

Figure 6.237   *Available electrical power in an AC fed intrinsically safe circuit versus frequency at the power matching condition for load resistances 50 ohms and 100 ohms (according to Gerlach et al., etz 18 (1998)).*

6.9.6.5   *Intrinsic safety for alternating current*

In the standards no criterion applicable in practice can be found for the assessment of the ignitability of alternating currents in intrinsically safe circuits. More recent investigations by PTB Physikalisch-Technische Bundesanstalt, Germany, have shown that the power transformed will increase significantly by application of alternating currents in intrinsically safe circuits in comparison with direct current. In the frequency range 10–300 kcps, a multiple power (see Fig. 6.237) is available.[*] In experiments, PTB has determined reference curves for resistive AC circuits and analysed the time dependence of electrical spark parameters. It could be demonstrated that the half-wave spark energy value is of importance for the assessment of the ignitability.[*]

In a dissertation at Technische Universität Braunschweig/Germany, theoretical and experimental contributions have been made for an extension of the application domain of fieldbus systems in areas endangered by potentially explosive atmospheres by use of alternating currents in a higher frequency range.[**] A significantly increased power in intrinsically safe circuits is available at frequencies higher than 20 kcps (Fig. 6.238). The results arising

---

[*] Published in: Gerlach, U., Wagner, S., Wehinger, H. and Varchmin, J.-U.: Eigensichere Wechselstromschnittstelle für Feldbusanwendungen, *etz Elektrotechnik +Automation* 18, 1998, 34–37
[**] Published in: Gerlach, U. and Wagner, S.: Feldbusanschaltung mit Wechselstromspeisung für die Zündschutzart Eigensicherheit, Dissertation, Technische Universität Braunschweig/Germany

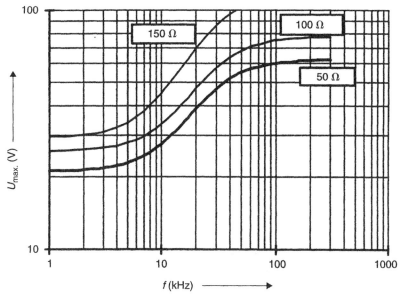

Figure 6.238 *Ignition voltage versus AC frequency for different resistances, 21% (v/v) hydrogen in air, ignition probability = 10^{-3} (according to Gerlach and Wagner, Technische Universität Braunschweig).*

from these projects can be applied not only for fieldbus systems but also for other intrinsically safe AC systems.

Complicated power distributions may be caused by reflections on bus systems with long cable lengths (see Fig. 6.239):

- local, e.g. at branch connections
- time dependent
- electrical, e.g. an apparatus has been disconnected.

An assessment for certification can be performed on the basis of simulations only, due to the complexity of the relations pointed out above. The software developed for it enables qualified statements concerning the power distribution on cables including bus connections. Here, concerns of explosion protection are taken into account. The aim of these investigations is a statement valid in general for the intrinsic safety at different bus configurations.

The ES bus system (in German ES = 'Eigensicherer Feldbus' = intrinsically safe fieldbus) is supposed to comply with minimum requirements, which are of importance in industrial practice, as follows:

- total available power in intrinsically safe circuits:
  5 W (at least) for a total cable length of 400 m
  8 W (at least) for a total cable length of 200 m
  10 W (at least) for a total cable length of 100 m

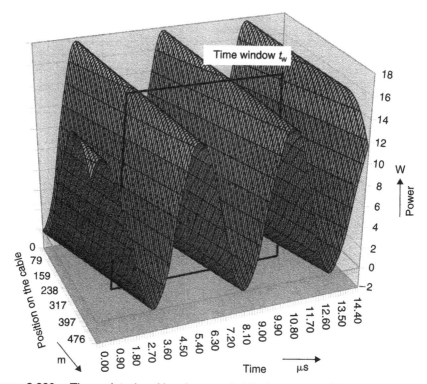

Figure 6.239　*Time-related and local power distribution on an AC fed cable. The integral of power over the time window* t_w *must not exceed a certain limiting value in order to ensure the 'intrinsic safety' of this system.*

- supply voltage: $U_{max.} = 50\,\text{V}$ at $f = 80\,\text{kcps}$
- bus cable length: $0\,\text{m}$ up to $400\,\text{m}$
- type of protection: EEx ib (ia)
- data transmission rate: $1.5\,\text{Mbit/s}$
- voltage available for a 'user': $U_{max.} = 30\,\text{V} \ldots 50\,\text{V}$
- 2-wire cable with a simple and inexpensive connection technique.

Compared with DC, particular difficulties arise from reflections on the cable in the mismatched state by application of alternating current in a higher frequency range (Fig. 6.240). This fact results in functional and safety-related problems, which may be controlled by restriction to relatively short cables and by applying the AC safety control circuit as established by PTB Physikalisch-Technische Bundesanstalt, Germany.

6.9.6.6　Intrinsic safety combined with other types of explosion protection

For some years, it has increasingly been thought that 'intrinsic safety' must not be practised in an 'unadulterated' way all the time. On the contrary, it depends on product design, so that the features typical for intrinsic safety

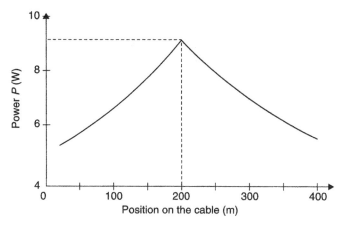

Figure 6.240   *Power distribution on a cable (power versus position).*
Cable length: 400 m; Resistance matching condition: (100 ohms); Inductive
termination: $Z_A = 200\,\mu H$.

and appreciated by the users will be preserved, e.g. the replacement of a
defective apparatus or module during normal operation.

Manufacturers succeed increasingly to transform this idea into marketable
products, in particular remote I/O systems and fieldbus products.

Component assemblies with an enhanced power consumption can be protected according to, e.g., 'flameproof enclosure – d –' or 'encapsulation – m –'.
Suitable small and lightweighted flameproof enclosures made of plastics are
'state of the art'. Plug-and-socket connections, which can be disconnected in
the live state, can be put into practice even in miniature versions: each pin is
enclosed individually to form a flameproof enclosure when plugged into its
counterpart.

For the connection of a e.g., non-intrinsically safe auxiliary power source,
type of protection 'increased safety – e –' has been available for a long time.
Among other things, the trick is to apply standardized types of protection in
an intelligent way and to combine them accordingly.

Two examples arising from topical fieldbus and remote I/O technology
will demonstrate the following (Fig. 6.241):

- overpassing the power limitations of classical intrinsic safety
- intrinsically safe field devices
- intrinsically safe fieldbus or according to 'increased safety – e –'
- 'hot swap', i.e. a defective module can be replaced in the live state.

The first example (Fig. 6.242) is a remote I/O to be installed in zone 1.
The auxiliary power is connected to the carrier complying with 'increased
safety – e –'.

Fieldbus (data transmission only) and terminal connections of the field
devices are intrinsically safe and designed to be pluggable. The power supply is fitted into a flameproof enclosure ('d') made of plastics. Its connection

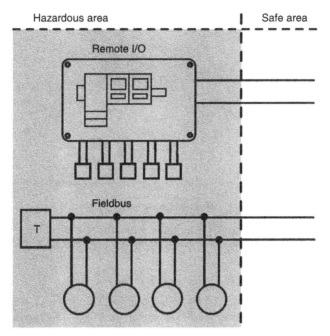

Figure 6.241 *Remote I/O and fieldbus.*

Figure 6.242 *Remote I/O system for zone 1.*

is made by individually enclosed pins as a part of a flameproof enclosure. All the other modules are intrinsically safe. All active modules can be replaced in the live state.

In the second example (Figs 6.243 and 6.244) a fieldbus is shown which has been developed in Norway for sub-sea applications. Here, the fieldbus

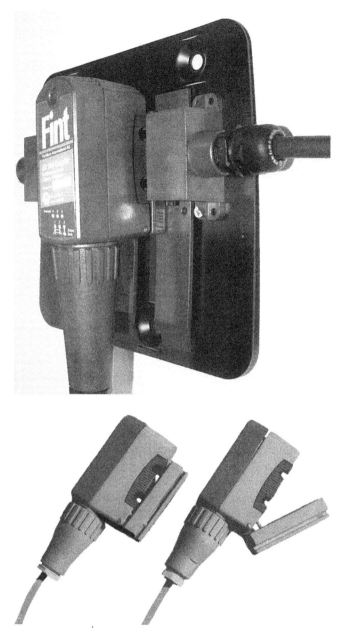

Figures 6.243 and 6.244   *AC fed fieldbus system with inductive coupling for zone 1.*

consists of two conductors running in parallel and complying with 'increased safety – e –'. Data transmission and auxiliary power supply with alternating current are combined on this 'pair of wires'. Intrinsically safe field devices are connected to the bus via an inductive coupler (non-sparking!) which is powered from the bus. The coupler itself is encapsulated ('m') and contains an isolating stage to operate the field device intrinsically safe. A special power supply unit (not shown here) provides AC auxiliary power to the bus and separates data flow and auxiliary power.

6.9.6.7   Intrinsically safe electric light?

Optoelectronic components for remote control and monitoring and for data transmission are being installed in hazardous areas more and more. In particular, the optical-fibre transmission technique enables numerous modern applications. Apart from the great advantages of this technology, the risk of an ignition caused by optical radiation at a power level sufficiently high is to be taken into account (see Section 1.2.7).

Of course, the concepts of 'intrinsic safety' cannot be transferred to optical systems. The basic principle to limit power or energy, however, can be applied for optical radiation.*

The lowest value for optical radiation power capable of igniting a gas–air mixture has been measured for CS_2–air mixtures with 50 mW (Nd: YAG-Laser, $\lambda = 1064$ nm). Including a margin of safety of $\times 1.5$, the optical radiation power shall not exceed 35 mW.

In practice, based upon this, optical data transmission systems may be developed and certified:

The following testing practice for continuous wave optical radiation sources as a part of explosion protected electrical apparatus has been introduced by PTB:

- **proof of compliance with the limitation of radiation intensity, based on electrical values recognized to be safe, for each point of the optical path under ambient temperature conditions to be taken into account,**

  or

- **proof, that the radiation source is not capable of exceeding the radiating power limitation under all ambient temperature conditions to be taken into account and under assumption of supply currents arbitrarily high,**

  or

- **proof that a safety-related countable limitation of the electrical input values will ensure compliance with the limitation of radiation power under all ambient temperature conditions.**

---

* Investigations in this field have been published in: Schenk, S.: Entzündung explosionsfähiger Atmosphäre durch gepulste optische Strahlung, PTB-Report ThEx-17, Physikalisch-Technische Bundesanstalt, Braunschweig and Berlin/Germany, 2001 and Welzel, M. M.: Entzündung von explosionsfähigen Dampf/Luft- und Gas/Luft-Gemischen durch kontinuierliche optische Strahlung, PTB-Report W-67, Physikalisch-Technische Bundesanstalt, Braunschweig and Berlin/Germany, 1996

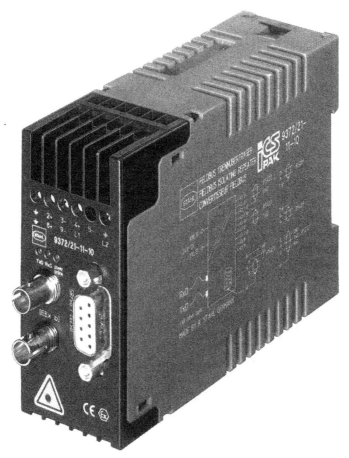

Figure 6.245 *Fieldbus isolating repeater as an RS 485 electrical optical link. The maximum optical radiation power is limited to values <35 mW.*

The result of such an assessment can be documented by PTB in an EC-type examination certificate, in an Expert's Statement (in German: 'Gutachtliche Stellungnahme') or in a Test Report according to apparatus configuration.
(from PTB: www.explosionsschutz.ptb.de)

In Figs 6.245 and 6.246, fieldbus isolating repeaters as electro-optical links are shown:

- Figure 6.245: electrical-to-optical link
  RS 485 to fibre optics
  Installation in the safe area or in zone 2
  Marking, type of protection: Ex II 3(2) G EEx nA II T6
  Fibre optics transmission in zone 1
  Optical radiation power <35 mW
  (according to certificate).

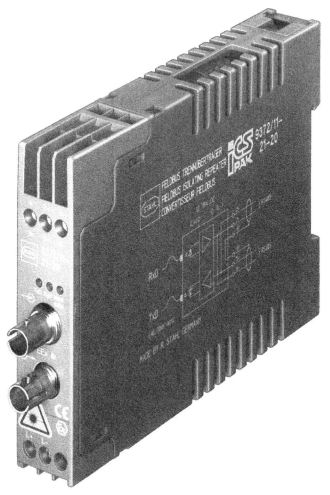

Figure 6.246 *Fieldbus isolating repeater as RS 485 electrical optical links. The maximum optical radiation power is limited to values <35 mW.*

- Figure 6.246: optical-to-electrical link
  Fibre optics to RS 485
  Installation in zone 1
  Marking, type of protection: Ex II 2G EEx ib IIC T4
  Auxiliary power supply intrinsically safe
  Optical radiation power <35 mW
  (according to certificate).

Chapter 7

Analysers and analyser rooms

7.1 Pressurized enclosures with an internal release of flammable substances

Pressurization (abbreviation: – p –) is a protection technique based upon a positive pressure differential within an enclosure referring to the environmental atmosphere. This overpressure prevents the ingress of a surrounding combustible atmosphere and by this way enables the operation of sparking or arcing components (contactors, circuit breakers, sliprings, commutators) and of components with surface temperatures exceeding the ignition temperature of combustible substances (e.g. the oven of a gas chromatograph, the rotor of induction motors and their stator windings, especially for 'low' temperature classes T4–T6). The protective gas may be air delivered by a pipeline with pressurized air or by a fan operating in a non-hazardous area or an inert gas, e.g. nitrogen, argon, carbon dioxide or sulphur hexafluoride (SF_6). An essential safety requirement is a purging procedure before energizing the apparatus. The continuous flow of protective gas in the purging phase sweeps out any combustible atmosphere which may have entered the apparatus interior in a non-operating period. In case of a failure of pressurization (i.e. the positive pressure differential falls below a given limiting value), the components inside a p-enclosure shall be de-energized or shall be explosion protected if they continue operation. As an example, a zone 1 p-enclosure may contain intrinsically safe circuits for data transmission and monitoring or flameproof heating elements remain operational as standstill heaters in a p-motor.

Requirements for pressurization are given in IEC 60079-2 and (especially for zone 1) in EN 50016.

Compared with EN 50016/1st edition, the 2nd edition of this standard opens an extended field of application: it incorporates protection techniques for internal sources releasing combustible substances (which may be added by parts of oxygen) into the pressurized enclosure. By this way, gas chromatographs and other analysers (e.g. mass spectrometers) for combustible substances have found an adequate protection technique and can be certified according to article 100 directives (see Table 3.7(b), in this case Directive 97/53/EC) by an 'E'-generation certificate of conformity or by an approval according to the ATEX 100a Directive (94/9/EC) for an EU (European Union)-wide acceptance, resulting in free commissioning and trading.

The 'solution' for a release of combustible substances has been found in the 'dilution area' (Fig. 7.1). In this part of the pressurized room surrounding the

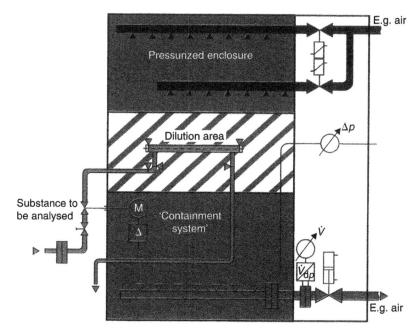

Figure 7.1   *Pressurized analyser for zone 1 application. Some parts of the containment system may show a release of a combustible substance which can be handled in a safe way in a dilution area.*

source of release, a continuous flow of protective gas (air, inert gas) reduces the content of combustibles to a value below the LEL or reduces the oxygen content to a 'safe' level ($\leq$2% (v/v)) excluding any possibility of starting an explosion (Figs 7.2(a) and (b)).

Depending on the mode of operation:

• leakage compensation, i.e. a safety system monitoring and regulating the pressure differential to a given value by an (undefined) flow of protective gas due to the fact that in practice the leakage of an enclosure never equals zero

• continuous flow, i.e. a defined continuous flow of protective gas is maintained by a safety system monitoring and regulating this flow. The pressure differential 'floats' within a given bandwidth above the rated minimum overpressure.

Air or inert gases can be used both as protective gas and for dilution of any release of combustible substances.

The containment system (this includes the piping, analyser cell and all related parts such as valves, flanges etc.), as the constructional part of an analyser which is permanently connected to the process to be monitored and which conducts the gas to be analysed, sometimes in combination with a carrier gas, shall be assessed with respect to its leakage rate (Table 7.1). If there is no oxygen present in this gas, air may be used for dilution. The air flow is

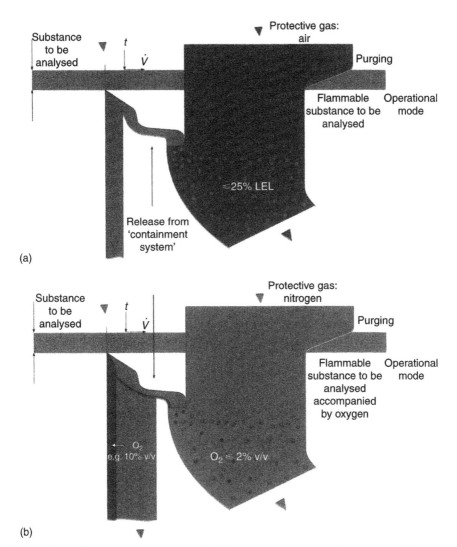

Figure 7.2 *Schematic diagram of the 'dilution concept', including purging. Time axis* t = *top to bottom; protective gas flow* $\dot{V}$ = *left to right. (a) Containment system as a source of release of a flammable substance, protective gas is air. After purging, the protective gas flow is reduced to ensure an adequate dilution of the flammable release down to* ≤25% LEL. *(b) Containment system as a source of release of a flammable substance accompanied by oxygen, protective gas is an inert gas (e.g. nitrogen, N_2). After purging, the protective gas flow is reduced to ensure a safe oxygen level* ≤2% (v/v) O_2.

determined by the fact that the substance released by the containment system shall be diluted to a concentration not exceeding 0.25 × LEL. If oxygen is present in the containment system, the dilution process can operate with inert gas only to ensure an oxygen concentration not exceeding 2% (v/v) within the dilution area.

Table 7.1 Pressurization – p – according to EN 50016/2nd edition and IEC 60079-2/4th edition. Conditions of release for combustible substances and oxygen from a containment system into the pressurized enclosure

No release	Limited release	Unlimited release
• c.s. is infallible and/or • gas concentration inside the c.s. is always lower than LEL and/or • pressure in the p-room exceeds the pressure inside the c.s. by 50 Pa at least	• rate of release shall be predictable due to flow limiting devices, e.g. orifices, nozzles, valves, secured joints, if c.s. is *not* infallible	• rate of release is *not* predictable (e.g. in the case of liquids inside the c.s.) • if there is an oxygen content inside the c.s., then its maximum flow rate shall be predictable

Note: c.s. = containment system
LEL = Lower Explosive Limit
UEL = Upper Explosive Limit
An 'infallible' containment system is defined in EN 50016/2nd edition in clause 10.1; the appropriate type tests are stated in clause 14.6. The corresponding clauses in IEC 60079-2/4th edition are 12.2 and 16.6

Table 7.2 Pressurization – p – according to EN 50016/2nd edition and IEC 60079-2/4th edition. Conditions of release and appropriate protection techniques

Protection technique	No release	Limited release	Unlimited release
1 Leakage compensation	Protective gas • air or/as a choice • inert gas	Protective gas • inert gas if: • oxygen concentration inside c.s. $\leq 2\%$ (v/v) • no deliberate release in normal operation • combustible substance inside c.s. with UEL $\leq 80\%$ (v/v)	Protective gas • inert gas if: • oxygen concentration inside c.s. $\leq 2\%$ (v/v) • no deliberate release in normal operation • combustible substance inside c.s. with UEL $\leq 80\%$ (v/v)
2 Continuous flow of	Protective gas	Dilution required, Protective gas	Dilution required, Protective gas

(continued)

Table 7.2 (*continued*)

Protection technique	No release	Limited release	Unlimited release
protective gas	• air or/as a choice • inert gas	• air substance from c.s. shall be diluted to $\leqslant 0.25 \times$ LEL • inert gas oxygen release from c.s. shall be diluted to $\leqslant 2\%$ (v/v)*,**	• inert gas only oxygen release from c.s. shall be diluted to $\leqslant 2\%$ (v/v)* no deliberate release in normal operation

Note: c.s. = containment system
LEL = Lower Explosive Limit
UEL = Upper Explosive Limit
*With inert gas as protective gas and an oxygen content inside the containment system $\leqslant 2\%$ (v/v), the flow rate of protective gas after the purging phase may be reduced to that necessary for maintaining the minimum pressure differential (this may be considered as a 'shift towards leakage compensation')
**Thermodynamic instability of flammable substances, e.g. an UEL exceeding 80% (v/v), does not allow the use of this protection technique. In this case, continuous dilution with air is required

Table 7.2 indicates that a release of combustible substances out of the containment system can be handled in a safe way, even in the case of an undefined flow of protective gas using 'leakage compensation' and 'continuous flow of protective gas' as protective techniques.

7.2   Pressurized enclosures in zone 2 and for use in the presence of combustible dusts

Pressurized apparatus with air or with an inert gas as protective gas succeeded in finding an extended field of application. Group II apparatus (in chemical plants, oil and gas industry, refineries) are installed as high power motor drives, as analysers and process control and monitoring units mainly in zone 1. In coal mines, Group I pressurized apparatus as 'power supply units' with all integrated high voltage switchgear, transformers, low voltage switchgear, frequency convertors and remote controlling and monitoring installations, mainly with intrinsically safe data transmission lines outside the p-enclosure, show a considerable reliability in operation [31].

In zone 2, the pressurization technique presents the greatest allowable bandwidth in selecting the electrical components to be installed inside an explosion protected enclosure in comparison with other types of explosion protection techniques as stated in IEC 60079-15, EN 50021 or VDE 0165. As a special feature, pressurization enables intervention and access of personnel via doors (sometimes in combination with an airlock as in zone 1) during continuous operation without any need for an operational interrupt. Thus, adjustment of parameters or replacement of components is simplified and results in reduced expenses for operation and maintenance of complex systems for power distribution and remote monitoring and signalling [21].

High operational reliability when surrounded by an environment formed mainly by combustible dusts and a very high degree of dust ingress protection – this is the users' experience over long periods and opens the field of application for p-enclosures in zones 11, or 21 and 22.

7.2.1   Pressurized enclosures with simplified control units – a concept for zone 2

Requirements for zone 2 p-apparatus are given in IEC 60079-14 (identical with EN 60079-14), overtaken as German Standard VDE 0165 Teil 1/1998-08 and IEC 60079-2/4th edition. In addition, clause 23 of EN 50021/1999-04, overtaken as German Standard VDE 0170/0171 Teil 16/2000-02, is under consideration. EN 50021 refers to electrical apparatus, Group II, category 3G.

The advantage of pressurization – compared with other protection techniques, e.g. 'restricted breathing enclosure' – is given by a positive pressure differential permanently maintained in the apparatus interior. Load cycles of the p-apparatus do not affect this pressure differential, and by this way, a cooling-down period caused by reduced or no-load conditions cannot cause an inverted pressure differential (the pressure of the surrounding atmosphere is somewhat higher compared with that in the apparatus interior) which may force the 'backstream' of environmental flammable atmosphere into the interior of an apparatus. Pressurization guarantees safe operational conditions completely independent of load cycles.

Compared with pressurized apparatus for zone 1 application, some benefits are given for zone 2 application:

- according to IEC 60079-14 and EN 60079-14 (VDE 0165), purging (which is obligatory for zone 1 apparatus to ensure that after purging with air the concentration of flammable substances is $\leqslant 0.25 \times$ LEL or, after purging with inert gas, the oxygen concentration is $\leqslant 2\%$ (v/v)) may be omitted, provided that the atmosphere inside the p-enclosure and associated ducting is below $0.25 \times$ LEL of flammable substances
- according to IEC 60079-14 and EN 60079-14 (VDE 0165), an acoustic or optical alarm is appropriate in case of failure of pressurization (internal overpressure and/or flow of protective gas fall below their minimum limiting values) for zone 2 equipment (for zone 1 application, an alarm and an automatic de-energizing of internal components are required).

These facts enable a more simple layout and construction of the safety device and related parts for controlling and monitoring the pressurization.

The discussions concerning internal sources of release of flammable substances into the pressurized enclosure and the technical solutions found to handle this problem for zone 1 application have forced the implementation of these concepts into the standards for zone 2 – pressurized apparatus, e.g. in IEC 60079-2/4th edition. Compared with other zone 2 protection concepts, e.g. 'restricted breathing enclosure,' which generally exclude any internal sources of release of flammable substances, the pressurization technique obtained a considerable advantage.

7.2.2   Pressurized enclosures and combustible dusts

'Clean rooms' with a very restricted amount of particles or gaseous components in air (other than oxygen and nitrogen) are often used in technical production processes, e.g. in semiconductor manufacturing. Generally, an overpressure inside prevents the ingress of impurities into the clean room. Following this philosophy, it is only a small step to create a pressurized enclosure in an environment hazardous due to combustible dusts as an explosion protection technique.

As explained in Section 2.5 and Table 2.5, the relevant standards ask for a certain tightness of the enclosure against the ingress of dust and a surface temperature ensuring a certain margin of safety related to the glow temperature and ignition temperature of the combustible dust referred to.

Maintaining a surface temperature limit normally does not impose severe problems. Electronic components (semiconductors) and switchgear are restricted to a maximum temperature in operation seldom exceeding +60°C to +70°C. In addition, convective heat transfer to the surface of the enclosure may be replaced by a closed circulation of the protective gas inside the enclosure passing a water-cooled heat exchanger to remove heat dissipated in the interior of the pressurized apparatus. Thus, fin tubes or air-cooling ducts on the surface of an explosion protected apparatus can be avoided. A smooth surface is not deemed to be susceptible to dust deposits, which may be stirred up during operation. 'Primary type of dust explosion protection' is closely related to a strict avoidance of dust deposits, i.e. a high degree of cleanliness.

Gaskets as sealing joints for covers and enclosure modules show high reliability during long operational periods due to the fact that a great part of combustible dusts does not react with gasket materials. A good solution for sealing elements at doors (e.g. doors for an airlock) has been found in hollow-body rubber gaskets pressurized (by compressed protective gas) in the closed position of the door and relieved automatically when operating the door opener.

Attention should be directed to the airlock concerning the dragging of dust deposits on clothes or shoes to the airlock's interior. Cleaning procedures may be required, e.g. by an air shower, to avoid dust that may be transported into the pressurized rooms.

Pressurization techniques are similar to those used in zones 1 and 2, with the exception that purging shall not be applied. The high flow rate of protective gas during the purging period accompanied by a high velocity of gas may stir up dust layers inside the p-room on components and thus adversely affect the safety of this type of protection. Static pressurization, leakage compensation and continuous flow technique may be used. Standard IEC 61241-4/1st edition as a standard for pressurization in zones 21 and 22 does not consider any sources of release of flammable substances into the p-room. This standard (whose key letters are 'pD', 'D' = Dust) asks for automatic de-energizing of internal components and/or for initiating an optical/acoustical alarm in the case of failure of pressurization.

7.3   Analyser rooms – manned pressurized enclosures?

In the previous chapters pressurized enclosures with airlocks for 'walking in' have been described. What standards shall be referred to? The scope of EN 50016 and IEC 60079-2 does not cover 'pressurized rooms or analyser houses'. In a similar way, standard IEC 61241-4/1st edition – type of protection 'pD' – does not cover 'the requirements for pressurized rooms'. But these clauses may need an interpretation: an explosion protected (pressurized) apparatus may be considered as a product regardless of its size or volume, which is subject to trading or commissioning in its entirety. This point of view covers p-enclosures intended for 'walking in' or manned apparatus respectively. Referring to EN 50016 and the subsequent need for a certifying procedure or an approval as described in Chapter 3, the majority of European certifying or notified bodies considers manned p-enclosures to be outside the scope of EN 50016. An important reason for this interpretation must be based on the lack of requirements concerning health and safety of the crew inside such an apparatus, e.g. ensuring a sufficient content of oxygen as an essential part of the protective gas.

Overlooking the IEC standardization work, two standards exist for rooms or buildings respectively and analyser houses:

- IEC 60079-13: Construction and use of rooms or buildings protected by pressurization
- IEC 60079-16: Artificial ventilation for the protection of analyser houses.

These standards cover the aspects of explosion protection for the most part. In contrast to this is IEC 61285 (1994-09): 'Industrial process control – safety of analyser houses' (overtaken by CENELEC as a European Standard EN 61285/1994-11 and overtaken as a German National Standard VDE 0400, Teil 100/1995-12) dealing with the protection of personnel, environmental protection and energy saving aspects. Due to the fact that toxic and combustible gases may be present permanently at a workplace only in a very small concentration for the purpose of the health and safety of personnel, and this concentration remaining below LEL by some orders of magnitude, aspects of explosion protection do not govern the technical content of this standard.

Starting some years ago, a working group of CENELEC subcommittee SC 31-7 established a (draft) standard pr EN 50381 for TVRs (transportable ventilated rooms with and without an internal source of release). This standard will cover the requirements for 'walking in' pressurized rooms (as manned explosion protected equipment) including both the aspects of explosion protection and the health and safety of personnel. It is intended that this draft will be incorporated in the variety of types of explosion protection given in EN 50014, forming a new European Standard with the abbreviation 'v' (=ventilated). Fields of application will be:

- coal mines (Group I, category M2, comparable with the 'classic' firedamp-proof electrical equipment)
- chemical plants (Group II, categories 2G and 3G or for zone 1 and zone 2 respectively)
- general application in a non-hazardous environment, but with internal sources of release of flammable substances.

Discussions concerning the technical content of this draft have been closed. It is expected that this draft standard will be published in the near future.

Chapter 8

Testing explosion protected electrical equipment

Within the framework of type verifications for and testing of electrical apparatus in general, technical tests shall furnish proof of mechanical, thermal or electrical properties as specified in the relevant standards. Some examples for type tests are given in the following:

A Mechanical tests

- ascertaining the tightness of an enclosure against the ingress of solid foreign bodies, dust and water
- impact tests
- testing the ability of a flameproof enclosure to withstand internal explosion pressure and tests for non-transmission of flames
- torque test for bushings
- tests of clamping of cables.

B Material tests

- test of erosion by flame
- test to ascertain the resistance to tracking of insulating materials
- resistance to light (for enclosures made of plastic materials not protected from light)
- resistance to chemical agents for plastic enclosures and plastic parts of enclosures.

C Electrical tests

- heating test (including temperature measurements) to ascertain that overtemperatures caused by ohmic, eddy current and hysteresis losses and/or dielectric losses comply with the relevant standards
- voltage tests
- partial discharge measurements as a non-destructive quality test for insulating materials subject to electrical stresses
- short-circuit current tests to ascertain the mechanical robustness and thermal resistance of electrical conductors and their insulating parts
- spark test apparatus for intrinsically safe electrical circuits.

The majority of the type tests listed above refers to electrical apparatus in general, i.e. these tests are valid for non-explosion protected electrical equipment as well.

In the following, type tests which are applied exclusively to firedamp-proof or explosion protected apparatus or which represent an enormous potential in material quality verification will be considered in more detail:

1 Tests of the ability of an enclosure to withstand an internal overpressure and for non-transmission of an internal ignition, applied to flameproof – d – enclosures
2 Partial discharge (PD) measurements applied to components or apparatus subject to electrical stresses, e.g. components according to 'encapsulation – m –' or complying with 'increased safety – e –'
3 Spark test apparatus applied to intrinsically safe electrical circuits.

8.1   Tests for flameproof enclosures

Remembering the basic principles of flameproof enclosures, two main characteristics of such enclosures shall be proved by type testing: they shall withstand the internal overpressure caused by the ignition of a combustible gas–air mixture inside, and they shall prevent flame transmission to the environmental hazardous atmosphere. So, such a test cycle is divided into three parts:

- determination of the internal explosion pressure (reference pressure)
- overpressure test
- test for non-transmission of an internal ignition.

  The internal overpressure of an ignited combustible mixture depends on:

- gas type and gas concentration
- enclosure shape and internal parts (e.g. the rotor of an asynchronous motor and its speed)
- location of ignition source
- area of doors or covers combined with the dimensions of the joints at shafts, spindles, doors and covers which act as 'venting systems'
- pressure and temperature of the combustible mixture.

For the different Groups and subgroups (I, IIA/B/C), representative gases have been selected with defined volumetric concentrations in air to obtain the maximum peak pressure as a reference pressure. These concentrations are somewhat higher than the stoichiometric point (see Section 1.2.1 and Table 1.2) due to dissociation of gas molecules, which 'captures' a part of the combustion heat and decreases the explosion pressure at the stoichiometric point. To compensate an additional amount of combustible substance is required – this explains the 'shift to the rich mixtures'. This shift increases with higher combustion temperatures and increasing dissociation levels resulting hereof. For acetylene, C_2H_2, showing combustion temperatures exceeding 3000 K (Tables 6.34 and 6.35) the 'maximum explosion pressure' concentration (in air) is 14% (v/v) whereas the stoichiometric point is at 7.75% (v/v), see Table 1.2.

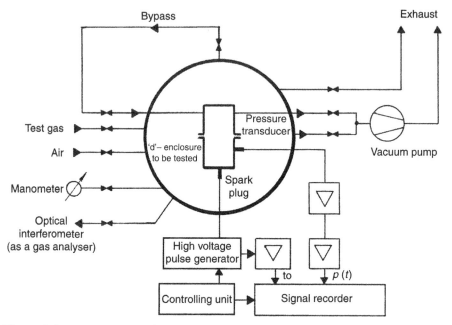

Figure 8.1   *Explosion test facility for flameproof enclosures.*

In spite of the fact that the enclosure to be tested can be filled with the test gas in a straightforward way, a test setup according to Fig. 8.1 with a closed testing vessel is more advantageous. The gas filling and mixing process is identical with that for the flame transmission test, and the closed vessel forms a 'barrier of safety' against enclosure damage or even burst with enclosure parts generated hereby, forming projectiles with a considerable kinetic energy. The test gas to determine the reference pressure, a gas–air mixture according to Table 8.1, Part A, is prepared in the testing vessel according to the partial pressures of the gas and air (see Section 1.2) controlled by a manometer (in practice with an additional pressure transducer to generate an electrical signal) and controlled in an independent way with a gas analyser, e.g. an optical interferometer. The enclosure to be tested is fitted with a spark plug to ignite the test gas in its interior and with one (or even more) pressure transducer(s), whose output is amplified and filtered (bandwidth from DC to, e.g., 5 kcps as required in EN 50018 and IEC 60079-1/4th edition) in order to suppress pressure oscillations (in a higher frequency range), ringing or resonance effects. A controlling unit, a high voltage pulse generator to supply the spark plug and a signal recorder complete the signal recording equipment. After gas mixing and ascertaining the correct mixture, the test gas enters the interior of the enclosure via a bypass, supported by means of a vacuum pump (in practice a water seal pump combined with an air jet pump and a rotating mechanical pump). When the gas pressure equals the atmospheric pressure, and with all valves in the closed position, the test gas is ignited and pressure recording starts (measured at different points).

Table 8.1 Test conditions for determination of the reference pressure and overpressure test (according to EN 50018 and IEC 60079-1/4th edition)

Group	I	IIA	IIB	IIC
Test gas, concentration in air	CH_4 $9.8 \pm 0.5\%$ (v/v)	C_3H_8 $4.6 \pm 0.3\%$ (v/v)	C_2H_4*** $8.0 \pm 0.5\%$ (v/v)	C_2H_2 $14.0 \pm 1\%$ (v/v) H_2 $31.0 \pm 1\%$ (v/v)
Part A Determination of the reference pressure				
Precompression of gas–air mixture	atmospheric pressure, $1.013\,25 \cdot 10^5$ Pa ($1.013\,25$ bar abs.)			
Number of tests**'	3	3	3	3 with C_2H_2–air 3 with H_2–air
Part B Overpressure test				
Precompression of gas–air mixture	floating, overpressure = $1.5 \times$ reference pressure, minimum overpressure = 3.5 bar			
Number of tests	1	1	1	3 with C_2H_2 3 with H_2

*Detachable gaskets shall be fitted to the enclosure under test

**In cases where pressure piling (the pressure rise time is shorter than 5 ms, or the peak pressure values obtained deviate from one to another by a factor ≥1.5) occurs during the tests, the number of tests shall be at least five (with each gas)

***In case of pressure piling, for IIB enclosures an additional test sequence is required:
test gas: MG 15/85 (i.e. 15% (v/v) CH_4 and 85% (v/v) H_2)
test gas concentration: ±24% (v/v) in air
precompression: atmospheric pressure
number of tests: 5 (at least)

Figure 8.2   *Testing vessel for large flameproof enclosures (according to [17]), the biggest testing facility worldwide.*
Volume: 56 m^3; Max. permissible internal overpressure: 2 MPa (=20 bar);
Inner diameter: 3.2 m; Length of the cylindrical part: 6 m; Payload: 20 tons.
In the photo, the test vessel is shown at the manufacturer's site.

Rotating electrical machines shall be tested at standstill and running with $0.9\times \cdots 1.0\times$ their rated speed. The rotor causes a certain turbulence of the test gas, resulting mainly in a considerably increased dp/dt value [2] and enhanced peak pressure. Cage induction motors may be fed from outside with an adjustable voltage and frequency via a frequency convertor, motors with sliprings or commutators need an auxiliary drive, e.g. a small cage induction motor powered by said frequency convertor.

The photo of Fig. 8.2 may give an impression of a testing vessel for large motors or the machine bodies of shearer-loaders for coal mines [17].

Regarding the overpressure test, an overpressure of 1.5 times the reference pressure is required by the d-standards, with a minimum of 3.5 bar (referring to atmospheric pressure). This test may be a static test using a liquid (oil, water) or compressed air (in this case, to avoid parts of an enclosure which fails becoming dangerous projectiles, the overpressure test shall be made with a 'barrier' – e.g. within a closed testing vessel) or a dynamic test based upon the same procedure as described for the determination of the reference pressure. The test conditions are summarized in Table 8.1, Part B. The main difference may be seen in the 'floating' value of precompression of the test gas. Theoretically, the peak pressure of an exploding gas–air (or gas–oxygen) mixture is proportional to the initial pressure, see Section 6.8.1. But an increased internal pressure and the increased peak pressure generated hereby cause stronger (elastic) deformations of enclosure parts showing joints, which act as 'pressure vents' increasing their cross-sectional areas in an (in most cases) unpredictable manner. So, the precompression of the test gas may be adjusted according to peak pressure measurements. An additional complication in this procedure is the heating of the enclosure walls to be tested 'shot by shot', and a subsequent undefined heating of the test gas inside the enclosure which results in a decreasing density of the test gas. So, correct overpressure values will be obtained in a few-step testing cycle with a well-skilled testing plant staff only.

The tests for non-transmission of an internal ignition are carried out according to the method of determining the reference pressure described in Fig. 8.1; however, with test gases specified in Table 8.2. The preconditions for these tests are:

- all gaskets are to be removed
- the gaps i_E of the enclosure to be tested shall be measured, and they shall be at least equal to 0.9 times the maximum constructional gap i_C specified by the manufacturer.

The reason for applying these preconditions is that gaskets may 'seal' the flameproof joints to be tested against flame transmission, and that gaps near the lower limit of the tolerance specified by the manufacturer may assume a safety against flame transmission which cannot be maintained with a gap near the upper limit of the specified tolerance. For enclosures produced in small series only, the condition $0.9 i_C \leqslant i_E \leqslant i_C$ is difficult to meet. For $i_E < 0.9 i_C$ (but within the tolerance!), the precompression of the test gas may be increased (see Table 8.2) or test gases with reduced gas concentrations may be used (see Table 8.3). Increased precompression of a gas–air mixture causes lower MESG values in general, see Figs 1.5–1.7. For hydrogen within a concentration c, 10% (v/v) $\leqslant c \leqslant$ 50% (v/v) in air, and with precompression values in the range $3 \cdot 10^4$ Pa (=0.3 bar abs.) to $8 \cdot 10^5$ Pa (=8.0 bar abs.), the MESG decreases with increasing precompression p_o according to a hyperbola. With MESG $= a \cdot p_o^b$ (a, b being constants), a double-logarithmic representation shows a straight line

$$\lg \text{MESG} = \lg a + b \cdot \lg p_o$$

The exponent b results in $-0.65 \geqslant b \geqslant -1.0$ [18].

At $p_o = 1 \cdot 10^5$ Pa (=1.0 bar abs.), the gradient $\Delta \text{ MESG}/\Delta p_o = 4$ nm/Pa (=0.4 mm/bar).

The diagrams MESG versus gas concentration (see Figs 1.3, 1.5, 1.6) are parabola shaped. So, for each MESG value (with the exception of the vertex)

Table 8.2 Test for non-transmission of an internal ignition (according to EN 50018 and IEC 60079-1/4th edition)

Group	I	IIA	IIB	IIC
MESG classification mm	for CH_4 only[1] 1.14 (1.17)	MESG > 0.9	$0.5 \leq$ MESG ≤ 0.9	MESG < 0.5[2]
Test gas, concentration in air	MG 58/42[3] 12.5 ± 0.5% (v/v)	H_2 55.0 ± 0.5% (v/v)	H_2 37.0 ± 0.5% (v/v)	C_2H_2 7.5 ± 1% (v/v) H_2 28 ± 1% (v/v)[4]
Precompression of gas–air mixture	atmospheric pressure $1.013\,25 \cdot 10^5$Pa (1.013 25 bar abs.)			1.5 bar (abs.)
MESG of test mixture	0.80 mm	0.65 mm	0.35 mm	0.24 mm[5]
Gap i_E[6]	$0.9\, i_C \leq i_E \leq i_C$ for all Groups (I, IIA/B/C)			
Safety factor χ[7]	1.42 (1.46)[1]	1.38	1.43	1.21[8]
Number of tests	5	5	5	5 with C_2H_2-air 5 with H_2-air

1 Different values for MESG of CH_4 are given in the literature
2 The lowest known MESG value is MESG = 0.29 mm for H_2– air mixtures
3 MG 58/42 = 58 ± 1% (v/v) CH_4 and 42 ± 1% (v/v) H_2
4 IEC 60079-1/4th edition asks for 27.5 ± 1.5% (v/v) H_2
5 This MESG value refers to the hydrogen–air mixture. Acetylene–air mixtures show somewhat higher MESG values, see Table 1.6
6 Gap i_E = gap of the enclosure to be tested. Gap i_C = maximum constructional gap as specified by the manufacturer (in his drawings). For values $i_E < 0.9 i_C$ test gases with a lower MESG shall be used. This may be achieved with precompression of the test gases given above according to
$P_K = i_C / i_E \times 0.9$ for Groups I, IIA and IIB
$P_K = i_C / i_E \times 1.35$ for Group IIC
P_K = precompression factor, referring to atmospheric pressure. A good approximation is to read P_K in bar (abs.). An alternative way for Groups I, IIA and IIB is the use of test gases with deviating volumetric concentrations (see Table 8.3)
7 The safety factor χ is defined as χ = lower MESG limit according to classification × (MESG of test gas)$^{-1}$
For Group I, χ = MESG of CH_4 × (MESG of test gas)$^{-1}$
and for Group IIC, χ = MESG of H_2 × (MESG of test gas)$^{-1}$
Note: EN 50018 and IEC 60079-1 define a margin of safety, K, with a similar denotation. K refers to a representative gas of the group concerned, e.g. to C_2H_4 for Group IIB
8 This safety factor refers to hydrogen, χ = 0.29 mm/0.24 mm = 1.21

Table 8.3 Test for non-transmission of an internal ignition (for enclosure gaps $i_E < 0.9 \times$ maximum constructional gap i_C, according to EN 50018 and IEC 60079-1/4th edition)

Group	I		IIA		IIB	
i_E/i_C ratio	$\geqslant 0.6$ <0.75	$\geqslant 0.75$	$\geqslant 0.6$ <0.75	$\geqslant 0.75$	$\geqslant 0.6$ <0.75	$\geqslant 0.75$
Test gas, concentration in air % (v/v)	H_2 50 ± 0.5	55 ± 0.5	H_2 45 ± 0.5	50 ± 0.5	H_2 28 ± 1	28 ± 1
Precompression of gas–air mixture	atmospheric pressure 1.013 25 · 10^5 Pa (=1.013 25 bar abs.)				1.5 bar (abs.)	atmospheric pressure
Number of tests	5		5		5	

two gas concentrations can be adjusted: one at the 'poor mixture branch' – left side in these diagrams – and one at the 'rich mixture branch', right side. As explained above, the explosion pressure in the 'rich mixture branch' is somewhat higher than in the 'poor mixture branch' due to dissociation. This is the reason why Tables 8.2 and 8.3 specify hydrogen–air mixtures as test gases with H_2 concentrations exceeding the stoichiometric point for Groups IIA and IIB, and Table 8.3 for Group I. To ensure a margin of safety in these tests, all test gases show MESG values below the lower limit of the group classification according to the MESG values of all gases (vapours and mists) belonging to this group. The safety factor χ (see Table 8.2) is near 1.4 for Groups I, IIA and IIB, whereas $\chi = 1.21$ for Group IIC.

The pressure dependence on time obtained in all these tests (assuming that they have been passed successfully) is similar to that given in Figs 6.78 and 6.79. In case of a flame transmission, e.g. during the tests according to Table 8.2, the first pressure peak (caused by the combustion of the gas–air mixture inside the enclosure to be tested) is followed by a second pressure peak after a certain time delay (see Fig. 8.3) [17]. The flame transmission starts combustion of the gas–air mixture inside the testing vessel. Thus, the pressure in the vessel exceeds the pressure inside the enclosure, and so a 'backstream' of gas finds its way into the enclosure via the joints. The pressure decay obtained in this special case is characteristic of the testing vessel (i.e. its volume), whereas, in general, the pressure decay is characteristic of the enclosure (shape, volume, sectional area of joints) under test. In Fig. 8.3 the precompression has been adjusted to $1.3 \cdot 10^5$ Pa (=1.3 bar abs.), which results in a somewhat higher MESG value compared with the precompression of $1.5 \cdot 10^5$ Pa (=1.5 bar abs.) required for Group IIC enclosures, see Table 8.2. So, the diagram in Fig. 8.3 indicates a flame transmission test failed in an unambiguous way.

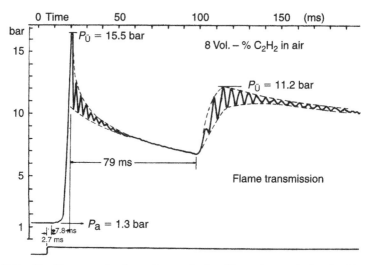

Figure 8.3   *Oscillogram of a flame transmission (flame transmission test has been failed), according to [17]. Pressure (within the test specimen) versus time. EEx dII C motor, without stator iron core and windings.*
Rated power: 535 kW; Internal free volume: 0.930 m³; Test gas: 8% (v/v) acetylene in air; Test gas precompression: $1.3 \cdot 10^5$ Pa (=1.3 bar (abs.))

8.2   Partial discharge (PD) measurements

In electrical apparatus (mainly in the high voltage range, but in small-sized components for low voltages, too), high electric field strengths in restricted areas cannot be excluded in general. Such areas are, e.g., the vicinity of screwed connection joints of electrical conductors or busbars, cable terminations, gas-filled voids in solid dielectrics or lowered levels of insulating fluids (oil) in the bushings of transformers. High electric field strengths in gases cause ionization, i.e. a statistical flow of electrical charges, or, in other words, an electrical current showing a broad-band frequency spectrum. In general, partial discharges adversely affect reliability and operational lifetime of electrical components and apparatus [20], [30]. Partial discharges can generate ozone, O_3, in air, or they decompose hydrocarbons as insulating materials into gases such as hydrogen, H_2, methane, CH_4, and an electrical conducting carbon skeleton.

So, partial discharge measurements have become a well-established method for type and routine testing as a non-destructive material test and quality control [23], [26], [53]. Apart from optical and acoustical detection methods of partial discharges, the high frequency detection and measuring method preferably covers a wide range of applications. It will be considered in more detail in the following.

The standards referred to are IEC 60270/2000 and EN 60270/2001-03 (overtaken as a German National Standard VDE 0434/2001-08).

The basic principle of partial discharge measurement may be explained with a rather simple model (Fig. 8.4). A coaxial cable with an intact dielectric

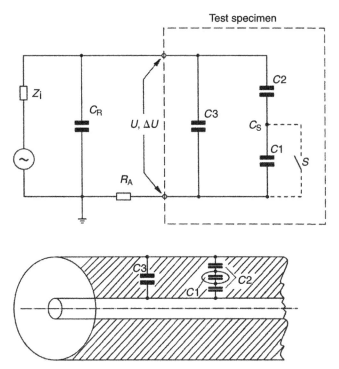

Figure 8.4 *Equivalent network circuit of a partial discharge in a solid dielectric.*

forming the capacitance $C3$ is connected to an AC voltage source with an internal impedance Z_i via the resistance R_A. The capacitor C_R acts as a 'shorting link' in the high frequency range (i.e. the measuring frequency of the partial discharge testing equipment is short-circuited by the reactance of C_R). A fault in the insulating material of the cable may be represented by a 'void' with capacitance $C1$ in series with the intact part of the insulation at this point, forming the capacitance $C2$. At low voltages U or low electric field strengths, the resulting (series) capacitance is:

$$C_S = C1 \times C2 \times (C1 + C2)^{-1}$$

and the total capacitance of the cable is:

$$C = C3 + C_S$$
$$= C3 + C1 \times C2 \times (C1 + C2)^{-1}$$

the total electrical charge in the cable:

$$Q = C \times U$$

With increasing U, the partial discharge in $C1$ is 'ignited', showing a constant low voltage on $C1$. As a model, a 'switch' S across $C1$ may be closed. So, the total capacitance of the cable is:

$$C = C2 + C3$$

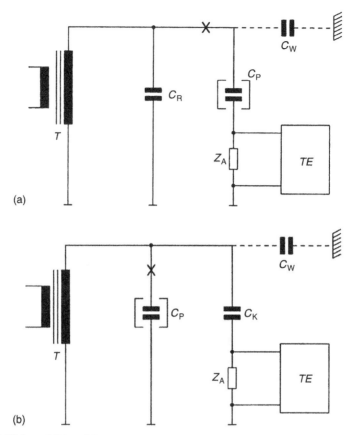

(a)

(b)

Figure 8.5(a) and (b)   *PD testing circuits.*

with the difference to the PD-free state

$$\Delta C = C2 - C1 \times C2 \times (C1 + C2)^{-1}$$

$$= C2^2 \times (C1 + C2)^{-1}$$

At the cable terminals, a compensating charge due to ΔC

$$q = U \times \Delta C$$

$$q = U \times C2^2 \times (C1 + C2)^{-1}$$

can be detected. So, a short-current pulse is to be measured across R_A, or, a voltage pulse ΔU across the cable. The charge q is called the 'apparent charge' – the electrical charge flowing to the series connection C1 and C2 when S is closed cannot be measured directly due to C3 (and other capacitances of the test circuit). So, q has found its name, indicating that q is based on a measurement across the test object terminals – usually expressed in picocoulombs, $1\,\mathrm{pC} = 10^{-12}\mathrm{C}$, or in nanocoulombs, $1\,\mathrm{nC} = 10^{-9}\mathrm{C}$.

The testing circuits are shown in Figs 8.5(a) and (b).

A transformer T as a high voltage source with C_R in parallel as a 'shorting link' is connected to the test object showing a capacitance C_P at point X (Fig. 8.5(a)). The earth terminal of C_P is connected in series with a coupling device with impedance Z_A. The PD signals pass Z_A whose impedance is real in the PD measurement frequency bandwidth, and so a voltage signal proportional to the PD current is generated for the PD measurement system *TE*. For the test voltage frequency, Z_A shall be as low as possible due to the capacitive current via C_P. In addition, an efficient overvoltage protection should be integrated into Z_A in order to limit the input voltage into *TE* in case of a flashover at C_P. Due to environmental high frequency 'noise', PD measurements, especially type tests, are often carried out in a shielded testing facility (forming a 'Faraday cage'). Test objects showing large dimensions (e.g. current and voltage transformers in the high voltage range) show a capacitance C_w which cannot be neglected compared with C_R, C_P and C_K and formed by the metallic walls of the testing facility, resulting in a bypass for the PD pulses. In Fig. 8.5(b), the positions of test object C_P and the 'shorting link' capacitor have been interchanged. The coupling device is connected to the high voltage line via a coupling capacitor C_K. In this testing circuit, Z_A is not endangered to potential flashovers at C_P.

The frequencies for PD measurements are recommended as follows:

1 Broadband equipment:
   lower frequency limit   30...100 kcps
   upper frequency limit   500 kcps
   bandwidth   100...400 kcps
2 Narrow band equipment:
   frequency range   50 kcps...1 Mcps
   bandwidth   9...30 kcps

The transformer T shall be rated to comply with the capacitive current demand of the test object.

In order to limit size and weight of T, resonant circuits or low frequency (e.g. 0.1 cps) test voltage sources may be used [20] instead of the 'conventional' transformer.

Test objects acting as a high voltage source, e.g. a voltage transformer, may be used as their own testing voltage source (see Fig. 8.6).

The secondary windings are excited according to the test voltage required, and the primary windings are earthed or connected to Z_A via a coupling capacitor C_K. For test voltages exceeding the rated primary voltage U_N, e.g. 1.2 times U_N, the exciting voltage shall be increased in its frequency to achieve a constant voltage/frequency ratio. For a test voltage of 1.2 times U_N, the supply frequency of a 50 cps voltage transformer shall be increased by a factor of 1.2, resulting in 60 cps. Thus, the saturation state of the transformer core can be evaded.

For a correct interpretation of PD measurement results it is very helpful to obtain information in addition to a pure PD intensity indicator, e.g. the phase angle (referring to the test voltage) of the PD pulses. A more conventional method is to record the test voltage superposed with the PD pulses.

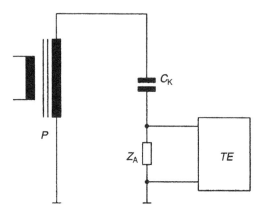

Figure 8.6   *PD testing circuit for (voltage) transformers.*

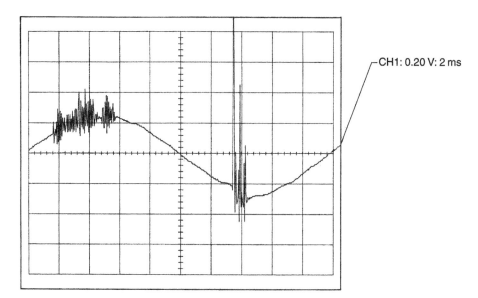

Figure 8.7   *50 cps AC test voltage with superposed PD signals, according to [20].
Sphere electrode–needle electrode in air.*
Test voltage: 4.2 kV rms; PD intensity: 2500 pC.

In Fig. 8.7, partial discharges in air have been detected: the PD pulses appear near the peak voltage, i.e. at the moment of maximum electric field strength. The PD pulses are 'asymmetrical' with respect to the test voltage polarity. This indicates an asymmetrical field configuration based on the fact that electrons show a considerably higher velocity than (positive) ions. Indeed, a sphere electrode opposite to a needle electrode has been used. Modern PD measuring systems present a sophisticated analysis of PD parameters, resulting in so-called 'fingerprints' (see Fig. 8.8). So, PD measurements can be interpreted in detail.

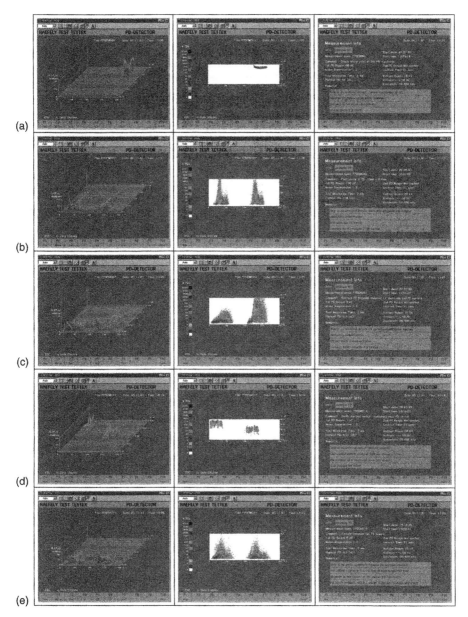

Figure 8.8   *'Fingerprints' of PD. (a) Corona discharge in air. Range: 50 pC; (b) Dielectric bounded flat cavity (polyethylene). Range: 100 pC; (c) Cylinder in polyethylene plane system. Range: 3 nC; (d) Floating object (ungrounded metallic plate). Range: 1 nC; (e) Fissure PD between polyethylene foils; Range: 6 nC.*

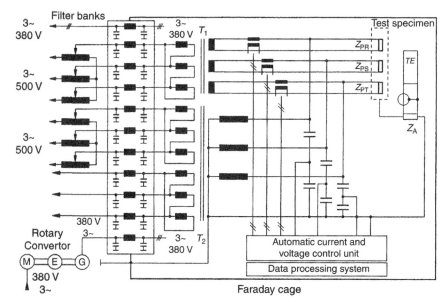

Figure 8.9 *Schematic circuit diagram of a 104 MVA 3 AC 'synthetic' PD test facility (according to [17], [20]).*

The PD test circuits given in Figs 8.5(a) and (b) do comply with the standards IEC 60270/2000, EN 60270/2001-03 and VDE 0434/2001-08. But in the field of research and development, they show two evident disadvantages:

- during the PD measurement, the test object remains at ambient temperature. The dielectric losses in insulating materials caused by the test voltage are too low as to increase the temperature of the test object to values attained in the operational mode due to ohmic losses in conductors, eddy current and hysteresis losses in conductors and metallic parts
- apparatus with a three-phase system can be tested with a one-phase circuit only step by step (conductor n versus conductor m, conductor n versus earthed parts, $n, m = 1 \dots 3$). The PD test sequence is interrupted frequently by different circuit interconnection wirings and does not allow continuous PD measurements over long periods.

So, for three-phase apparatus representing the majority in public and industrial power supply systems, a three-phased 'synthetic' PD test circuit has been developed and put into operation [17], [20].

All conductors are part of three closed current loops with impressed currents which are kept constant, and a three-phase high voltage system enables simultaneous PD measurements in the three conductors (Fig. 8.9). The complete test circuit is installed in a shielded cabin as a 'Faraday cage', all power lines enter this cage via filter banks to suppress unwanted signal transmission to the internal test circuit. The main parts of the test circuit are:

- the 'current' transformer T_1, which supplies the current into the closed current loops with impedances Z_{PR}, Z_{PS} and Z_{PT}

Table 8.4 Main technical data of the 'synthetic' test circuit

Test voltage	0–60 kV rms (phase to phase)
	0–35 kV rms (phase to earth)
Frequency	50 cps
Maximum continuous current of voltage transformer	0.4 A
Rated power of voltage transformer	41 kVA
Test current	1000 A rms per phase
Frequency	50 cps
Maximum voltage of current transformer	10 V
Rated power of current transformer	30 kVA
Insulation voltage of current transformer (windings to earth)	45 kV
Maximum continuous simulated throughput power	$60\,kV \times 1\,kA \times \sqrt{3} = 104\,MVA$

- the 'voltage' transformer T_2 delivering the three-phase high voltage system
- the test specimen, e.g. cable joint boxes, terminal compartments, busbar boxes or complete switchboards according to 'increased safety – e –'
- the PD measurement system TE with coupling device Z_A
- the (external) regulating transformers for T_1 and T_2
- the voltage and current control unit with three current transformers and three capacitive voltage dividers which, in addition, act as 'shorting links' as described in Fig. 8.5(a) (C_R)
- a bidirectional signal transmission system via optical fibres for the voltage and control unit, for television, telephone and signal transmission.

The main technical data of this 3 AC PD test circuit are summarized in Table 8.4. This test facility enables continuous PD measurements for electrical 3 AC apparatus up to 11 kV and 1000 A, the voltage limitation complying with EN 50019 and IEC 60079-7, but with a 'margin of safety' in the rating of the test voltage. As a total, a throughput power of 104 MVA can be simulated.

To close this section, two main aspects of PD measuring technique shall be explained in more detail. The first point is to determine the maximum permissible PD intensities in an adequate way. The general aim is to ensure an appropriate quality of dielectrics in electrical apparatus and components or to predict the residual operational lifetime of such parts after a certain time on duty. There is absolutely no 'algorithm' for this, i.e. all values for maximum permissible PD intensities are based on experience in practice. So, the values given in Table 8.5 are 'near to practice', but may be reduced accordingly for special applications.

The second point is to compare the appropriateness of PD measurements with that of more conventional loss factor (tan δ) measurements. In general, tan δ measurements show an 'integral' over the insulating material in its entirety. Deviations from tan δ due to an insulation impairment restricted to a small (local) area cannot be detected to have been caused by the limited

Table 8.5 Maximum permissible PD intensities (valid in general, not restricted to explosion protected apparatus, according to [30])

Apparatus/ component/cable	PD test voltage*	PD intensity**	according to	Comments
Bushings (>1kV)	$1.05 \times U/\sqrt{3}$	<10 pC (o) ≤10 pC (cr) ≤100 pC (lp)	IEC 60137	o = oil impregnated cr = cast resin lp = laminated paper
	$1.5\ U/\sqrt{3}$	<10 pC (o) ≤250 pC (lp)		
Capacitors (coupling and voltage divider s)	$1.1 \times U_m/\sqrt{3}$ $1.1 \times U_m$	≤10 pC ≤100 pC	IEC 60358, VDE 0560	AC systems with isolated neutral
Cables	$2 \times U_o$ $2 \times U_o$ $1.5 \times U_o$ $1.5 \times U_o$ $1.5 \times U_o$	≤5 pC (VPE) ≤20 pC (PVC) ≤20 pC (PE, VPE) ≤40 pC (PVC) ≤5 pC (PE, VPE)	VDE 0273 VDE 0271 IEC 60502 IEC 60502 IEC 60840	PE = polyethylene VPE = cross-linked polyethylene PVC = polyvinyl chloride
Cable fittings	$2 \times U_o$ $2 \times U_o$	≤20 pC (VPE) ≤40 pC (PVC) ≤20 pC	VDE 0278, Teil 2 VDE 0278, Teil 6	cable sleeves cable sleeves pluggable/screw type enclosed cable connections
Current transformers	$1.1 \times U_m$ $1.1 \times U_m$ $1.1 \times U_m/\sqrt{3}$ $1.1 \times U_m/\sqrt{3}$	≤100 pC (li) ≤250 pC (si) ≤10 pC (li) ≤50 pC (si)	IEC 60044-4, VDE 0414 Teil 10	3 AC systems with isolated neutral li = liquid insulated si = solid insulating material

Voltage transformers (2 poles, isolated from neutral)	$1.1 \times U_m$ $1.1 \times U_m$	$\leqslant 10\,pC$ (li) $\leqslant 50\,pC$ (si)		–
Power transformers	$1.3 \times U_m/\sqrt{3}$ $1.5 \times U_m/\sqrt{3}$	$\leqslant 300\,pC$ $\leqslant 500\,pC$	IEC 60076-3 and VDE 0532, Teil 3	–
Power transformers, dry type	$1.1 \times U_m/\sqrt{3}$	$\leqslant 20\,pC$ (cr)	VDE 0532, Teil 6	cr = cast resin
Stepping switches (for power transformers)	$1.5 \times U_m/\sqrt{3}$	$\leqslant 50\,pC$	IEC 60214 and VDE 0532, Teil 30	–
Optocouplers (for safe electrical isolation)	$1.6 \times U_{IORM}$ $(1.2 \times U_{IORM}$ $1.0 \times U_{IORM})$***	$\leqslant 5\,pC$	VDE 0884	

---

* U = rated voltage of apparatus

U_m = rated maximum permissible voltage of apparatus

U_o = rated cable voltage between conductor and earth or between conductor and screening

U_{IORM} = insulation voltage (peak value) to be applied for 1 minute

** $1\,pC = 1 \cdot 10^{-12}\,C$

  $= (1 \cdot 10^{-12}\,As)$

*** Optocouplers are type tested in different lots with PD test voltages specified accordingly

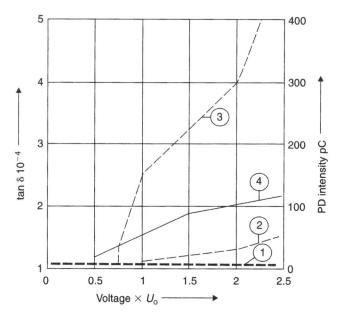

Figure 8.10   *PD intensities and tan δ values versus test voltage for a cross-linked polyethylene insulated cable (according to [30]).*
Curve 1: PD intensity, intact insulation; Curve 2: PD intensity, damaged area 1 cm²;
Curve 3: PD intensity, damaged area 6 cm²; Curve 4: tan δ for cases 1–3; U_o = rated cable voltage between conductor and earth or between conductor and screening.

measuring accuracy. On the contrary, even the smallest defects in an insulation can be detected by applying the PD measuring technique, assuming that the very small PD signals are free of superposing environmental disturbances.

Figure 8.10 may illustrate this fact: a cable insulated with cross-linked polyethylene shows a constant PD intensity versus test voltage (curve 1) with an intact insulation. Small local damages of the screening (curve 2 for 1 cm² (!) and curve 3 for 6 cm² area of damaged screening) show significantly increased PD intensities, whereas tan δ measurements give identical results (curve 4) for all three cases.

8.3   Testing intrinsic safety of electrical circuits

The standards for 'intrinsic safety – i –', EN 50020 and IEC 60079-11, define the scope of type verifications and tests as well as routine verifications and tests (in clauses 10 and 11). Among the type tests:

- spark ignition test
- temperature tests
- voltage tests
- small component ignition test
- determination of parameters of loosely specified components

- tests for cells and batteries
- mechanical tests
- tests for apparatus containing piezoelectric devices
- tests for diode safety barriers and safety shunts
- cable pull tests

the spark ignition test shall be elucidated in more detail. As explained in Section 6.9.1, one basic philosophy of 'intrinsic safety' is to limit the electrical parameters current, voltage, power and stored energy in capacitors $(0.5 \times CU^2)$ and in inductances $(0.5 \times LI^2)$ in such a way that make or break in an electrical circuit cannot create ignitable sparks. Make and break are simulated applying a spark test apparatus (in a certain sense an 'interrupter') at the points of make and break. This spark test apparatus as now standardized (in EN 50020, IEC 60079-11) has been proved to be the most sensitive in numerous trials over some decades. It is the reference for all ignition tests in electrical circuits as far as 'intrinsic safety' is concerned.

The spark test apparatus (Fig. 8.11) consists of a contact arrangement in an explosion chamber (1) with a volume of $250 \, cm^3$ at least. It produces make – sparks and break – sparks in an explosive test mixture. The contact arrangement consists of a rotating cadmium disc (5) with two slots (4) and a counter-rotating electrode holder (2) with four tungsten contact wires (3). In operation, the electrode holder (2) rotates so that the tungsten contact wires (3) slide over the slotted cadmium disc (4, 5). The shafts of the electrode holder and the cadmium disc are geared together by a non-conductive gearbox with a ratio of 50 (cadmium disc) to 12 (electrode holder), so that the electrode holder rotates (roughly) four times faster than the cadmium disc. The electrode holder is rotated at $80 \, min^{-1}$ by an electric drive, and so the cadmium disc rotates with roughly $20 \, min^{-1}$ in the opposite direction. The current to the contact arrangement is fed to the shafts of electrode holder and cadmium disc via sliprings (6).

The shafts shall be tightened when entering the explosion chamber (1). This chamber shall withstand an explosion pressure of 1.5 MPa (=15 bar). Figure 8.12 shows a longitudinal section drawing of the spark test apparatus and Fig. 8.13 this apparatus in its entirety (cover of explosion chamber removed, however).

Due to its construction, there are some limitations for the spark test apparatus:

- the test current shall not exceed 3 A
- for resistive or capacitive circuits the operating voltage shall not exceed 300 V
- for inductive circuits the inductance shall not exceed 1 H
- the frequency of the circuit shall not exceed 1.5 Mcps.

The spark test apparatus must not influence the circuit to be tested, e.g. the circuit inductance must not be increased by the apparatus inductance. So, the parameters of the apparatus are:

- self-capacitance $\leqslant 30 \, pF$
- self-inductance $\leqslant 3 \, \mu H$
- resistance (at 1 ADC) $\leqslant 0.15 \, ohm$.

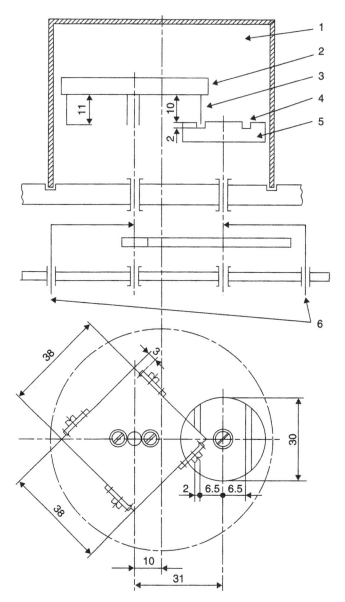

Figure 8.11   *Spark test apparatus. Functional scheme. Dimensions are given in mm.*

With this spark test apparatus, the reference curves (minimum ignition curves) of Figs 6.162 to 6.167 (showing price ratio ex-protected to non-ex-protected cage induction motors versus rated power: types of protection – (E) Ex e II T_3; (E) Ex d IIC T_4; and (E) Ex d IIB T_4) have been established with reference gas–air mixtures according to Table 6.38.

In practice, 'simple' circuits may be assessed with respect to the reference curves, i.e. there is no need to apply the spark test apparatus. However, more

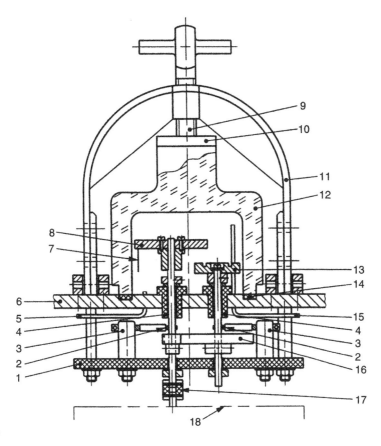

Key

1. Insulating plate
2. Current connection
3. Insulated bolt
4. Insulated bearing
5. Gas outlet
6. Base plate
7. Contact wire
8. Contact holder
9. Clamping screw
10. Pressure plate
11. Clamp
12. Chamber
13. Cadmium contact disc
14. Rubber seal
15. Gas inlet
16. Gear wheel drive 50:12
17. Insulated coupling
18. Drive motor with reduction gears 80 r/min

Figure 8.12   *Spark test apparatus longitudinal section drawing, according to EN 50020 and IEC 60079-11.*

complicated circuits, e.g. with a non-linear characteristic or an AC fed circuit, ask for a 'practical' test with the spark test apparatus.

> *The spark test apparatus shall be inserted in the circuit under test at each point where it is considered that an interruption or interconnection may occur. Tests shall be made with the circuit in normal operation, and also with one or two faults, as appropriate to the category of electrical apparatus ..., and with the maximum values of the external capacitance (C_o) and inductance (L_o) or inductance to resistance ratio (L_o/R_o) for which the apparatus is designed.*

(from EN 50020 and IEC 60079-11 respectively)

Figure 8.13 *Spark test apparatus, total view (cover of explosion chamber removed).*

The number of rotations of the electrode holder is specified as follows:

- for DC circuits 400 revolutions as a total, 200 revolutions at each polarity. With respect to the nominal speed of $80\,min^{-1}$ for the electrode holder, the test cycle takes 5 minutes
- for AC circuits 1000 revolutions (12.5 minutes)
- for capacitive circuits 400 revolutions as a total, 200 revolutions at each polarity.

No ignition shall occur in any test series at any of the chosen test points.

Figure 6.236 may give an impression of the test points to be selected in an intrinsically safe circuit.

When using the spark test apparatus, the safety factor of 1.5 (if required) shall be obtained by one of the following methods:

- for inductive and resistive circuits, the current shall be increased to 1.5 times the fault current by decreasing the current limiting resistance, or, if the factor of 1.5 cannot be obtained, by increasing the voltage additionally
- for capacitive circuits, the voltage shall be increased to 1.5 times the fault voltage
- instead of applying an increased current or voltage, use test mixtures with an MIC (**minimum ignition current**) decreased accordingly. These test

Table 8.6 Test mixtures to be used in the spark test apparatus to ensure a safety factor of 1.5 (at $T = 293\,\text{K}$ and $p = 1.013 \cdot 10^5\,\text{Pa}$), according to EN 50020 and IEC 60079-11

For Group	Composition of test mixtures, in % (v/v)				
	O_2–H_2–air mixtures			O_2–H_2 mixtures	
	H_2	Air	O_2	H_2	O_2
I	52	48	–	85	15
IIA	48	52	–	81	19
IIB	38	62	–	75	25
IIC	30	53	17	60	40

mixtures (oxygen–hydrogen–air mixtures or pure oxygen–hydrogen mixtures) are listed in Table 8.6.

They may be compared with those of Table 6.38, which do not include a safety factor.

Chapter 9

Financial considerations – selecting explosion protected electrical equipment

When selecting explosion protected apparatus, there are technical (or physical) and financial aspects to be considered.

Technical and physical aspects were discussed in Chapter 5, first of all by considering the appropriate choice of apparatus according to safety related data, e.g. MESG, MIC or temperature class. Financial aspects are concerned as far as 'low' temperature classes (T5, T6) and will result in a rather uneconomic design of motors in 'increased safety – e –' or, in a weakened fashion, according to 'flameproof enclosure – d –' (with the exception of totally water-cooled engines). In the same way, the need for a Group IIC flameproof switchgear does restrict considerably the selection from various apparatus available on the market.

In this chapter, financial aspects – independent of safety related data – are given priority, e.g. the comparison between 'e'- and 'd'-motors or light fittings.

9.1 Mains-operated light fittings in zone 1

In Table 6.23, a survey of lamps is given for general lighting in industrial plants. Sodium vapour low pressure lamps (point 5.1 in Table 6.23) are not within the scope of the standards for explosion protected luminaires. Incandescent lamps (1.1 in Table 6.23) and blended lamps (2 in Table 6.23) are unimportant for general lighting applications in industry due to their poor luminous efficiency. Therefore, the following luminaire–lamp combinations are 'the state of the art':

- tubular fluorescent lamps in luminaires according to 'increased safety – e –' or 'flameproof enclosure – d –', the latter of importance for low temperature classes (T5, T6) only
- high pressure (mercury vapour, metal-halide, sodium vapour, sodium–xenon) lamps in flameproof – d – luminaires
- induction lamps in e- or d-luminaires
- halogen or high pressure discharge lamps in flameproof – d – floodlights.

Table 9.1 gives an overview of capital investment for zone 1 luminaires. Obviously, flameproof luminaires with high pressure discharge lamps show

Table 9.1 Capital investment* for mains operated luminaires in zone 1 (prices in arbitrary units)

Type of luminaire	Luminaire Type of protection/ temp. class	Luminaire Price per luminaire	Lamps Number/type	Lamps Luminous flux lm	Lamps Price total	Operating efficiency of the luminaire	Total luminous flux lm	Investment luminaire and lamps	Luminous flux per price unit lm	Operational lifetime of lamp(s) hours
1 Light fitting	e, T4	493	2 × tub. fluoresc. 2 × 65 W	2 × 4800	21	0.72	6900	514	13.4	6000
2 Luminaire for induction lamp	e, T4	1430	1 × induction lamp 1 × 150 W	1 × 12000	153	0.7	8400	1583	5.3	60000
3 Floodlight	d, T3									
3.1		2773	1 × HST 600** 1 × 600 W	1 × 90000	46	0.64	57600	2829	20.4	24000
3.2		2203	1 × HST 400** 1 × 400 W	1 × 48000	19	0.64	30700	2222	13.8	24000
3.3		2203	1 × HIT 400** 1 × 400 W	1 × 32000	35	0.64	20480	2238	9.2	6000
4 Pendant light fitting	d, T4	1307								
4.1			1 × HST 400** 1 × 400 W	1 × 48000	19	0.71	34080	1326	25.7	24000
4.2			1 × HIT 400** 1 × 400 W	1 × 32000	35	0.71	22720	1342	16.9	6000

*Capital investment costs do not include costs for lamp replacement and luminaire cleaning. The latter depend on the specific situation in the plant and may not be specified in general

** Key for lamp type marking see Section 6.8.5

a higher total luminous flux (22720 lm to 57600 lm) compared with
e-luminaires (6900 lm to 8400 lm), a fact which indicates that a lighting system with d-luminaires requires a smaller amount of luminaires compared
with a system of e-luminaires. This may be a considerable advantage for
situations with high mounting levels (above floor level) of luminaires and
increased cleaning and lamp replacement costs caused hereby.

The specific luminous flux – luminous flux per price unit – however, does
not differ significantly between e-luminaires with tubular fluorescent lamps
(13.4 lm) and d-luminaires (9.2 lm to 25.7 lm).

Luminaires for induction lamps play a special part with a specific luminous
flux of 5.3 lm only, but they are compensated with a multiple operational lifetime and decreased lamp replacement costs.

For a lighting system, the total costs per year (including costs for electric
energy) can be calculated according to:

$$c_{total} = c_c + c_e + c_l + c_m$$

c_{total} = total costs per year
c_c = constructional costs
c_e = energy costs
c_l = lamp costs
c_m = maintenance costs

with:

$$c_c = n_1/n_2(k_1c_1 + k_2c_2) \cdot 1/100$$

$$c_e = n_1 \cdot t_o \cdot a \cdot P$$

$$c_l = n_1 \cdot c_3 \cdot t_o/t_L$$

$$c_m = n_1 \cdot c_4 \cdot t_o/t_L + w \cdot n_1/n_2$$

Here, a lighting system with n_1 lamps and n_1/n_2 luminaires is assumed
(n_2 = number of lamps per luminaire).

c_1 = costs per single luminaire
k_1 = rate for c_1 (in per cent)
c_2 = installation costs per single luminaire
k_2 = rate for c_2 (in per cent)
t_o = operational time of the lighting system per year
t_L = operational lifetime of the lamp(s)
a = costs of electrical energy (per kWh), including fixed costs
P = power consumption of a lamp (in kW), including the ballast (if any)
c_3 = costs per single lamp
c_4 = replacement costs per single lamp
w = cleaning costs per luminaire

The formula for the total costs can be separated into two terms, one determined by the amount of luminaires, n_1/n_2, and the other by the number of lamps, n_1:

$$c_{total} = n_1/n_2[k_1 \cdot c_1/100 + k_2 \cdot c_2/100 + w] + n_1 \cdot t_o[aP + 1/t_L(c_3 + c_4)]$$

For n luminaires with one single lamp (e.g. flameproof luminaires with a high pressure discharge lamp), $n_2 = 1, n_1 = n$, it is:

$$c_{total} = n[k_1 \cdot c_1/100 + k_2 \cdot c_2/100 + w + t_oaP + t_o/t_L(c_3 + c_4)]$$

and for n luminaires with two lamps (e.g. 'e' luminaires with two tubular fluorescent lamps):

$$n_2 = 2, n_1/n_2 = n, n_1 = 2n, \text{ it follows:}$$
$$c_{total} = n[k_1 \cdot c_1/100 + k_2 \cdot c_2/100 + w + 2t_oaP + 2t_o/t_L(c_3 + c_4)]$$

9.2 Motors and generators

For zone 1 applications, three types of protection enter into competition:

- flameproof enclosure – d –
- increased safety – e –
- pressurization – p –

Advantages and disadvantages are discussed in Sections 6.4.6 for 'p', 6.7.3 for 'e', and 6.8.3 and 6.8.11 for 'd'. As an overview, the following facts are of importance:

- 'e'-motors are designed 'near to normal industrial equipment', with additional requirements for the radial air gap between rotor and stator iron core and stringent temperature limitations for rotor and stator windings resulting in a poor economy for 'low' temperature classes (T5, T6). Asynchronous and synchronous motors comply with the 'e'-standards (sliprings or diode wheels shall be protected according to, e.g., 'flameproof enclosure – d –')
- 'd'-motors ask for a special enclosure with machined parts withstanding an internal overpressure of 1 MPa (10 bar) as an order of magnitude, and they are subject to restrictions for the radial clearance of bearings
- 'p'-motors are near to non-ex-protected motors (with the exception of their mains connection facilities in some cases), but they ask for additional purging and pressurizing installations and control and monitoring units.

So, the main domains of application can be described:

- in the 'small' power range – some 10^0 to 10^1 kW – 'e'-motors are more economical than 'd'-motors for T1 to – say – T3 or T4 temperature classification (see Figs 9.1–9.4)

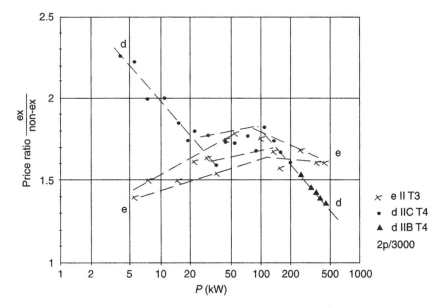

Figure 9.1 *Two-pole motors, synchronous speed: 3000 min⁻¹ for 50 cps. Price ratio ex-protected to non-ex-protected cage induction motors versus rated power. Types of protection: (E) Ex e II T3; (E) Ex d IIC T4; (E) Ex d IIB T4.*

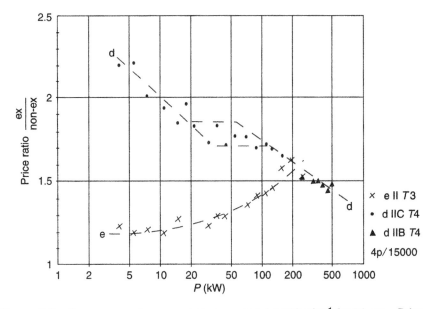

Figure 9.2 *Four-pole motors, synchronous speed: 1500 min⁻¹ for 50 cps. Price ratio ex-protected to non-ex-protected cage induction motors versus rated power. Types of protection: (E) Ex e II T3; (E) Ex d IIC T4; (E) Ex d IIB T4.*

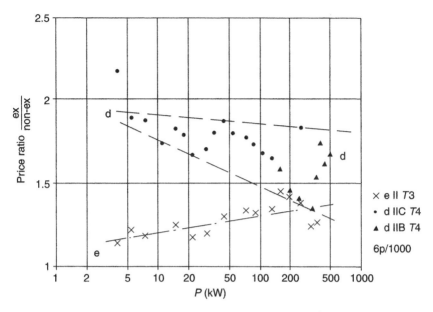

Figure 9.3   *Six-pole motors, synchronous speed: 1000 min^{-1} for 50 cps. Price ratio ex-protected to non-ex-protected cage induction motors versus rated power. Types of protection: (E) Ex e II T3; (E) Ex d IIC T4; (E) Ex d IIB T4.*

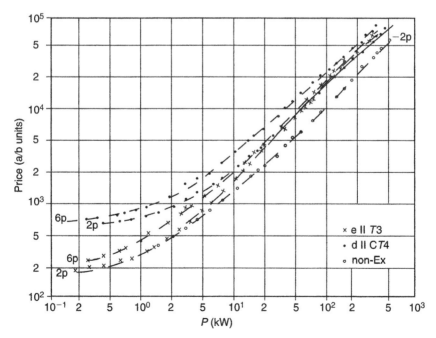

Figure 9.4   *Price versus rated power for ex-protected and non-ex-protected cage induction motors. Types of protection: (E) Ex e II T3; (E) Ex d IIC T4; 2p = two-pole motors; 6p = six-pole motors.*

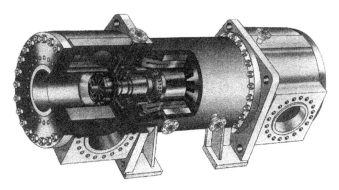

Figure 9.5 *Sectional view of an integrated two-stage compressor–motor unit (one centrifugal compressor stage at each rotor side) for natural gas.*

- in the 'medium' power range – some 10^1 to 10^2 kW – 'e'- and 'd'-motors are at the same price level, with an increasing advantage for 'd'-motors at increasing rated power, especially for dII B-motors
- in the 'high' power range – some 10^3 kW – 'p'-motors start to compete with 'e'- and 'd'-motors, and they dominate the power range of 10^4 kW (see Figs 6.2, 6.6 and 6.30–6.33).

It should be emphasized that special duty cycles for motors, e.g. arduous starting conditions with high moments of inertia or intermittent operation with short duty cycles, restrict the field of application for 'e'-motors (or generally exclude 'e'-motors).

Variable-speed motor drives with a frequency convertor show high efficiency – the losses caused by the convertor are some per cents only – good economy and high reliability. Flameproof – d – motors as well as 'p'-motors allow a 'free choice of the frequency convertor' in a given convertor frequency range and with an appropriate thermal control of the motor. For 'e'-motors with rotor and stator windings exposed to the environmental explosive atmosphere (caused by the thermal cycles of the motor, see Section 6.7.1) and temperature limitations resulting hereof, a frequency convertor supply shall be considered carefully. An additional complication arises from the non-sinusoidal voltage at the convertor output and an appropriate sizing of clearances and creepage distances in 'e'-terminal compartments (the same is valid for 'e'-terminal compartments of flameproof or pressurized motors).

For (process) gas compressors, there is a tendency to integrate the cage induction motor and the compressor into a single housing. In this conception, rotor and stator windings are exposed to the gas flow (see Fig. 9.5) which shall be free of impurities and non-corrosive and which is used – in parts – as a coolant. This motor–compressor design evades sealing and leakage problems of bearings and combines small dimensions and low weight with zero emission and considerably reduced expenditures for base plates and foundations.

Table 9.2 summarizes typical ratings and sizes of integrated compressor–motor units.

Table 9.2   Integrated compressor–motor units – main technical data (by courtesy of ALSTOM Power Conversion GmbH, Essen/Germany)

Rated power kW	Max. speed min^{-1}	Total length mm	Unit diameter mm	Unit weight kg
4200	12 700	2700	1000	9000
7500	12 000	3150	1350	11 300
9000	10 000	3400	1350	12 500
14 600	9500	4200	1650	19 000
23 000	7500	4800	2100	31 000

Figure 9.6   *Motor section of a compressor–motor unit according to Fig. 9.5 for a natural gas compressor station.*
Peterstow/UK, British Gas; Rated power: 8000 kW; Rated speed: 10 000 min^{-1}.

The motor part of such a compressor–motor unit is shown in Fig. 9.6.

Gas expansion processes (e.g. in a public natural gas supply system) are traditionally equipped with pressure relief valves (see Fig. 9.7, left branch in this diagram). More recent expansion plants convert the energy obtained by gas expansion into electric energy by means of an 'expander' engine and a generator (Fig. 9.7, right branch). The 'expander' can be designed as a reciprocating engine (Fig. 6.35) or as an expansion turbine (Figs 6.34 and 9.8) coupled with an explosion protected generator. For generators installed in this domain of application, the same 'tendency towards integration' as for motors can be observed for gas expansion units [22, 52].

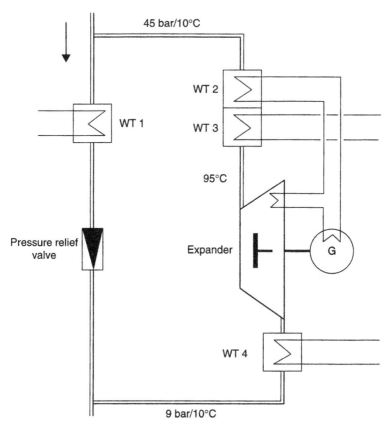

Figure 9.7   *Schematic diagram of a natural gas expansion station.*
WT 1:   preheater for pressure relief valve
WT 2:   preheater (1st stage) for gas flow through expander with waste heat
             from expander–generator set
WT 3:   main preheater (2nd stage)
WT 4:   heat exchanger at the cold gas side for, e.g., air conditioning
G: generator
45 (9) bar: 45 (9) $\times 10^5$ Pa

Figure 9.9 shows a natural gas expansion unit with an integrated turbine–generator set.

9.3   Switchgear assemblies and frequency convertors

Low voltage switchgear assemblies for installation in zone 1 are designed purely flameproof – d – (Figs 6.105 and 6.107) or flampeproof – d – combined with terminal compartments and busbar housings complying with 'increased safety – e –', see Figs 6.102 and 6.103. Parallel to this more conventional design, switchgear assemblies in 'increased safety – e –' have been developed

Figure 9.8   *Gas expansion turbine (left side) in a natural gas expansion station, coupled with an explosion protected 300 kW asynchronous generator. The 'classic' pressure relief valve is shown at the right side (according to [17] and [22]).*

(Figs 6.72 and 6.73). Sparking components (e.g. fuses, contactors, circuit breakers) inside or parts not complying with the temperature class of the assembly are individually protected by an independent type of protection (for the most part in a flameproof housing). More recently, attempts have been made to introduce pressurized switchgear assemblies onto the market (Fig. 6.19). In the USA, the 'classic solution' has been a flameproof enclosure with the 'conduit technique', see Fig. 6.83.

In the high voltage range (i.e. 3–11 kV as a guideline for this field of application), flameproof switchgear assemblies dominate (Figs 6.108, 6.109 and 6.111). In German coal mines, high voltage power distributions in 'increased safety – e –' with oil-blast circuit breakers (typical 6 kV/630 A and 200 MVA breaking capacity), type of protection (Sch)eod according to VDE 0170/0171/1969-01 served for a long time [17].

Frequency convertors – in the low voltage range – are designed according to 'flameproof enclosure – d –' (Fig. 6.104) or to 'pressurization – p –' (Fig. 6.20), combined with 'e'-terminal compartments for the most part.

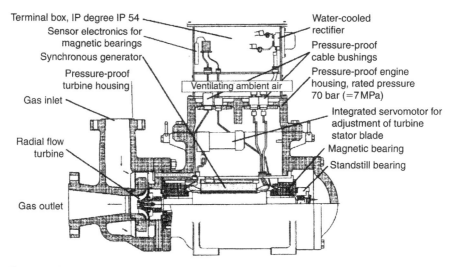

Figure 9.9 *Longitudinal section drawing of a natural gas expansion unit (integrated turbine–generator set).*

Rated power: 450 kW
Generator: 4 pole-synchronous type with permanent magnetic rotor, magnetic bearings
Speed: max. 32 000 min^{-1}
Output voltage: 500–550 V DC via integrated rectifier
Type of protection: EEx pe II T4
Gas inlet pressure: max. 70 bar (=7 MPa)
Expansion ratio: 4.5
Volume flow: 15 000 m^3/h (referring to 1×10^5 Pa)

Large-scale integration of power distribution units, i.e. the combination of high voltage switchgear, transformers, low voltage switchgear assemblies and frequency convertors, ask for a considerably increased space for installation in a single unit, which is compensated by multiple interconnection cables between individual housings and cable entries to be dropped in this case. In order to guarantee a large choice of components for installation, the enclosure shall be either flameproof (d) or pressurized (p). Fig. 9.10 gives guidance for price per unit volume versus internal volume for d- and p-enclosures [31]. Volumes up to 1 m^3 can be put into practice with a d-enclosure. Bigger volumes are the domain of p-enclosures, with a 'field of competition' in the 0.1 m^3 range. Smaller p-enclosures seem to be uneconomic due to the expenditures for purging and pressurizing installations and the control unit. The next step of integration has been achieved by p-enclosures which enable quick access for personnel via an airlock for maintenance, repair and inspection. Such 'manned' pressurized power distribution systems (Figs 6.23, 6.24 and 9.11) show a maximum level of integration and concentration of components combined with an excellent economy and high reliability in service [31]. The handling of these systems is identical with those installed in a safe area.

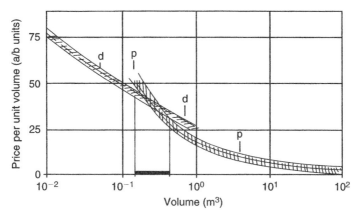

Figure 9.10 *Price per unit volume versus internal volume for flameproof – d – and pressurized – p – enclosures (according to [31]).*

Figure 9.11 *Interior view of a pressurized – p – power distribution unit: control panels (foreground) and 1 kV switchgear (background).*
Type of protection: EEx p I

Attention shall be given to national installation rules when planning a pressurized power distribution system. They may contain additional requirements for p-apparatus (e.g. the German standard VDE 0118 Teil 1/2001-11 for Group I p-enclosures in coal mines) or they may give restrictions to the field of application. Additional requirements and restrictions can both affect financial calculations as a basis for investment activities.

9.4   Remote controlling and monitoring, data transmission and communication

The description of the different types of protection in Chapter 6 indicates that there are two very different ways to solve this problem – if an electrical transmission is required at all. One way is to use intrinsically safe circuits, the other one applies industrial equipment as usual, additionally explosion protected by an enclosure as appropriate, e.g. flameproof housings for small-sized devices. In the history of process instrumentation, the appearance of semiconductors and integrated circuits has drastically reduced the power consumption of field devices. So, intrinsically safe circuits dominate this field today.

Advantages and disadvantages of 'intrinsic safety – i –' can be summarized as follows:

Advantages
- small-sized and lightweight compact construction (without heavy enclosures)
- value for money and very economic design of apparatus
- calibration, maintenance and repair activities in the live state. No interruptions of duty cycles. This is a special advantage for the so-called 'conti-plants', i.e. plants which are operational to some extent over some years without any interruption
- the standards for 'intrinsic safety – i –' are very similar worldwide (European Union, USA, Canada, Russia, Australia etc.), in contrast to, e.g., 'flameproof enclosure – d –'. Thus, apparatus and their installation techniques are applicable worldwide in a similar way
- cables with intrinsically safe circuits are a part of explosion protection, i.e. a mechanical cable damage or an electrical cable fault does not adversely affect the safety of the system. (In Chapter 12 methods are described to protect power cables adequately under severe environmental conditions, e.g. in coal mines)
- if certain 'rules of the game' are taken into account, interconnections can be put into practice to a given degree (see Section 6.9.5)
- 'intrinsic safety – i –' is suitable for zone 0 applications.

Disadvantages
- 'intrinsic safety – i –' asks for a special design of the circuits. This will be 'near to market' for great apparatus series only, whereas apparatus which are installed in hazardous areas in a small number of items shall be protected by, e.g., a flameproof enclosure (Figs 6.120 and 9.12) or a pressurized housing
- suitable for low power equipment only
- when planning intrinsically safe systems, there shall be a proof for 'intrinsic safety – i –'.

Optical data transmission systems have captured an increasing field of application. In order to avoid the ignition of a gas–air mixture, the optical

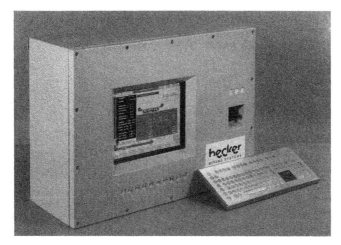

Figure 9.12 *Firedamp-proof PC workstation for installation in coal mines. The main components are fitted into a flameproof – d – Group I housing. The interfaces for PC keyboard and mouse are intrinsically safe as well as interfaces for 4–20 mA analogous inputs, RS 485 and FSK-PROFIBUS. An optical–electrical interface (Ethernet 10/100 Mbit) is available additionally.*

Power input:	230 V 1 AC 50 cps
Processor:	Pentium 266 Mcps
Working memory:	128 Mbyte, dynamical RAM
Mass memory:	2.5″ HD, 4 Gbyte
Display:	15″ active TFT-LCD display, resolution 1024 × 768
Dimensions (W × H × D):	815 × 575 × 310 mm^3
Weight:	approx. 70–80 kg, depending on internal components
Operating system:	Windows 98, Windows NT 4.0, MSDOS (alternatively).

radiation power is limited to 35 mW for Group IIC (see Section 6.9.6.7). The general advantages are:

- no equipotential bonding required, no transmission of overvoltages into a hazardous area
- high degree of safety from interception
- broadband system including audio and video transmission.

An optical data transmission system for coal mines is characteristic of:

- optical waveguides form a double ring in the system as closed loops
- up to 200 nodal points in the system as electrical-to-optical/optical-to-electrical interfaces
- distance between two nodal points up to 80 km
- transfer rate: 600 Mbit/s
- time division multiplex
- 'optical cable' with 24 fibres
- electrical inputs/outputs of the interfaces intrinsically safe
- nodal points located in hazardous areas installed in a flameproof – d – enclosure (Fig. 9.13).

(a)

(b)

Figure 9.13   *Interior view into the nodal point of an optical data transmission system for coal mines. All components are fitted into an EEx dl-enclosure.*
(a) Module with (left to right)
   4 low speed slots
   optical transceivers
   4 low or high speed slots
   power supply modules.
(b) Modules with interfaces for intrinsically safe circuits (2 modules, up to 12 printed
   circuit boards per module).

Investment decisions concerning optical data transmission systems are mainly determined by:

- the relatively high costs of the nodal points with their flameproof enclosures (where installed in a hazardous area)
- the transfer rate required as a matter of fact compared with the transfer rate as specified (in this example 600 Mbit/s).

In applications where a transfer rate increasing year by year is to be expected with investments step by step, the 'electrical solution' with an intrinsically safe data transmission may be favourable. The optical data transmission system shows investment costs c_{opt}, as a total in the first year, and the intrinsically safe system investment costs c_i in the year number i ($i = 1, \ldots n$). So, the total costs are c_{opt} for the optical system and:

$$c_{tot} = \sum_{i=1}^{n} c_i \text{ over } n \text{ years for the electrical system.}$$

With a rate k (in %) to be constant over n years, the annual costs are $c_{opt} \times k/100$ for the optical and:

$$k/100 \times \sum_{i=1}^{j} c_i$$

for the intrinsically safe system in year j ($j = 1, \ldots n$).
This will result in a 'gap' (decreasing for the most part, e.g. in the first years) of annual costs (year j):

$$\delta_j = k/100 \times \left(c_{opt} - \sum_{i=1}^{j} c_i \right)$$

between the optical and intrinsically safe system. Summarized over n years,

$$\Delta = \sum_{j=1}^{n} \delta_j$$

may result in a financial benefit for the intrinsically safe transmission system in cases where step by step installations are adequate (and a financial procedure is preferred to save money at the first steps).

Chapter 10

Inspection, maintenance and repair of explosion protected equipment

As with industrial electrical equipment in chemical plants, explosion protected apparatus and systems are subject to inspection, maintenance and repair. These apparatus and systems show special features in order to comply with the requirements of the relevant standards (see Chapter 6) to guarantee a safe operation in hazardous areas. Inspection, maintenance and repair shall ensure this safe operation during the total lifetime of the apparatus and systems installed. For Group II applications (installations in chemical plants, oil and gas industry), requirements are given in EN 60079-17 (German Standard: VDE 0165 Teil 10) and IEC 60079-17. (In Germany, additional regulations are given in ElexV [8]. At the time of writing, a 'successor' is under consideration to transform Directive 1999/92/EC into national legislation ('Betriebssicherheitsverordnung – BetrSichV').)

To clarify things: neither type tests and verifications nor routine tests are the point of this chapter, but inspection and maintenance procedures after installation and before or during service. These procedures shall be done – and this is a general rule – by qualified personnel, who are experienced in the various types of explosion protection, installation practices and rules and the principles of area classification.

After installation of an explosion protected apparatus and before it is brought into service, it shall be given an initial inspection. During service, regular periodic inspections or continuous supervision shall guarantee the safe operation of the apparatus or system. If necessary, maintenance or repair shall be carried out. The standards define different grades of inspection:

- visual inspection: an inspection which identifies, without the use of access equipment or tools, those defects, such as missing bolts, which will be apparent to the eye
- close inspection: an inspection which encompasses those aspects covered by a visual inspection and, in addition, identifies those defects, such as loose bolts, which will be apparent only by the use of access equipment, e.g. steps (where necessary), and tools
- detailed inspection: an inspection which encompasses those aspects covered by a close inspection, and, in addition, identifies those defects,

such as loose terminations, which will only be apparent by opening the enclosure, and/or using, where necessary, tools and test equipment

(these definitions are quoted from EN 60079-17 and IEC 60079-17). Inspection schedules are summarized in Tables 10.1–10.3.

Table 10.1 Inspection schedule for (E) Ex d, (E) Ex e and (E) Ex n installations D = Detailed, C = Close, V = Visual (according to EN 60079-17, IEC 60079-17)

Check that:		Ex 'd'			Ex 'e'			Ex 'n'		
		Grade of inspection								
		D	C	V	D	C	V	D	C	V
A	APPARATUS									
1	Apparatus is appropriate to area classification	X	X	X	X	X	X	X	X	X
2	Apparatus group is correct	X	X		X	X		X	X	
3	Apparatus temperature class is correct	X	X		X	X		X	X	
4	Apparatus circuit identification is correct	X			X			X		
5	Apparatus circuit identification is available	X	X	X	X	X	X	X	X	X
6	Enclosure, glass parts and glass-to-metal sealing gaskets and/or compounds are satisfactory	X	X	X	X	X	X	X	X	X
7	There are no unauthorized modifications	X			X			X		
8	There are no visible unauthorized modifications		X	X		X	X		X	X
9	Bolts, cable entry devices (direct and indirect) and blanking elements are of the correct type and are complete and tight									
	– physical check	X	X		X	X		X	X	
	– visual check			X			X			X
10	Flange faces are clean and undamaged and gaskets, if any, are satisfactory	X								
11	Flange gap dimensions are within maximal values permitted	X	X							
12	Lamp rating, type and position are correct	X			X			X		
13	Electrical connections are tight					X			X	

(*continued*)

Table 10.1   (*continued*)

Check that:	Ex 'd'			Ex 'e'			Ex 'n'		
	Grade of inspection								
	D	C	V	D	C	V	D	C	V
14 Condition of enclosure gaskets is satisfactory					X			X	
15 Enclosed-break and hermetically sealed devices are undamaged								X	
16 Restricted breathing enclosure is satisfactory								X	
17 Motor fans have sufficient clearance to enclosure and/or covers	X			X			X		
18 Breathing and draining devices are satisfactory	X	X		X	X		X	X	
B INSTALLATION									
1 Type of cable is appropriate	X			X			X		
2 There is no obvious damage to cables	X	X	X	X	X	X	X	X	X
3 Sealing of trunking, ducts, pipes and/or conduits is satisfactory	X	X	X	X	X	X	X	X	X
4 Stopping boxes and cable boxes are correctly filled	X								
5 Integrity of conduit system and interface with mixed system is maintained	X			X			X		
6 Earthing connections, including any supplementary earthing bonding connections are satisfactory (for example connections are tight and conductors are of sufficient cross-section)									
– physical check	X			X			X		
– visual check		X	X		X	X		X	X
7 Fault loop impedance (TN systems) or earthing resistance (IT systems) is satisfactory	X			X			X		
8 Insulation resistance is satisfactory	X			X			X		
9 Automatic electrical protective devices operate within permitted limits	X			X			X		
10 Automatic electrical protective devices are set correctly (auto-reset not possible)	X			X			X		
11 Special conditions of use (if applicable) are complied with	X			X			X		
12 Cables not in use are correctly terminated	X			X			X		

(*continued*)

Table 10.1 (*continued*)

Check that:	Ex 'd'			Ex 'e'			Ex 'n'		
	Grade of inspection								
	D	C	V	D	C	V	D	C	V
13 Obstructions adjacent to flameproof flanged joints are in accordance with EN 60079-14/IEC 60079-14	X	X	X						
14 Variable voltage/frequency installation in accordance with documentation	X	X		X	X		X	X	
C ENVIRONMENT									
1 Apparatus is adequately protected against corrosion, weather, vibration and other adverse factors	X	X	X	X	X	X	X	X	X
2 No undue accumulation of dust and dirt	X	X	X	X	X	X	X	X	X
3 Electrical insulation is clean and dry					X			X	

Note: 1 General: the checks used for apparatus using both types of protection 'e' and 'd' will be a combination of both columns

2 Items B7 and B8: account should be taken of the possibility of an explosive atmosphere in the vicinity of the apparatus when using electrical test equipment

Table 10.2 Inspection schedule for (E) Ex i installations (according to EN 60079-17, IEC 60079-17)

Check that:	Grade of inspection		
	Detailed	Close	Visual
A APPARATUS			
1 Circuit and/or apparatus documentation is appropriate to area classification	X	X	X
2 Apparatus installed is that specified in the documentation – fixed apparatus only	X	X	
3 Circuit and/or apparatus category and group correct	X	X	
4 Apparatus temperature class is correct	X	X	
5 Installation is clearly labelled	X	X	
6 There are no unauthorized modifications	X		
7 There are no visible unauthorized modifications		X	X
8 Safety barrier units, relays and other energy limiting devices are of the approved type,	X	X	X

(continued)

Table 10.2   (*continued*)

Check that:	Grade of inspection		
	Detailed	Close	Visual
installed in accordance with the certification requirements and securely earthed where required			
9 Electrical connections are tight	X		
10 Printed circuit boards are clean and undamaged	X		
B INSTALLATION			
1 Cables are installed in accordance with the documentation	X		
2 Cable screens are earthed in accordance with the documentation	X		
3 There is no obvious damage to cables	X	X	X
4 Sealing of trunking, ducts, pipes and/or conduits is satisfactory	X	X	X
5 Point-to-point connections are all correct	X		
6 Earth continuity is satisfactory (for example connections are tight and conductors are of sufficient cross-section)	X		
7 Earth connections maintain the integrity of the type of protection	X	X	X
8 The intrinsically safe circuit is isolated from earth or earthed at one point only (refer to documentation)	X		
9 Separation is maintained between intrinsically safe and non-intrinsically safe circuits in common distribution boxes or relay cubicles	X		
10 As applicable, short-circuit protection of the power supply is in accordance with the documentation	X		
11 Special conditions of use (if applicable) are complied with	X		
12 Cables not in use are correctly terminated	X	X	X
C ENVIRONMENT			
1 Apparatus is adequately protected against corrosion, weather, vibration and other adverse factors	X	X	X
2 No undue external accumulation of dust and dirt	X	X	X

Table 10.3   Inspection schedule for (E) Ex p installations (according to EN 60079-17, IEC 60079-17)

Check that:	Grade of inspection		
	Detailed	Close	Visual
A APPARATUS			
1 Apparatus is appropriate to area classification	X	X	X
2 Apparatus group is correct	X	X	
3 Apparatus temperature class is correct	X	X	
4 Apparatus circuit identification is correct	X		
5 Apparatus circuit identification is available	X	X	X
6 Enclosure, glass parts and glass-to-metal sealing gaskets and/or compounds are satisfactory	X	X	X
7 There are no unauthorized modifications	X		
8 There are no visible unauthorized modifications		X	X
9 Lamp rating, type and position are correct	X		
B INSTALLATION			
1 Type of cable is appropriate	X		
2 There is no obvious damage to cables	X	X	X
3 Earthing connections, including any supplementary earthing bonding connections are satisfactory, for example connections are tight and conductors are of sufficient cross-section			
– physical check	X		
– visual check		X	X
4 Fault loop impedance (TN systems) or earthing resistance (IT systems) is satisfactory	X		
5 Automatic electrical protective devices operate within permitted limits	X		
6 Automatic electrical protective devices are set correctly	X		
7 Protective gas inlet temperature is below maximum specified	X		
8 Ducts, pipes and enclosures are in good condition	X	X	X
9 Protective gas is substantially free from contaminants	X	X	X
10 Protective gas pressure and/or flow is adequate	X	X	X
11 Pressure and/or flow indicators, alarms and interlocks function correctly	X		

(continued)

Table 10.3   *(continued)*

Check that:	Grade of inspection		
	Detailed	Close	Visual
12 Pre-energizing purge period is adequate	X		
13 Conditions of spark and particle barriers of ducts for exhausting the gas in hazardous area are satisfactory	X		
14 Special conditions of use (if applicable) are complied with	X		
C ENVIRONMENT			
1 Apparatus is adequately protected against corrosion, weather, vibration and other adverse factors	X	X	X
2 No undue accumulation of dust and dirt	X	X	X

The initial inspection shall ensure that the type of protection of the apparatus and its installation are appropriate. As a rule, initial inspections shall be detailed inspections. Periodic inspections are of the visual or close type. They shall 'detect' all the factors effecting the deterioration of apparatus, e.g. corrosion, exposure to chemicals or solvents, accumulation of dust or dirt, water ingress, exposure to excessive ambient temperature, mechanical damage, unauthorized modifications or adjustments and inappropriate maintenance.

The interval between periodic inspections shall not exceed three years (with the exception that an expert may extend this interval). A typical inspection procedure is given in Fig. 10.1. Sample inspections (inspections of a proportion of the electrical apparatus, systems and installations) of different grades shall 'adjust' the interval between periodic inspections to an appropriate value.

For maintenance or repair of explosion protected apparatus, three 'categories' may be differentiated:

A: replacement of a part or component which is not of relevance for the safety of the apparatus, i.e. this part or component does not affect the type of protection of the apparatus
B: replacement of a part or component which does affect the type of protection of the apparatus. The part or component itself complies with the relevant standard and/or is an original replacement part or may be fitted to the apparatus as an optional part or component as defined in the annex of the certificate
C: replacement of a part or component which does affect the type of protection of the apparatus. The part or component itself is neither defined in the annex of the certificate nor an original replacement part. Such

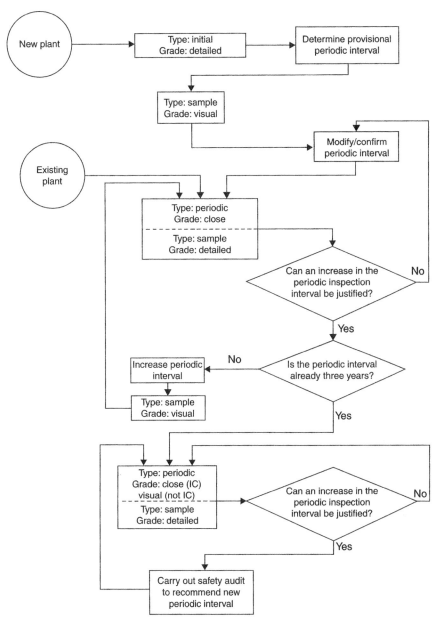

Figure 10.1   *Typical inspection procedure for periodic inspections (according to EN 60079-17 and IEC 60079-17). IG: ignition capable internal components.*

replacements will be out of the scope of the certificate and they abolish the explosion protection of the apparatus.

In Tables 10.4–10.6, typical maintenance or repair activities are listed. To give guidance, they are related to one of the categories A, B or C as defined above.

Table 10.4 Maintenance and repair of flameproof – d – apparatus

Operation	Category
Reasons for classification A/B/C	
1 Replacement of a lamp (of an identical type and power rating) in a luminaire	A
Identical type and power rating complies with the certificate and temperature classification	
2 Replacement of a transformer or ballast (with identical technical data)	A
Identical technical data comply with the certificate	
3 Replacement of a capacitor for a new one of a different type but with identical electrical data	C
Impregnating fluids in capacitors may be decomposed by fault current arcing into hydrogen and other constituents which may adversely affect the safety of a flameproof enclosure (see Section 6.8.1)	
4 Replacement of rods, shafts and spindles (original parts)	B
Dimensions of joints are an essential characteristic of a flameproof enclosure	
5 Replacement of conductor bushings (original parts)	B
Current and voltage ratings and thermal stability shall comply with certificate. Unchanged dimensions of joints	
6 Fitting of components not listed in the certificate	C
Additional components inside a flameproof enclosure may affect the temperature class of the apparatus. They may contain liquids subject to thermal decomposition or insulating materials with poor CTI values (see Section 6.8.1). Aluminium may be used as a conductor material (see Section 6.8.11)	
7 Fitting of cable entries listed in the certificate	B
These cable entries comply with the requirements given in 'General requirements' and 'Flameproof enclosure – d –', e.g. EN 50014 and EN 50018	
8 Machining of commutators or sliprings of motors/generators	A
There is no influence on the safety of a flameproof enclosure	
9 Rewiring the stator windings of a cage induction motor, voltage/power/frequency/number of poles according to the certificate	A
There are no changes of mechanical or electrical data	
10 Rewiring the stator windings of a cage induction motor, voltage and frequency not in accordance with the certificate	C

(continued)

Table 10.4 *(continued)*

Operation	Category
Reasons for classification A/B/C	

An increased motor speed (at a higher frequency) may cause an increased explosion pressure in a flameproof enclosure (see Section 8.1). Bearings may show an increased temperature and exceed the temperature class of the motor where fitted outside the flameproof enclosure (Fig. 6.85(b)). In terminal compartments according to 'increased safety – e –', clearances and creepage distances at an increased voltage may not comply with the requirements

11 Replacement of a rolling bearing inside the flameproof enclosure   A

Not of influence on the safety of a flameproof enclosure

12 Replacement of an end shield of a motor (original part)   B

Dimensions of the joints are unchanged and in accordance with the certificate, as well as the material quality to withstand the explosion pressure

13 Machining the surfaces of a flameproof joint, gap and width of joint according to the certificate   B

Unchanged dimensions of joints are in accordance with the certificate

14 Machining the surfaces of a joint considered to be flameproof, gap and width of joint not in accordance with the certificate   C

Dimensions of joint may be outside the requirements of the relevant standard. Joints with changed dimensions shall pass successfully a type test (see Section 8.1)

15 Replacement of inspection windows (original parts)   B

With an identical material quality and dimensions of the joint they comply with the certificate

16 Replacement of inspection windows; type not listed in the certificate   C

Material quality and joint dimensions shall comply with the relevant standard. All type tests shall be passed successfully

Table 10.5   Maintenance and repair of apparatus according to 'increased safety – e –'

Operation	Category
Reasons for classification A/B/C	
1 Replacement of a lamp of an identical type and a reduced/ identical power rating in a luminaire	B
An identical lamp type complies with the certificate. Lamp power rating is of influence on luminaire temperature classification	
2 Replacement of a lamp in a luminaire for a new one of a different type (e.g. a compact fluorescent for an incandescent lamp) to increase efficiency	C
The lamp type shall comply with the standard's requirement. An industrial-type compact fluorescent lamp contains electronics not in accordance with 'increased safety – e –'	
3 Replacement of a gasket (original part)	B
Material quality and dimensions shall comply with those given in the certificate to ensure an appropriate IP degree of the apparatus	
4 Replacement of a ballast (original part)	B
Mechanical, thermal and electrical properties in accordance with certificate, as well as clearances and creepage distances	
5 Replacement of a ballast for a new one not listed in the certificate	C
Mechanical, thermal and electrical properties may deviate from those given in the certificate. Clearances and creepage distances may not comply with the relevant standard	
6 Fitting a new wiring in a luminaire	B
Conductor cross-sections and insulation quality shall comply with the standard's requirements	
7 Replacement of terminal blocks (original parts)	B
Mechanical, thermal and electrical properties in accordance with certificate as well as clearances and creepage distances	
8 Assembling of individual 'e'-enclosures to form a complete low voltage power distribution unit, within the limiting values given in the certificate and using parts listed in the certificate	B
Mechanical, thermal and electrical data comply with certificate. If fitted according to the manufacturer's instructions, the IP degree will comply with the certificate	

(continued)

Table 10.5　*(continued)*

Operation	Category
Reasons for classification A/B/C	
9　Fitting of additional cable entries, exceeding the scope of the certificate in number and/or size	C
Mechanical and thermal properties of the 'e'-enclosure may be deteriorated, its IP degree may be adversely affected	
10　Replacement of screws to fit the cover	A
Not of influence on the safety of 'increased safety – e –'	
11　Replacement of the external ventilating fan of a motor (original part)	B
Material quality, mechanical properties and surface resistance in accordance with certificate	
12　Replacement of the external ventilating fan of a motor for a new one made of an undefined material	C
Dimensions, mechanical strength, content of light metals and surface resistance may not comply with the standard's requirements	
13　Replacement of a measuring instrument for a new one (with a component certificate) but of a different housing design	C
The IP degree of an enclosure with removable parts strongly depends on surface quality, gasket material quality and dimensions of packing strip (if any)	

Table 10.6　Other maintenance and repair operations

Operation	Category
Reasons for classification A/B/C	
1　Replacement of the analyser part of a pressurized gas chromatograph for a new one of a different type	C
The analyser part forms the 'containment system' of the p-apparatus. Its properties (infallible or not, leakage flow) determine dilution requirements (see Chapter 7)	
2　Use of an industrial-type rubber jacket cord instead of a heat-resistant special cable for an apparatus marked 'X' (the 'X' asking for such heat-resistant cables)	C
Thermal properties at cable entries may require special cables able to withstand higher overtemperatures than 30 K for the entry point and 40 K for the branching point (see Section 6.7.2)	

(continued)

Table 10.6   (*continued*)

Operation	Category
Reasons for classification A/B/C	

3 Replacement of a colourless protecting glass of an (E)Ex eII   C
   T ... luminaire with an incandescent lamp for a new blue-
   coloured one with identical dimensions, not listed in the
   certificate

   *T class rating of an 'e'-luminaire with an incandescent lamp is
   determined by the power rating of the lamp and the part of radiation
   which 'escapes' from the luminaire, mainly visible light and infrared
   radiation. A blue glass cuts down the radiation in the visible part
   and raises the temperature profile. The original T class rating
   may be passed*

Chapter 11

Explosion protected apparatus for zone 0 and zone 2

To complete Chapters 6 and 7 describing zone 1 equipment, this chapter is focused on electrical apparatus for zone 0 and zone 2 application, i.e. hazardous areas endangered by combustible gases, vapours or mist. The relevant standards and their technical content have been covered in Sections 2.3 for zone 0 and 2.4 for zone 2.

An overview of these standards is given in Table 11.1.

11.1   Apparatus for zone 0

For the purpose of visually inspecting the interior of a vessel, tank or barrel which is classified as zone 0, light generated by a zone 1 luminaire, which is located outside the vessel in zone 1, is brought into zone 0 via a fibre optics bundle. In the luminaire, the light is focused on the aperture of the fibre optics bundle by means of a condensor lens (Figs 11.1 and 11.2).

Previously, luminaires powered by compressed air have been used for lighting purposes in zone 0 during inspections [17]. The luminaire contains a synchronous generator with a permanent magnetic rotor coupled with a single-stage turbine to supply a high pressure mercury vapour lamp. The compressed air enters the lamp housing via a pressure governor and an annular-shaped nozzle arrangement. After purging and pressurization of the lamp housing, the compressed air expands in the turbine and leaves the luminaire via an exhaust hose. A typical luminaire marked (Ex)s d3n G5 according to VDE 0170/0171/1969-01 shows a luminous flux of 2.800 lm with an 80 W lamp. The turbine–generator set rotates at a speed of 12.000 min^{-1} and supplies 120 V at 600 cps within a pressure range of 3–7 bar ($=3 \times 10^5 Pa$–$7 \times 10^5 Pa$). The air consumption is 15 m^3/hour, referring to atmospheric pressure.

The air flow arrangement as described ensures that in case of damage to or destruction of the protecting glass the air flow to the turbine–generator set is stopped automatically and that with decreasing generator speed the lamp circuit falls below the limit of ignitability. The lamp housing remains 'pressurized' with air.

Very different from these 'indirect' ways into zone 0 is instrumentation for inductive conductivity or concentration measurement in a pipeline or

Table 11.1   Standards for electrical apparatus for use in zone 0 and zone 2

Hazardous area	International Standard	Overtaken as a German Standard
Zone 0	EN 50284/1999-04	VDE 0170/0171
	Special requirements for construction, test and marking of electrical apparatus of equipment Group II, category 1G	Teil 12-1/2000-02
	Draft Standard:	
	pr IEC 60079-26/2002	–
	Electrical apparatus for explosive gas atmospheres	
	Part 26:	
	Special requirements for construction, test and marking of electrical apparatus for use in zone 0	
For coal mines*:	EN 50303/2000-07	VDE 0170/0171
	Group I, category M1 equipment intended to remain functional in atmospheres endangered by firedamp and/or coal dust	Teil 12-2/2001-05
Zone 2	EN 50021/1999-04	VDE 0170/0171
	Electrical apparatus for potentially explosive atmospheres – Type of protection 'n'	Teil 16/2000-02
	IEC 60079-15/2nd edition, 2001-02	–
	Electrical apparatus for explosive gas atmospheres – Part 15 Type of protection 'n'	

*Coal mines are not subject to an area classification. In most countries, firedamp-proof electrical equipment which is – as a general rule – comparable with zone 1 equipment (or, in the wording of the ATEX Directives, Group I, category M2 apparatus comparable with Group II, category 2G equipment) shall be disconnected from the grid when a given level of firedamp (methane, CH_4) is exceeded, e.g. 1 or 1.5% (v/v). Nevertheless, installations for safety devices and gas monitoring shall be operational in this case. This equipment has been classified as Group I, category M1 apparatus

vessel (Fig. 11.3). The sensor head is located in zone 0, whereas the two-wire transmitter is located outside in zone 1. Instead of a flange-mounted transmitter with integrated sensor head, the latter may be fitted separately in zone 0.

An example for pH measurement instrumentation is the pH immersion assembly according to Fig. 11.4. Here, the electrical values are lower than

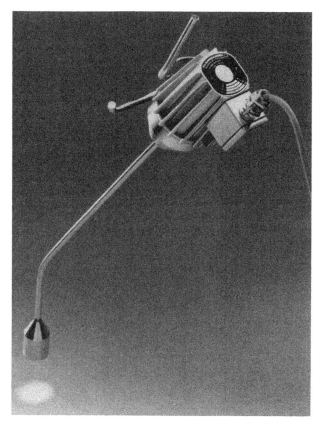

Figure 11.1   *Luminaire for inspection of vessels, tanks, barrels or pipes classified as zone 0. The luminaire body shall remain outside, i.e. in zone 1. The light is brought into zone 0 via a fibre optics bundle.*
Type of protection: EEx de IIC T4
Certificate: PTB Ex-93.C.2087
Lamp: Halogen lamp, 12 V, 12 W
Power supply: 12 V AC/DC, or 230 V 1 AC 50 cps with internal transformer
Weight: approx. 3.3 kg (12 V type)
                      approx. 3.8 kg (230 V type)

1.5 V, 100 mA and 25 mW. Thus, a certificate is not required (see Section 6.9.3.1). The transmitter is intented for zone 1 application (see Fig. 6.177).

Apart from this zone 0 equipment, a methane transmitter to monitor the CH_4 concentration in coal mines is shown in Fig. 6.174 as a Group I, category M1 apparatus.

11.2   Apparatus for zone 2

In the history of the development of explosion protection techniques for electrical apparatus located in zone 2, the 'non-sparking concept' (see Section 2.4)

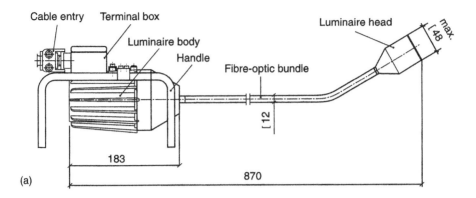

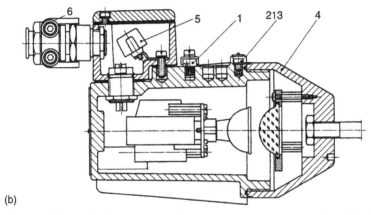

(b)

Figure 11.2   *(a) Dimensions of the luminaire according to Fig. 11.1. All dimensions given in millimetres.*
(b) Longitudinal section drawing of the luminaire according to Fig. 11.1 (luminaire body).
1   terminal for earthing conductor
2, 3 special fastener and securing clip
4   threaded part of the flameproof luminaire body
5   terminal box and terminal block according to 'increased safety – e –'
6   cable entry

has found its main domain of application in motors (cage induction motors), mains operated luminaires, control panels (in 'nP', simple pressurization) and switchgear assemblies (in 'nR', restricted breathing enclosure).

In more recent times, the n-concept has been extended to applications in the field of process control. In Table 11.2, a comparison is given between two CPU and power modules for remote I/O systems, one intended for zone 1 installation, the second for zone 2 application. Obviously, the zone 2 concept with a special protection technique offers some financial and technical benefits compared with a zone 1 apparatus which may be used in zone 2 as a matter of course.

Figure 11.3   *Sensor for inductive conductivity or concentration measurement in a pipeline or vessel classified as zone 0 with integrated flange-mounted 2-wire transmitter (in zone 1).*

Conductivity measurements

Measuring range:	5 μS/cm–2000 mS/cm
Max. resolution:	0.1 μS/cm
Measuring frequency:	2 kcps

Sensor

Marking:	Ex II 1G EEx ia IIC T6/T4
Type of protection:	EEx ia IIC T6/T4

Medium temperature range:
- for T4 classification $-20°C \leqslant T_{med} \leqslant +125°C$
- for T6 classification $-20°C \leqslant T_{med} \leqslant +75°C$

Certificate: DMT 99 ATEX E 075 X

Figure 11.4   *Sensor head for pH measurements in vessels or tanks classified as zone 0.*

Measuring range:	pH −2.00–+16.00
Resolution:	pH 0.01
Depth of immersion:	500–2500 mm
Max. allowable pressure:	10 bar (=1 MPa)

pH-signal input

Input resistance:	>1 × 10^{12} ohms
Input current:	<1.6 × 10^{-12} A

Table 11.2 CPU and power modules for remote I/O systems (by courtesy of R. STAHL Schaltgeräte GmbH, Waldenburg/Germany)

Model	For zone 1 (see Fig. 11.5)	For zone 2 (see Fig. 11.6)
Function	power supply and fieldbus interface	
Dimensions	both modules show an identical width	
Type of protection	EEx de [ia/ib] IIC/IIB T4	EEx nA II T4
Marking	Ex II 2G EEx de [ia/ib] IIC/IIB T4	Ex II 3G EEx nA II T4
Housing	flameproof enclosure – d – plastic material	housing with ventilating apertures
Auxiliary power	24 V DC	24 V DC
Dissipated power	approx. max. 12 W	approx. max. 20 W
Number of pluggable I/O modules	max. 8	max. 16
Connecting terminals		plug-and-socket connectors, fixed by screws to avoid any unintentional loosening
Auxiliary power	EEx e-terminals	
Others	EEx i – plug-and-socket connectors	
Connection between module and socket	1 × EEx d plug-and-socket connector	–
	1 × EEx i plug-and-socket connector	
Price (arbitrary units)	912,–	702,–

Figure 11.5   *CPU and power module for remote I/O systems, zone 1 application (for technical data see Table 11.2)*

Figure 11.6   *CPU and power module for remote I/O systems, zone 2 application (for technical data see Table 11.2)*

Chapter 12

Cable protection in coal mines and other areas hazardous due to combustibles

12.1 Principles of cable fault detection

In Chapters 6, 7 and 11, explosion protected apparatus for hazardous areas was described. But what about the cables? In industrial plants cables are laid on cable trays or ducts where a negligible risk arises for mechanical cable damage. Cable-fed mobile equipment, e.g. cranes and mechanical shovels, are subject to cable damage. However, a very different situation is given in coal mines: for the most part, electrical equipment is 'mobile', i.e. frequent changes of place are required following the wall face. A special risk exists for the cables of shearer-loaders (see Figs 6.3 and 6.137). Besides mechanical damage, cable faults are caused by frequent bending stresses over long periods, tumbling of conductor insulating material and breaking of conductor wires. In the case of a short-circuit arc it will decompose the insulating material and blow out within some milliseconds supported by enormous dynamical forces due to the short-circuit current. Even in the absence of methane a coal dust layer is stirred up and ignited by such a cable fault.

Cables with intrinsically safe circuits (see Section 6.9) are 'explosion protected' and cannot act as an ignition source in case of mechanical damage or cable fault. However, the power transferred per circuit is limited to some 10^0 W or 10^1 W.

When cable damage by foreign body penetration is considered, the velocity of the penetrating part shall be known. In the literature [35] values between 2 m/s (for damage by machinery) and 15 m/s (for damage caused by hand tools, e.g. an axe) have been reported. Overlooking the causes of short-circuits in cables in mines, 50% of the damage is due to caving, 40% results from damage by winding machinery and 10% is due to cables in dangerous locations which are not protected appropriately.

With the assumption that a three-phase AC grid with 'industrial' frequency (50–60 cps) represents the 'state of the art' of a power grid in a coal mine, two features shall be considered in parallel to establish a network and

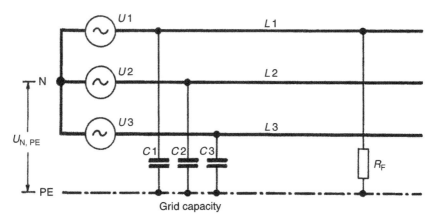

Figure 12.1   *Schematic diagram of a three-phase AC grid with isolated neutral N (type IT).*
L1, L2, L3: phase conductors; U1, U2, U3: voltages phase to neutral; C1, C2, C3: capacitances phase to earth; PE: protective earthed conductor; R_F: earth fault resistance; $U_{N, PE}$: voltage between neutral N and PE in case of an earth fault

network protection systems of adequate reliability and safety:

- the handling of the neutral point
- the design of cables (low voltage, high voltage up to – say – 11 kV) for installation in coal mines.

A comprehensive treatment of these points is given in [35]. So, some focal points only are discussed here.

As a basic rule, an IT three-phase network should be brought into hazardous areas (coal mines, zone 1 installations in industrial plants) with an isolated neutral. This network shows (nearly identical) capacitances C1, C2 and C3 (see Fig. 12.1) between the phase conductors L1, L2 and L3 and the cable or conductor screen at protective earthed (PE) potential. In its 'healthy state', such a network shows symmetrical phase voltages with respect to the neutral and PE (Fig. 12.2(a)). In case of an earth fault (which is in general the result of cable damage or insulation deterioration) a resistance R_F (time dependent for the most part) generates a fault current and a 'shift' of the neutral point away from PE (voltage $U_{N, PE}$), see Fig. 12.2(b). The fault current is determined by the voltage of the system, the capacitances C1–C3 and R_F. Using cables with individually screened phase conductors (and these three screens forming the PE conductor or a pilot conductor) ensures that an earth fault will be the 'predecessor' of a short-circuit in general. Only a double earth fault results in an immediate short-circuit condition.

With the exception of terminal compartments in 'e' or 'd' or within an enclosure in case of direct cable entries into 'd' or 'p', such an IT system shows totally screened phase conductors which are not subject to an immediate short-circuit condition in case of cable damage or an internal insulation fault. Even in the earth fault condition the network may remain operational. And this is the great advantage of these systems compared with, e.g., a TN system

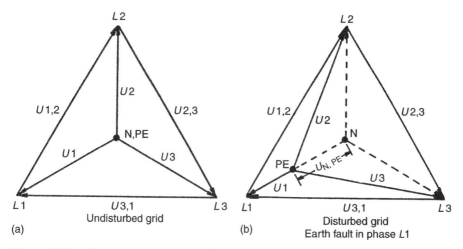

Figure 12.2 *Vector representation of voltages in an IT grid.*
L1, L2, L3: phase conductors; N: neutral; PE: protective earthed conductor;
U1, U2, U3: voltages phase to neutral; U1,2, U2,3, U3,1: voltages phase to phase
(a) Undisturbed grid.
(b) Disturbed grid. Earth-fault in phase L1. $U_{N, PE}$: voltage between neutral N and PE
in case of an earth fault.

with an earthed neutral. In addition, the 'weak point' of unscreened partitions of the phase conductors in terminal compartments and cable joint boxes
may be dropped by a connecting technique as shown in Figs 6.54 and 6.55.
So, two methods are available to monitor an IT grid:

- the insulation resistance between earth and grid is monitored continuously
- or the voltage between earth and neutral is monitored (see Fig. 12.2(b)),
  in other words, the 'shift' of PE away from the neutral in case of an earth
  fault.

In order to detect a cable fault or damage as early as possible, cables for coal
mines show a special design. In low voltage cables (Figs 12.3(a)–(d)), the
three-phase conductors are individually covered with a copper braiding (forming the protective earthed conductor in its entirety) or with a semi-conductive
layer which includes one (or two or three) non-insulated protective earthed
conductors to enable an appropriately low longitudinal resistance. One or
two protective earthed conductors may be replaced by (insulated) auxiliary
conductors for remote control. In addition, the cable may be armoured (Fig.
12.3(d)), normally with steel wires combined with copper wires, the entirety
forming the protective earthed conductor, or, alternatively, a pilot or monitoring conductor.

In German coal mines, two different protective circuits are established:
method A uses a common concentric conductor (kz and t in Fig. 12.3(d)) as a
protective earthed (PE) conductor. The semiconductive layer (pp) is used as
a monitoring conductor (ÜL) with (z) an uninsulated copper conductor to
lower the longitudinal resistance. Between PE and ÜL, a sinusoidal monitoring

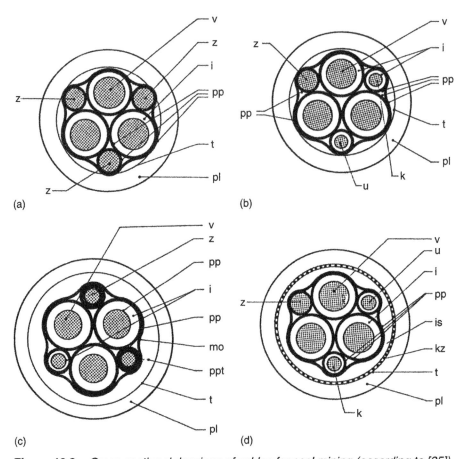

Figure 12.3   *Cross-sectional drawings of cables for coal mining (according to [35]).*
(a) Low voltage cable. v: phase conductors; i: phase conductor insulation;
pp: semi-conductive layer; z: protective earthed conductors, non-insulated; t: fabric
tape (braiding); pl: cable sheathing. (b) Low voltage cable. v: phase conductors;
i: phase conductor and auxiliary conductor insulation; pp: semi-conductive layer;
z: protective earthed conductor, non-insulated; u, k: auxiliary conductors (insulated),
e.g. for control and monitoring; t: fabric tape (braiding); pl: cable sheathing. (c) Low and
high voltage cable (key letters for high voltage type in parentheses). v: phase conductors;
i: phase conductor and auxiliary conductor insulation; (pp: inner conducting layer
between phase conductor and insulation); pp: semi-conductive layer; (mo: phase and
auxiliary conductor metal braiding); z: protective earthed conductors, non-insulated;
–: auxiliary conductor, insulated; ppt: inner cable sheathing; (ppt: semi-conductive cable
core tape, braiding); t: fabric tape; pl: cable sheathing. (d) Low voltage cable. This cable
may show two different designs. Design I – v: phase conductors; i: phase conductor
and auxiliary conductor insulation; pp: semi-conductive layer; z: protective earthed
conductor, non-insulated; u, k: auxiliary conductors (insulated); is: inner cable
sheathing; kz: cable armouring (steel wires, steel–copper braiding); t: fabric tape;
pl: cable sheathing. Design II – v: phase conductors; i: phase conductor and auxiliary
conductor insulation; pp: protective conductor as an individual screen for each phase
conductor, copper braiding; z, u, k: three auxiliary conductors (insulated); is: inner
cable sheathing; kz, t: cable braiding (steel and copper wires); pl: cable sheathing.

voltage is applied. As a circuit terminating equipment a resistor connected in series with a diode 'terminates' the PE–ÜL circuit. This PE–ÜL circuit shows 'ripped' DC in its undisturbed condition. A break or short-circuit causes a sinusoidal current in the PE–ÜL circuit.

Additionally, an 'inverted' design of this monitoring circuit may be put into practice: the common concentric conductor acts as monitoring conductor (ÜL), whereas the protective earthed (PE) conductor is composed of metallic screens covering each phase conductor individually.

With method B, the PE and ÜL conductors are at identical PE potential, i.e. the protection method is reduced solely to earth fault monitoring. An insulated auxiliary conductor is used as a monitoring conductor to ensure a protective earthed conductor correctly connected all over the total system length.

What about insulation resistances and fault detecting times?

To give guidance, in [35] insulation resistances for 3 AC grids up to 1000 V have been reported (Table 12.1). The fault detection times for methods A and B as described above are summarized in Table 12.2. To give an example: with method A, 25% of the detection times are within the interval 0–50 ms or 0–75 ms, whereas 20% only can be found within a time interval 0–100 ms for method B. The 100% limit is within 150 ms for method A and within 200 ms for method B. It is of interest that the monitoring circuit (PE–ÜL) of method A is somewhat faster than the insulation resistance monitoring module (Table 12.2, Part B) used additionally in method A. The PE–ÜL circuit detects all cable faults (=100%) within the time interval 0–100 ms, whereas the insulation resistance monitoring module needs 200 ms to cover the 100% limit.

Table 12.1 Insulation resistances in low voltage 3 AC grids up to 1000 V (according to [35])

Condition of insulation	Insulation resistance* ohms/V
Good	1000
Adequate	>100
Provisionally adequate	40–100
Inadequate	<40
Earth fault (dangerous)	⩽20

* For technical correctness, the insulation resistance R should be related to the grid voltage U and grid length l with a 'specific insulation resistance', R', as follows:

$R = R' \times U/l$
with $R' = R \times l/U$

The values given in this table are values for R'.
For values of $l < 1$ km, $l = 1$.
E.g. for $U = 500$ V, $l = 1$,
$R' = 1000$ ohms/V, it is
$R = 500$ kohm

Table 12.2   Cable fault detection times for methods A and B

Part A   Distribution of detection times into individual 'time intervals'

Time interval ms	Share of detection time %	
	Method A	Method B
0–10	0	0
0–20	12.5	10
0–50	25	10
0–75	25	20
0–100	87.5	20
0–150	100	60
0–200	100	100

Part B   Distribution of detection times into individual 'time intervals' for method A

Time interval ms	Share of detection time %	
	Monitoring the PE–ÜL circuit	Insulation resistance monitoring module
0–10	0	0
0–20	0	0
0–50	0	20
0–75	20	20
0–100	100	20
0–150	100	80
0–200	100	100

Tables 12.3 and 12.4 give a survey of the main technical data of low and high voltage cables for mining. The design of high voltage cables is shown in Figs 12.4(a) and (b).

12.2   Cable protection by rapid-acting switchgear with electrodynamic linear drives

In Section 12.1, detection (or response) times for cable monitoring equipment are summarized in Table 12.2. These times are in an order of magnitude of 10^1–10^2 milliseconds and are valid for 'classic' monitoring systems, whose aim is to indicate an earth fault, to start an alarm or to disconnect the faulty branch of a grid. Things change drastically if the aim of such a cable monitoring system is extended to a protection system, i.e. to detect faults in cables or apparatus and isolate the faulty partition so fast to avoid any short-circuit arcing in electrical apparatus or the ignition of coal dust or even methane in case of a cable fault. As described in Section 12.1, foreign bodies penetrate

Table 12.3 Low voltage cables for coal mines – main technical data

Cross-section of …					
• phase conductor mm²	25	35	50	70	95
• protective earthed conductor mm²	16	16	25	35	50
• monitoring/auxiliary conductor mm²	1.5	1.5	1.5	1.5	1.5
U_o/U^* kV			0.6/1		
Phase conductor:					
• type			fine wire		
• surface			bare		
• number of conductor wires	349	494	516	736	776
• diameter of conductor wire mm	0.31	0.31	0.36	0.36	0.41
• diameter of conductor mm	7.8	9.2	11.0	13.1	15.1
• thickness of insulation mm	1.4	1.4	1.6	1.6	1.6
Thickness of …					
• inner cable sheathing mm	2.8	2.8	3.0	3.0	3.4
• outer cable sheathing mm	3.5	3.5	3.5	3.5	4.5
Outer diameter of cable, max. value mm	50.5	50.5	51.7	57.9	64.4
Cable weight kg/m	3.01	3.25	4.18	5.3	7.1
Phase conductor:					
• resistance per unit length					
• for +90°C mohms/m	0.995	0.706	0.492	0.347	0.263
• for +20°C mohms/m	0.780	0.554	0.386	0.272	0.206
• inductance per unit length µH/m	0.250	0.240	0.240	0.230	0.230
• capacitance per unit length nF/m	0.600	0.690	0.720	0.840	0.860
• capacitive reactive current for 50 cps, $U = 1$ kV and 1000 m cable length A	0.110	0.125	0.131	0.152	0.156
• max. current load A at +30°C ambient temperature	131	162	202	250	301
• thermal short-circuit current for 1s kA	3.57	5.00	7.15	10.0	13.6
Cable design:					
• according to Fig. 12.3(d)					
• design I					

$^*U_o$ = rated cable voltage between phase conductor and conductor screening
U = rated cable voltage between phase conductors

into a cable with typical velocities between 2 m/s and 15 m/s, or, in other words, with 2 mm/ms and 15 mm/ms. Tables 12.3 and 12.4 indicate that the thickness of cable sheathings is in the order of some millimetres, and the same values are valid for the insulation thickness of a phase conductor.

Table 12.4   High voltage cables for coal mines – main technical data

Cross-section of ...					
• phase conductor mm^2	25	35	50	70	95
• protective earthed conductor mm^2	16	16	25	35	50
• monitoring/auxiliary conductor mm^2	2.5	2.5	2.5	2.5	2.5
U_o/U^* kV			3.6/6		
Phase conductor:					
• type			fine wire		
• surface			bare		
• number of conductor wires	196	276	396	580	740
• diameter of conductor wire mm	0.41	0.41	0.41	0.41	0.41
• diameter of conductor mm	7.8	9.2	11.0	13.1	15.1
• thickness of insulation mm	3.4	3.4	3.4	3.4	3.4
Thickness of ...					
• inner cable sheathing mm	3.0	3.0	3.0	3.0	3.0
• outer cable sheathing mm	3.0	3.0	3.0	3.0	3.0
Outer diameter of cable, max. value mm	53.0	57.0	61.0	65.0	68.0
Cable weight kg/m	3.9	4.6	5.5	6.7	7.9
Phase conductor:					
• resistance per unit length					
• for +70°C mohms/m	0.933	0.663	0.462	0.325	0.246
• for +20°C mohms/m	0.780	0.554	0.386	0.272	0.206
• inductance per unit length μH/m	0.310	0.300	0.280	0.270	0.260
• capacitance per unit length nF/m	0.620	0.700	0.810	0.930	1.05
• capacitive reactive current for 50 cps, $U = 6$ kV and 1000 m cable length A	0.676	0.763	0.883	1.014	1.145
• max. current load A at +30°C ambient temperature	101	126	153	196	238
• thermal short-circuit current for 1 s kA	2.88	4.02	5.75	8.05	10.9

Cable design:
• similar to Fig. 12.4(b)
• with three auxiliary conductors
• kz = pilot or monitoring conductor (ÜL)
• kk = armour (steel braiding)

$^*U_o$ = rated cable voltage between phase conductor and conductor screening
U = rated cable voltage between phase conductors

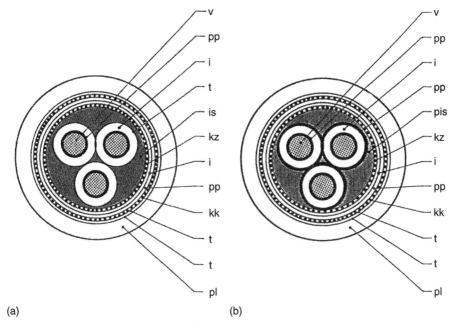

(a)   (b)

Figure 12.4   *Cross-sectional drawings of high voltage cables for coal mining (according to [35]).*
(a) v: phase conductors; pp: semi-conductive phase conductor screen; i: phase conductor insulation; t: fabric tape (braiding); is: core filler; kz: concentric protective conductor; i: insulation; pp: semi-conductive layer; kk: concentric pilot conductor; t: fabric tape (braiding); pl: cable sheathing. (b) v: phase conductors; pp: semi-conductive phase conductor screen or layer; i: phase conductor insulation; pp: semi-conductive layer *or* copper braiding as a protective conductor; pis: (semi-conductive) core filler; kz: concentric protective conductor *or* concentric pilot conductor; i: insulation; kk, t: concentric pilot conductor, with fabric tape *or* steel braiding (armour); pl: cable sheathing.
Note: For many mining applications, this cable may additionally contain three (insulated) auxiliary conductors fitted into the core filler

So, such a protection device has to detect a fault and to isolate within 1 millisecond at the latest.

In the history of electrical engineering with regard to mining equipment, attempts have been made to put such systems into practice [6, 27]. In the following, a cable protection system will be described in more detail. This system successfully demonstrates for the first time how to avoid the ignition of gas–air atmospheres in case of cable faults [57].

12.2.1   Damage risks – fault currents and arcs

In power supply networks, fault currents cause arcs with the release of a considerable amount of energy. An AC system with period T, peak current I_o, causing an arc with arc voltage U_B generates the energy:

$$W = (U_B \cdot I_o \cdot T)/\pi \text{ per half cycle (see Section 6.8.11)}$$

(In this case, the arc voltage U_B is assumed to be constant due to the invariable distance between the conductors.) As an example, in a low voltage system with $f = 50$ cps corresponding to $T = 20$ ms, $U_B = 300$ V and $I_o = 30$ kA, W results with $W = 57.3$ kJ. Quenching the arc after 10 half-cycles, i.e. 100 ms, results in an energy release of 573 kJ. For comparison: an exploding methane–air mixture in an enclosure with a volume of $1 \times 0.4 \times 0.6$ m^3 $= 0.24$ m^3 typical for low voltage power distributions causes 766 kJ. In an enclosure assumed to be completely gas-tight, this energy will generate an internal overpressure of 7.2 bar. In general, fault currents and their arcs will distort the enclosures of switchgear and therefore expose the personnel to danger, accompanied by a considerable destruction of internal components.

Often, overhaul and rebuilding costs combined with production losses come up to a multiple of the capital expended for the switchgear.

Similar considerations can be made for semiconductors in rectifiers and frequency convertors. The maximum load integral for semiconductors in power equipment can be assumed with values starting from 10^3 A^{2}s up to (and exceeding) 10^7 A^{2}s.

Even in times near to the zero passage of a short-circuit current these values may be transgressed. As a good approximation, a sine-shaped current can be represented by linearity near its zero passage.

With $I_o = 30$ kA, the maximum load integral follows with $3 \cdot 10^4$ A^{2}s after 1 ms and with $2.4 \cdot 10^5$ A^{2}s after 2 ms only. This fact indicates that even in the vicinity of zero passage of a short-circuit current semiconductors with maximum rms on-state currents in the range 100 A to 800 A are endangered after a few milliseconds only.

These considerations can be summarized as: a protective device for power cables in coal mines, low voltage switchgear, rectifiers and frequency convertors shall respond within one millisecond. 'To respond' means to detect faults and subsequent fault current cut-off with isolation of network areas to be protected. The maximum load integral shall be limited to 10^3 A^{2}s (as an order of magnitude).

Following this route, the dynamical stresses of switchgear, busbars and load terminals due to short-circuit currents can be massively reduced, if a protective device responds in an adequate time and prevents the short-circuit current rising to its prospective value. There is no longer a need for an additional reactance as a short-circuit current limiting device, which unfortunately impedes the run-up of large asynchronous motors.

12.2.2   Principles of fault-current switching

Speaking very generally, a faulty power system is fed by a power source (Fig. 12.5) with an internal impedance Z1, connected to a load, e.g. a motor with impedance Z2. The point of fault shows an impedance (in most cases

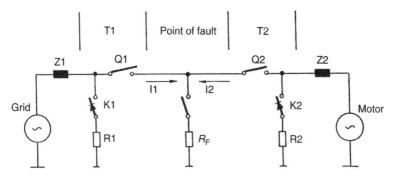

Figure 12.5   *Single-pole equivalent circuit of a faulty power system.*
$Z1$: internal grid impedance; $Z2$: load impedance; R_F: fault resistance; T1, T2:
protective devices; Q1, Q2: isolating switches; K1, K2: short-circuiting devices;
R1, R2: resistances of K1, K2; I1: fault current generated by the grid; I2: fault
current generated by a reactive load.

time dependent) with a dominant ohmic fraction R_F due to the arc. The protective device T is composed of:

- an isolating switch Q
- a short-circuiting device K with a resistance R (deviating from zero!).

In case of a fault, the current $I1$ – from the power source – feeds R_F. It should
be emphasized that an additional current $I2$ flows from the (reactive) load
to R_F, due to the stored energy at the load side. This aspect will become
important if cable protection is achieved. For other purposes, $I2$ can be
neglected.

Apart from cable protection, only one protective unit T1, composed of K1
(short-circuit device) and Q1 (isolating device) can meet the requirements as
stated above.

12.2.3   Electrodynamic linear drives – the way to sub-millisecond switching times

Two currents in parallel direction attract one another, in the opposite direction there is a repelling force. The second part of this finding is used as the
basic idea for a drive unit for switchgear. A disc-shaped coil as a part of a
thyristor controlled capacitor-fed oscillating circuit generates a time-variant
magnetic field. This, in turn, causes a current counter-rotating with respect to
the coil current in a moveable disc adjacent to the coil, made of electrical conducting material.

The current in an oscillating circuit with a capacitor C, a total inductance L
and ohmic losses R is time dependent due to:

$$I'' + (R/L)I' + (1/LC)I = 0$$

with the solution (assuming R to be time independent):

$$I = -(U_o/\omega L)e^{-\delta t}\sin \omega t$$
$$U_o = \text{charging voltage of capacitor } C$$
$$\omega = 2\pi f = 2\pi/T \text{ (angular frequency)}$$
$$\omega_o = (LC)^{-0.5} \text{ (undamped circuit)}$$
$$\delta = R/2L \text{ (damping constant)}$$
$$\omega = (\omega_o^2 - \delta^2)^{0.5} \text{ (damped circuit)}$$

An oscillating circuit with suitably selected U_o, C and L with:

$$f = 5.68 \text{ kcps}$$

and a stored energy $0.5CU^2 = 50\,\text{J}$ is used as an electrodynamic drive for a vacuum switching tube.

In the closed position of the tube, the atmospheric pressure supported by the spring tension of the vacuum bellows acts on the moveable contact. When firing the oscillating circuit, the disc is accelerated by the magnetic force due to the currents in the circuit (some 10^3 A) and in the disc (which can be considered as a single-turn short-circuited winding of an air core transformer). The disc acts directly on a connecting rod to the moveable contact and opens the contacts within the vacuum tube. Assuming a constant acceleration $b = 2s/t^2$ along the linear motion s (s: clearance between the contacts in off positon, $s = 4\,\text{mm}$) and limiting t to values smaller than/equal to one millisecond, b shall exceed $10^4\,\text{m/s}^2$.

Considering the mass of the moveable contact, vacuum bellows, rod and disc, the repelling force acting on the disc exceeds 20 kN (as an order of magnitude).

12.2.4   The ultra-rapid switch unit

These principles are the basic philosophy for design and construction of a triple-pole switch (Fig. 12.6). Three vacuum switching tubes Q1–Q3 with electrodynamic linear drives edA1–edA3 act as isolating devices in the power lines L1–L3. The linear drives are fed from capacitors C1–C3 via the thyristors V4–V6. Short times for the separation of the contacts within the switching tubes are in no way sufficient for an adequate time to isolate. Generally, arcing shall be avoided, or, in other words, the voltage drop across Q1–Q3 after contact separation has to be limited to some volts only.

In this voltage range an arc-free contact separation is possible. In order to achieve an adequate voltage drop, thyristors V1–V3 arranged at the source side in the power system are fired simultaneously with the drive circuits for edA1–edA3 via the transformers T1–T3. They act as a short-circuit device for – roughly speaking – half a cycle, time enough for an arc-free contact separation in Q1–Q3. After this time interval, V1–V3 return to the non-conducting state. Q1–Q3 are mechanically fixed in their off position.

Figure 12.7 shows the ultra-rapid switch in its triple-pole version. Technical data are given in Table 12.5.

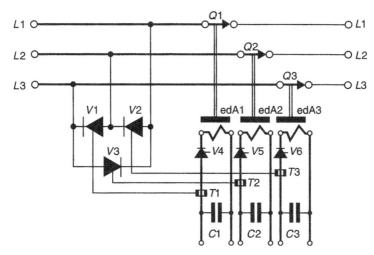

Figure 12.6 *Ultra-rapid switch with electrodynamic linear drives, basic circuit diagram.*
L1, L2, L3: phase conductors; Q1, Q2, Q3: vacuum switching tubes; edA1, edA2, edA3: electrodynamic linear drives for Q1, Q2, Q3; C1, C2, C3: capacitors for edA1, edA2, edA3; V4, V5, V6: thyristors to initiate edA1, edA2, edA3; V1, V2, V3: short-circuiting thyristors; T1, T2, T3: ignition transformers for V1, V2, V3.

Figure 12.7 *Ultra-rapid switch UR 1/400 – total view.*

Table 12.5   Ultra-rapid switch UR 1/400 – main technical data

Rated voltage	1000 V
Rated current	400 A
Current type	3 AC
Rated frequency	50 ... 60 cps
Rated instantaneous short-circuit current*	31.5 kA
Rated control voltage 1 AC 50–60 Hz	230 V
Dimensions	410 × 340 × 385 mm
Weight	45 kg
Mounting position of vacuum switching tubes	vertical ±15°
Reset to ON position	manually operated

*The thyristors V1–V3 shall meet the power system short-circuit conditions

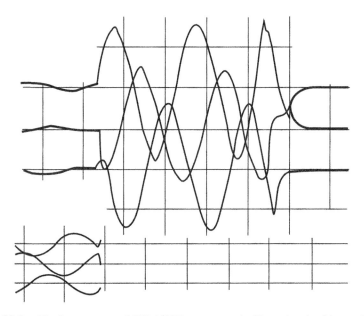

Figure 12.8   *Fault response of UR 1/400, compared with a standard low voltage circuit breaker. A three-phase AC current of 80 A rms to a load is followed by a short-circuit in a 50 cps grid.*
top: Response of the standard circuit breaker – current base/vertical: 1250 A/div.; time base/horizontal: 10 ms/div.
bottom: Response of UR 1/400 – current base/vertical: 200 A/div.; time base/horizontal: 10 ms/div.

Figure 12.8 demonstrates the efficiency of the UR 1/400 switch. A 3 AC current of 80 A to a load is followed by a short-circuit.

Two different switchgear are compared:

top of Fig. 12.8:	Standard low voltage circuit breaker
bottom of Fig. 12.8:	UR 1/400

Note: time base/horizontal 10 ms/div.
current base/vertical
top of Fig. 12.8   1250 A/div.
bottom of Fig. 12.8   200 A/div.

Within 1 millisecond, the short-circuit current has been switched off combined with the isolation of the faulty part of the 3 AC system using the UR 1/400 ultra-rapid switch.

12.2.5 Cable protection using the UR 1/400 switch

Due to the small voltage drop (some volts) across the fired (and conducting) thyristors V1–V3, the current flowing into the area protected by UR 1/400 is in the range of 10^3 A even in the case of short-circuits in an immediate vicinity to the protective device and low cable impedance. The switching-off time of 1 millisecond limits the maximum load integral for currents entering the protected area to $10^3 A^2 s$.

This fact enables UR 1/400 to act as a cable protection system in gassy coal mines endangered by firedamp, i.e. methane (CH_4).

Remembering Fig. 12.5, reactive loads (e.g. a motor) obviously cause a fault current *I*2 flowing to the point of fault of the power system. In order to limit the energy dissipated at this point to values unable to ignite a methane–air mixture, the current flow *I*2 from, e.g., the motor has to be cut off in the same manner as *I*1 coming from the power source. In consequence, two protective devices T1 and T2 have to be arranged, defining a protected area of the power system between T1 and T2, e.g. a cable to a shearer-loader. T2 shows an 'inverted' position of isolation device and short-circuit device with respect to T1 on the power source side. With respect to the flow of energy to the point of fault, the short-circuit devices shall be arranged ahead of the isolating devices.

For reasons of safety, the short-circuit device of UR 1/400 as given in Fig. 12.6 (V1–V3) may be replaced by a bridge-connected rectifier (Fig. 12.9) V1–V3 and V4–V6, followed by two thyristors V7 and V8 fired in parallel. The lack of any current zero passage at the output of V1–V3/V4–V6 ensures a short-circuit time not affected by the availability of firing pulses for the thyristors. They return to their non-conducting state with neglectable current.

A 'type test' for a cable protection system has been carried out (Fig. 12.10). A 250 kW cage induction motor M was fed via a standard circuit breaker Q, an ultra-rapid switch UR 1/400 at the source side, a testing device P and a second ultra-rapid switch UR 1/400 at the motor side. In P, a cable (for coal mines application) was cut in a surrounding $21 \pm 1\%(v/v)$ H_2–air atmosphere during the motor duty cycle using a pneumatically operated axe forming a 'cable guillotine'. This H_2–air mixture was used due to its minimum ignition energy of 19 µJ, which is much smaller than the corresponding value for methane–air mixtures (280 µJ, see Table 1.7). So, the choice of H_2–air mixtures guarantees a certain 'margin of safety' for these tests.

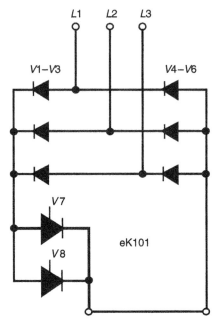

Figure 12.9   *Short-circuiting device with an 'inherent safety'.*
*L*1, *L*2, *L*3: phase conductors; *V*1–*V*3: positive branch of the rectifier bridge; *V*4–*V*6: negative branch of the rectifier bridge; *V*7, *V*8: short-circuiting thyristors to be fired in parallel.

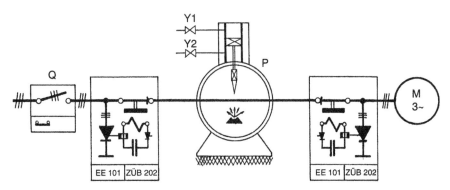

Figure 12.10   *Test set-up for cable protection systems.*
Q: standard low voltage circuit breaker at the grid side; M: two-pole cage induction motor, rated power 250 kW; P: 'cable guillotine' with a pneumatic drive to cut the cable under explosive hydrogen–air atmosphere
The protected area is located between the two ultra-rapid switches. EE 101 is a device to detect the voltage shift $U_{N, PE}$ between neutral and PE, ZÜB 202 a device to fire the thyristors of the electrodynamic drives (V4–V6, see Fig. 12.6). At the motor side, the short-circuiting device has been in accordance with Fig. 12.9.

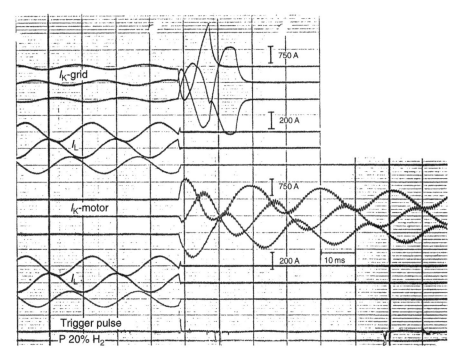

Figure 12.11 *Response of the cable protection system.*
Traces top to bottom – I_K-grid: current at the grid side: motor current, followed by the short-circuit current generated by *V1–V3* of the ultra-rapid switch at the grid side (the grid side current has been measured by non-integrating Rogowski coils [20]. Their output signals show a phase shift of $-90°$ compared with classical current transformers); I_L: cable current, behind the ultra-rapid switch at the grid side. The motor current is switched off within 1 ms; I_K-motor: current to the short-circuiting device of the ultra-rapid switch at the motor side. After firing *V7* and *V8* (see Fig. 12.9), the motor is short-circuited during its run down; I_L: cable current, ahead of the ultra-rapid switch at the motor side. The motor current is switched off within 1 ms; Trigger pulse: response of EE 101, caused by the cable cut and the voltage shift $U_{N, PE}$ generated hereby; P: output signal of pressure transducer, monitoring the pressure of a H_2–air mixture inside the 'cable guillotine' P housing: no ignition of the explosive mixture.

Moreover, the cable was cut at the vertex of a U-shaped loop to ensure gas entry to the point of cable damage as quick as possible.

The result is given in Fig. 12.11. During normal motor operation, all currents I_K-grid, I_L are identical. Cutting the cable fires the two ultra-rapid switches UR 1/400, and so short-circuit currents occur at the source side (I_K-grid) and, with a short-circuit device according to Fig. 12.9, at the motor side, I_K-motor.

The slow decay of I_K-motor indicates the flow of energy coming from the motor caused by the rotor's run out.

The track at the bottom shows the output signal of a pressure transducer monitoring the gas mixture pressure inside P. Obviously, there is no ignition of the H_2–air test mixture. To ensure its ignitability, the H_2–air mixture has

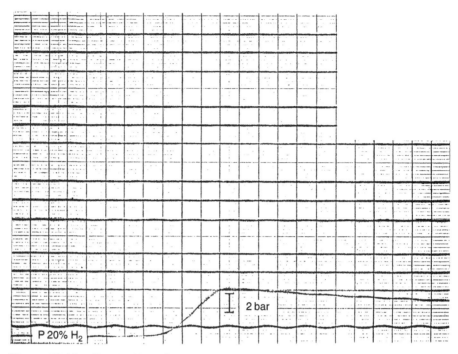

Figure 12.12   *Test ignition of the H₂–air mixture inside P immediately after the trial of Fig. 12.11, caused by a spark plug: the ignitability of the mixture has been demonstrated (peak pressure approx. 5.2 bar = 5.2 · 10⁵ Pa).*

been fired using a spark plug after cable cutting (Fig. 12.12). The track at the bottom indicates the pressure rise due to the exploding mixture. A test sequence of total 100 cutting procedures without any ignition of the test mixture demonstrates the reliability of this cable protection system for application in both gassy atmospheres and under severe operating conditions.

Bibliography

This bibliography does not claim to be complete. It deals with basic principles and essential subsections in the field of explosion protection.

1 Baker, W. E., Cox, P. A., Westine, P. S., Kulesz, J. J. and Strehlow, R. A.: Explosion hazards and evaluation, in: *Fundamental Studies in Engineering*, vol. 5. Elsevier, Oxford, 1983

2 Bartknecht, W.: *Explosionsschutz*, Springer, Berlin, 1993 (Original: German) *Explosion Protection*

3 Beyling, C.: Versuche zwecks Erprobung der Schlagwettersicherheit besonders geschützter elektrischer Motoren und Apparate sowie zur Ermittlung geeigneter Schutzvorrichtungen für solche Betriebsmittel, ausgeführt auf der Berggewerkschaftlichen Versuchsstrecke in Gelsenkirchen-Bismarck, *Glückauf*, 42, 1906 – Nr. 1, 1–9, Nr. 2, 34–42, Nr. 3, 70–74, Nr. 4, 93–99, Nr. 5, 129–138, Nr. 6, 165–171, Nr. 7, 201–206, Nr. 8, 237–244, Nr. 9, 273–278, Nr. 10, 301–306, Nr. 11, 338–346, Nr. 12, 373–383, Nr. 13, 400–418 (Original: German)
Experiments for testing the flameproofness of specially protected electrical motors and apparatus and for the determination of suitable safety devices for such apparatus, performed at the Berggewerkschaftliche Versuchsstrecke, Gelsenkirchen-Bismarck

4 Bradley, D. and Gupta, M. L.: Direct and alternating current coronas in flame gases, *Combustion and Flame* 40, 1981, 47–63

5 *Brenn- und Explosions-Kenngrößen von Stäuben*, Berufsgenossenschaftliches Institut für Arbeitssicherheit und Bergbau-Versuchsstrecke, Institut für Explosionsschutz und Sprengtechnik, Verlag Erich Schmidt, Bielefeld, 1987
(Original: German)
Combustion and Explosion Data of Dusts

6 Calis, A. and Goreux, G.: Développement d'un disjoncteur ultra rapide à semiconducteurs pour le réseau basse tension de mines souterraines, INIEX Institut National des Industries Extractives, Rapport final 7255-14/031/02, Colfontaine/B, 1983
(Original: French)
Development of a semiconductor-based protection switch for low voltage power lines in mines

7 ElBergV: Bergverordnung des Landesoberbergamtes NRW und des Oberbergamtes für das Saarland und das Land Rheinland-Pfalz (Ausgabe Saarland) für elektrische Anlagen, (Elektro-Bergverordnung), dated 1992-03-16, Verlag Glückauf, Essen, 1992
(Original: German)

Mining decree of the Landesoberbergamt – Supreme authority in mines – of the land Northrhine Westphalia and of the Oberbergamt – Supreme authority in mines – of the Saarland and Rheinland-Pfalz on electrical systems
and:
Verwaltungsanweisungen, Hinweise und Erläuterungen zur ElBergV, dated 1992-04-08, Verwaltungsvorschrift zu § 30 Abs. 2 ElBergV, dated 1992-05-06, Verlag Glückauf, Essen, 1992
(Original: German)
Administration orders, instructions and comments on ElBergV, Administration directive concerning clause 30.2 of ElBergV

8 ElexV: Verordnung über elektrische Anlagen in explosionsgefährdeten Räumen, Neufassung vom 13. Dezember 1996, Carl Heymanns Verlag, Köln, 1995, Bundesgesetzblatt Teil I, Nr. 65/1996
(Original: German)
Decree on electrical systems in areas hazardous due to explosive atmospheres

9 ElZulBergV: Bergverordnung über die allgemeine Zulassung schlagwettergeschützter und explosionsgeschützter elektrischer Betriebsmittel (Elektrozulassungs-Bergverordnung), Revised form 1993-03-10, Verlag Glückauf, Essen, 1993
(Original: German)
Mining decree on the general approval of firedamp-proof and explosion protected electrical apparatus

10 Enright, R. J.: *Sulfide Dust Explosions in Metalliferous Mines*, The Australasian Institute of Mining and Metallurgy, Parkville, Vic./Australia, 1984

11 ExVO: Explosionsschutzverordnung, Elfte Verordnung zum Gerätesicherheitsgesetz – Verordnung über das Inverkehrbringen von Geräten und Schutzsystemen für explosionsgefährdete Bereiche – 11. GSGV vom 12. Dezember 1996, Bundesgesetzblatt Teil I, Nr. 65/1996
(Original: German)
Decree on Explosion Protection, 11th decree on the Apparatus Safety Act – Decree on the circulation of apparatus and protection systems for areas hazardous due to explosive atmospheres

12 Förster, H. and Steen, H.: *Untersuchungen zum Ablauf turbulenter Gasexplosionen*, Bericht W-32, PTB Physikalisch-Technische Bundesanstalt, Braunschweig, 1986
(Original: German)
Studies on the Behaviour of Turbulent Gas Explosions

13 Fox, J. S. and Bertrand, C.: Measurement of local saturation current in flames, *Combustion and Flame* 43, 1981, 317–320

14 Groh, H.: Fotochemische Reaktionen in explosionsgeschützten Leuchten mit Entladungslampen, *Chem.-Ing.-Tech.* 52, 1980, Nr. 7, 590–592
(Original: German)
Photochemical reactions in explosion protected light fittings with discharge lamps

15 Groh, H.: Die Messung der Lichtbogenspannung an Schaltgeräten, *etz-Archiv* 2, 1980, 311–313
(Original: German)
Measurement of arcing voltage in circuit breakers and contactors

16 Groh, H.: Die Stickoxidproduktion der Schaltlichtbögen in Luftschalt-strecken, *etz-Archiv* 3, 1981, 255–264
(Original: German)
Nitrogen-oxide production of switching arcs in air

17 Groh, H.: *Der Schlagwetter- und Explosionsschutz elektrischer Betriebsmittel,* Verlag Glückauf, Essen, 1986
(Original: German)
Firedamp and Explosion Protection of Electrical Equipment

18 Groh, H.: Die zünddurchschlagsicheren Spaltweiten vorverdichteter Wasserstoff-Luft-Gemische, *Chem.-Ing.-Tech.* 59, 1987, 670–671
(Original: German)
The maximum experimental safe gaps of precompressed hydrogen–air mixtures

19 Groh, H.: Sintermetallbauteile als Zünddurchschlagsicherungen im Explosions- und Schlagwetterschutz, *Chem.-Ing.-Tech.* 59, 1987, 672–673
(Original: German)
Sintered metals as flame arrestors in chemistry and coal mining industry

20 Groh, H.: *Hochspannungsmeßtechnik,* 2nd edition, Expert-Verlag, Renningen-Malmsheim, 1994
(Original: German)
High Voltage Measuring Technology

21 Groh, H.: Überdruckkapseln in Zone 2 und im Staubexplosionsschutz, *etz* 19, 1995, 10–16
(Original: German)
Pressurized enclosures – application in zone 2 and for dust explosion protection

22 Van Gucht, A.: Entwurf, Werksprüfungen und industrielle Erprobung des ACEC-Turboentspanners, *ACEC-Zeitschrift* 1, 1983, 7–12, Ateliers de Constructions Électriques de Charleroi, Charleroi/B
(Original: German)
Design, plant testing and industrial trials of the ACEC turbo expander

23 Gulski, E.: *Computer-aided Recognition of Partial Discharges using Statistical Tools,* Delft University Press, 1991

24 Hall, J.: *Intrinsic Safety,* Marylebone Press Ltd/The Institution of Mining Electrical and Mining Mechanical Engineers, Manchester, 1985

25 Hertzberg, M.: *The Theory of Flammability Limits,* United States Department of the Interior, Bureau of Mines, Pittsburgh Mining and Safety Research Center, 1976

26 Hylten-Cavallius, N.: *High Voltage Laboratory Planning,* Emile Haefely & Co. Ltd, Basel, 1986

27 Killing, F. and Gruber, L.: Schutzeinrichtung zum schnellen Abschalten von Fehlerströmen in druckfesten Gehäusen, *Glückauf-Forschungshefte* 41, 1980, Nr. 5, 214–217

(Original: German)
Protection device for rapid breaking of fault currents in flameproof enclosures

28　Ko, Y., Anderson, R. W. and Arpaci, V. S.: Spark ignition of propane–air mixtures near the minimum ignition energy, Part I: An experimental study, *Combustion and Flame* 83, 1991, 75–87

29　Ko, Y., Arpaci, V. S. and Anderson, R. W.: Spark ignition of propane–air mixtures near the minimum ignition energy, Part II: A model development, *Combustion and Flame* 83, 1991, 88–105

30　König, D. and Narayana Rao, Y.: *Teilentladungen in Betriebsmitteln der Energietechnik*, VDE-Verlag, Berlin und Offenbach, 1993
(Original: German)
Partial Discharges in Electrical Power Apparatus

31　Lorenz, H.: Explosionsgeschützte Überdruckkapseln, *etz* 19, 1995, 18–24
(Original: German)
Explosion protected pressurized enclosures

32　Lunn, G. A.: An apparatus for the measurement of maximum experimental safe gaps at standard and elevated temperatures, *Journal of Hazardous Materials* 6, 1982, 329–340

33　Lunn, G. A.: The influence of chemical structure and combustion reactions on the maximum experimental safe gap of industrial gases and liquids, *Journal of Hazardous Materials* 6, 1982, 341–359

34　Lunn, G. A.: The maximum experimental safe gap: the effects of oxygen enrichment and the influence of reaction kinetics, *Journal of Hazardous Materials* 8, 1984, 261–270

35　Marinovic, N.: *Electrotechnology in Mining: Advances in Mining Science and Technology*, vol. 6, Elsevier, Amsterdam, 1990

36　Mathis, M. and Schnettler, A.: Optische Sensoren für Hochspannungs-schaltanlagen, *etz* 18, 1996, 40–43
(Original: German)
Optical sensors for high-voltage switchgear

37　Minard, A. and De Maneville, P.: *Systèmes et installations de sécurité intrinsèque*, LCIE Laboratoire Central des Industries Electriques, Fontenay-aux-Roses/F, 1979
(Original: French)
Intrinsically Safe Systems and their Installation

38　Nabert, K. and Schön, G.: *Sicherheitstechnische Kennzahlen brennbarer Gase und Dämpfe*, 2nd edition, Deutscher Eichverlag, Braunschweig, 1963
(Original: German)
and
Redeker, T. and Schön, G.: 6. Nachtrag, Deutscher Eichverlag, Braunschweig, 1990
(Original: German)
Safety Related Data of Combustible Gases and Vapours

39　Olenik, H., Rentzsch, H. and Wettstein, W.: *Handbuch für Explosionsschutz/ Explosion Protection Manual*, 2nd edition, W. Girardet, Essen, 1983

40 Phillips, H.: The mechanism of flameproof protection, Safety in Mines Research Establishment, Research Report 275, Buxton/UK, 1971

41 Phillips, H.: Maximum explosion pressure in flameproof enclosures: the effects of the vessel and the ambient temperature, *Journal of Hazardous Materials* 8, 1984, 251–259

42 Phillips, H.: *Safe Gap Revisited: Progress in Astronautics and Aeronautics*, vol. 114, AIAA, Washington, DC, 1988

43 Redding, R. J.: *Intrinsic Safety: The Safe Use of Electronics in Hazardous Locations*, McGraw-Hill, London, 1971

44 Redeker, T.: Classification of flammable gases and vapours by the flameproof safe gap and the incendivity of electrical sparks, Report W-18, PTB Physikalisch-Technische Bundesanstalt, Braunschweig, 1981

45 Richtlinien für die Vermeidung der Gefahren durch explosionsfähige Atmosphäre mit Beispielsammlung – Explosionsschutz-Richtlinien – (EX-RL), Carl Heymanns Verlag, Köln, 1994
(Original: German)
Directive concerning the avoidance of risks due to explosive atmospheres (aided by a collection of examples)

46 Schaefer, B. A.: The calculation of ion currents in hydrocarbon flames, *Combustion and Flame* 56, 1984, 43–49

47 Schwarz, K.-H.: Die eigensichere Variante – Profibus in der Verfahrenstechnik, *Elektronik* 14, 1995, 48–58
(Original: German)
The intrinsically safe variant – Profibus in process engineering

48 Scott, L. W. and Dolgos, J. G.: Electric arcing at high voltage during methane–air explosions inside explosion-proof enclosures, US Department of the Interior, Bureau of Mines, Technical Progress Report 115, 1982

49 Thater, R.: Beliebige Oberflächentemperatur, *etz* 19, 1995, 26–31
(Original: German)
Arbitrary surface temperature

50 VbF: Verordnung über Anlagen zur Lagerung, Abfüllung und Beförderung brennbarer Flüssigkeiten zu Lande vom 27. Februar 1980, Neufassung vom 13. Dezember 1996, TRbF: Technische Regeln für brennbare Flüssigkeiten, Carl Heymanns Verlag, Köln, 1980
Bundesgesetzblatt Teil I Nr. 65/1996
(Original: German)
Decree on installations for storage, racking and overland transportation of combustible fluids, revised form 1996-12-13: Technical rules concerning combustible fluids

51 Vogt, G.: Untersuchungen der Zündung von explosionsfähigen Methan-Luft-Gemischen durch elektrische Entladungen in induktiven Stromkreisen von eigensicheren Betriebsmitteln für den Steinkohlenbergbau, *Mitteilungen der Westfälischen Berggewerkschaftskasse* 42, 1980
(Original: German)
Studies on the ignition of explosible methane–air mixtures caused by electrical discharges in intrinsically safe inductive circuits for coal mines

52 Willmroth, G., Schmitz, H., Teermann, A., Fink, E. and Pauls, P.: Betriebserfahrungen mit der Erdgasexpansionsanlage der EWV Stolberg, *Gas – Erdgas – gwf* 138, 1997, vol. 9, 534–543
(Original: German)
Operation experience with the natural gas expansion plant of EWV Stolberg

53 Wittler, M.: Teilentladungen – Zündquellen in explosionsgeschützten Einrichtungen?, *etz* 14/15, 1993, 902–907
(Original: German)
Partial discharges – ignition sources in explosion protected apparatus?

54 Wörmann, W. and Groh, H.: Die Spannungsfestigkeit von Flammen in druckfesten explosionsgeschützten elektrischen Betriebsmitteln, *etz-Archiv* 5, 1983, 155–160
(Original: German)
Dielectric strength of flames in flameproof electrical equipment

55 Wörmann, W. and Groh, H.: *Die elektrische Leitfähigkeit abflammender Gas/Luft-Gemische*, ETZ-Report 17, VDE-Verlag, Berlin and Offenbach, 1983
(Original: German)
Electrical Conductance of Burning Gas/Air Mixtures

56 Wörmann, W. and Groh, H.: *Der Lichtbogenzünddurchschlag bei druckfesten explosionsgeschützten elektrischen Betriebsmitteln*, ETZ-Report 18, VDE-Verlag, Berlin and Offenbach, 1983
(Original: German)
Flame Transmission Caused by Arcing in Flameproof Electrical Equipment

57 Zupfer, H. and Groh, H.: Die schlagwettergeschützte Leitung – Leitungsschutz mit schnellwirkenden Schaltgeräten, *Glückauf-Forschungshefte* 49, 1988, 13–18
(Original: German)
The flameproof power cable – cable protection by rapid-acting switchgear

Index

Printed and bound by CPI Group (UK) Ltd, Croydon, CR0 4YY

03/10/2024

01040433-0015